Integration of Practice-Oriented Knowledge Technology: Trends and Prospectives

Madjid Fathi (Ed.)

Integration of Practice-Oriented Knowledge Technology: Trends and Prospectives

Springer

Editor
Madjid Fathi
Professor and Director
Institute for Knowledge Based Systems & Knowledge Management
Research Center for Knowledge Management and Intelligent Systems
University of Siegen
Siegen
Germany
fathi@informatik.uni-siegen.de

ISBN 978-3-642-34470-1 e-ISBN 978-3-642-34471-8
DOI 10.1007/978-3-642-34471-8
Springer Heidelberg New York Dordrecht London

Library of Congress Control Number: 2012950328

Printed on acid-free paper

Springer is part of Springer Science+Business Media (www.springer.com)

Preface

Scientific network of Integrated Systems, Design and Technology (ISDT) is an initiative that has been established to respond industrial needs for integration of **"Knowledge Technology" (KT)** with multi- and inter-disciplinary applications. In particular the objective of ISDT is to incorporate multilateral engineering disciplines i.e. Composite-, Automotive-, Industrial- , Control- and Micro-Electronics Engineering, and derive knowledge for design and development of innovative products and services. In this context, the discourse of KT is established to address effective use of Knowledge Management, Semantic Technologies, Information Systems and Software Engineering towards evolution of adaptive and intelligent systems for industrial applications.

Naturally, the point of view on any kind of integration confronts several obstacles on managing of Human Resource (HR), technical resources and Information Technology (IT) infrastructures. There is a need to a think-tank and cooperative research on the conceptual, systematical, and technological aspects of integration as well as feasibility study on potential mediums of application. In this era, ISDT is established since 2007 based on indication of a gap between KT scientists and industrial practitioners.

Furthermore, integration should be critically seen, especially for in-house development of robust and adaptive solutions and further adaptation. Thus, reusability and adaptivity are the major challenges. In this context, the primary aspect is to investigate possibilities which enable us to integrate multi-domain intellectual capitals as organizational, technological or human capitals in distinctive research fields. Interdisciplinary research encompasses identifying the potential of each field and also the needs for improvement. Such effort facilitates the entire process for development and transfer of innovation among inter- or intra-organizational stakeholders. In ISDT, we are cooperating on detection and finding adaptive solutions for bridging industrial challenges based on integration of computerized and intelligent methods. In today's SMEs the common required technology should be established particularly for reusing of existing values at HR and exploiting of available tools and methods raising production and industrial values. To succeed in this competition and achieve sustainable marketing strategies, selected KT components should be taken into use, namely knowledge modeling, representation, discovery, and delivery, as well as semantic and decision-making algorithms and data analytics. The common attribute in this matter is "*Knowledge*" for reasoning (i.e. reasoning under uncertainty), and decision-making. Reasoning under uncertainty requires handling uncertain knowledge and discovering hidden relations between knowledge attributes to increase the usability of decision-making. In sum, KT concentrates on conceptual solutions that enable the usage and further development of knowledge based techniques and methods to support human behavior like decision-making, learning, planning or controlling. In our understanding, the entire goal of integration is to sustain efficient management of HR and increase on return of investment in SMEs.

In the latest ISDT meeting, we have successfully discussed the mentioned issues and agreed for constituting and concretizing the discourse of KT integration with special involvement of leading researchers and industrial experts whose contributions are presented in the book chapters. This book consists of three main chapters, namely:

- **Chapter 1: Applied Knowledge Management in Practice**
- **Chapter 2: Semantic Technologies for Industrial Management and Process Controlling**
- **Chapter 3: Knowledge Driven Approaches for Product Engineering**

In addition, each article presents a unique in-progress research with respect to the target goal of integration. All articles have been double blind reviewed with a domain-specific expert, and published as a book chapter.

I honestly hope that this edition of ISDT publication leads to improve our common understanding of KT integration and promotes further researches and cooperation in future.

Siegen, September 2012 Madjid Fathi

Associated Reviewers for Blind Review Process

Prof. F. Sassani	University of British Columbia, Canada
Prof. A.G. Hessami	Vega Systems, U.K.
Prof. M.H. Abd Shukor	University of Malaya, Malaysia
Prof. M. Saif	University of Windsor, Canada
Prof. U. Kelter	University of Siegen, Germany
Prof. C. Bratianu	Academy of Economic Studies, Romania
Prof. M. Abramovici	Ruhr-Universität Bochum, Germany
Prof. H. Garmestani	Georgia Institute of Technology, U.S.A
Prof. R. Brück	University of Siegen, Germany
Prof. A. Gábor	Corvinno Technology Transfer Center, Hungary
Prof. S. Nahavandi	Deakin University, Australia
Prof. D. Gerhard	Vienna University of Technology, Austria
Prof. U. Seidenberg	University of Siegen, Germany
Prof. R. Talebi-Daryani	Cologne University of Applied Sciences, Germany
Dr. S.T. Mol	University of Amsterdam, The Netherlands
Dr. R. Tafreshi	Texas A&M University at Qatar
Dr. M. Saadat	University of Birmingham, U.K.
Dr. F. Schulz	SAP AG, Germany
Dr. D. Atlan	Phenosystems SA, Belgium
Dr. E. Fersini	University of Milano-Bicocca, Italy
Dr. D. Nestler	Chemnitz University of Technology, Germany
Dr. K. Hahn	University of Siegen, Germany
Dr. M. Baniasadi	University of Strasbourg, France
Dr. G. Kismihók	Budapest University of Economic Sciences and Public Administration (BUESPA), Hungary
Dr. A. Ghazavizadeh	University of Strasbourg, France
Dr. H.M. Navazi	Sharif University of Technology, Iran
Dr. R. Montino	ELMOS Central IT Services GmbH, Germany
Dr. S. Berlik	University of Siegen, Germany
Dr. U. Fischer	Deutsche Post AG, Germany

Contents

Chapter 1
Applied Knowledge Management in Practice

Nonlinear Integrators of the Organizational Intellectual Capital

Constantin Bratianu

Academy of Economic Studies, Bucharest
Piata Romana 6, Sector 1
010374 Bucharest, Romania
cbratianu@yahoo.com

Abstract. The purpose of this paper is to present a new perspective in interpreting the structure of the organizational intellectual capital by introducing the concept of *integrators*. The organizational intellectual capital is a conceptual system reflecting the fields of knowledge, emotions and values developed within a generic organization. The canonical theory of the organizational intellectual capital considers that its fundamental components are: human capital, structural capital and relationship capital. However, this theory is working with a static model unable to reflect the dynamic processes within a generic organization. The intellectual capital is conceived as a potential resulting from the linear aggregation of the input contributions of all employees. Human capital, structural capital and relationship capital are considered as fundamental components, but they are not independent entities. They overlap both conceptually and operationally. This paper presents a dynamic model of the organizational intellectual capital that is based on the new concept of integrators, and the new structure composed of: knowledge, intelligence, and values. Thus, the potential intellectual capital is transformed into operational intellectual capital due to the work of organizational integrators, especially of those that are nonlinear: leadership, management and organizational culture.

Keywords: Knowledge, Intellectual capital, Nonlinear integrators, Leadership, Management, Organizational culture.

1 Introduction

In the framework of Economics, a generic organization contains both tangible and intangible assets. Tangible assets are physical objects like buildings, infrastructures, technologies and equipments, and monetary assets. Intangible assets are knowledge, intelligences, talents, brands, reputation, intellectual properties, organizational values, traditions, and symbols. In a generic way, intangible assets may be called *knowledge assets*, that are the building blocks of the *organizational intellectual capital* [1-3]. For the knowledge intensive organizations, the intangible assets become more and more important. This is shown by the high value of the ratio between the market value and the book value of the company [4, 5]. For companies like Microsoft, Apple or IBM, there is a significant difference between total market capitalization and net fixed assets. According to Schiuma [6, p.291], "The value that can be seen and measured through the tangible components represents only a small part of the overall value of a firm that is hidden under the surface of an organization's intangible and knowledgeable assets. Knowledge assets, particularly for knowledge-intensive firms, define most of the value of an organization". Knowledge assets can be considered as value drivers that support the organizational value creation mechanisms. Creation, acquisition, deployment, sharing, dissemination and transfer processes of knowledge assets means actually to manage the organizational intellectual capital [7-10].

Due to various definitions and theories developed so far, the concept of *intellectual capital* is a fuzzy concept. Although it is a powerful concept, it is difficult to figure out how to measure it, due to its intangible nature. However, there are some definitions that cover the most part of the semantic domain. For instance, Klein and Prusak consider that intellectual capital is: "Intellectual material that has been formalized, captured, and leveraged to produce a higher-valued asset" [8, p.67]. Edvinsson and Sullivan consider that intellectual capital is: "Knowledge that can be converted into value" [11]. We appreciate that Roos et al. formulated a definition that is flexible enough to incorporate the contribution of the human resources: "Intellectual capital can be defined as all nonmonetary and nonphysical resources that are fully or partly controlled by the organization and that contribute to the organization's value creation" [4, p.19]. It is important to remark the fact that the organizational control may be fully or partially done. In the linear industrial management the control over the human resources is done directly as a coercive force. It can be a full control, especially if human resources are used mostly for physical activities. In the nonlinear creative management, that is characteristic for most knowledge intensive organizations, the control is done through motivation. Knowledge creation and business innovation cannot be a result of any coercive force. That is why we must change their approach when switching from tangible assets to intangible assets and intellectual capital management. The purpose of this paper is to present a new view of the structure of the intellectual capital, a structure that is dynamic and based on the concept of integrators. After defining the organizational intellectual capital, in the next sections of this paper we shall present the canonical approach on the structure of the intellectual capital, and then we shall introduce the concept of integrators and we shall discuss their role in transforming the potential of the organizational intellectual capital into its operational form. The contribution of this paper consists in this dynamic and integrative approach of the organizational intellectual capital, and in making the difference between the power of linear and nonlinear integrators in a given organizational context.

2 The Static Models of the Organizational Intellectual Capital

The most known models of the intellectual capital are: Sveiby's Intangible Asset Monitor (IAM), Skandia's Intellectual Capital Navigator, and the Canonical Intellectual Capital Model. We shall present the main features of each of these models, and their importance. They are static because there is no time variable, and intellectual capital is considered as a potential of the organization, resulting from the linear aggregation of all its components.

2.1 *Sveiby's Intangible Asset Monitor (IAM)*

The IAM model has been developed by Karl-Eric Sveiby, one of the most prominent personalities in the field of intellectual capital and knowledge management [12]. Sveiby classifies the intangible assets into three categories: external structure, internal structure and individual competence. The external structure refers to the customers, to the suppliers and to other stakeholders that are considered relevant to a specific company. This model takes into consideration the suppliers and other relevant stakeholders. Depending on the type of the organization, the external structure will be different from one to another.

Internal structure refers to all the systems, databases, patents, models, processes and routines that support the organization's operations and employees. The informal internal networks and the organizational culture are also components of the internal structure. All of these components are created by the employees and represent the company's ownership. That means that management has full control over these assets, and any decision made for developing the internal structure is almost risk free. Although internal structure may appear less interesting, it actually plays a very important role in transforming the human intellectual capital potential into a high level operational intellectual capital.

External structure consists of relationships with customers and suppliers, brand names, trademarks and reputation, or image of the company. Some of these cannot be controlled completely by the organization since they depend both on the internal business environment and the external business environment. The value of these assets is primarily influenced by how well the company solves the customers' problems, and thus there is always a degree of uncertainty in the decision making process. Since companies are not closed systems, their interaction with the external business environment is necessary and valuable. Improving the external structure of the intellectual capital top management may improve the company's performance and may achieve a strategic advantage.

Individual competence refers to individual experience, knowledge, competence, skills and ideas [1, 9]. It is about human resources and their capacity to find solutions for their working problems. People are the only true agents in business since they create knowledge and they process this knowledge with their intelligences, based on their cultural values. Actually, it is almost impossible to conceive a company without people.

2.2 Skandia's Intellectual Capital Navigator

This model is a result of the continuous efforts in the field of intellectual capital made by Leif Edvinsson and his team at Skandia, a Swedish Financial Services Company. The intellectual capital is composed of human capital and structural capital. The structural capital at its turn is made by customer capital and organizational capital. The organizational capital is composed of innovation capital and process capital [3, 7]. *Skandia's Intellectual Capital Navigator* has five areas of focus: customer, human, process, renewal and development or financial, providing a holistic view of the organization. The presumption behind this model is that the intellectual capital of a company is the difference between its market value and its book value. As Edvinsson remarks, the Navigator provides a 3D compass for charting a course towards tomorrow as well as a map of yesterday. It is a versatile leadership tool for any company. Also, it can be used as a diagnosis tool. "What the Navigator helps us understand is that intellectual capital is not only a way of assessing intangible assets. A course of action rather than a store of knowledge. A flow" [3, p.84]. This model tries to get out of the static structure of all the pioneering models, but it does not have dynamic forces to achieve a dynamic structure.

2.3 Canonical Intellectual Capital Model

The basic assumption of this model is that organizational intellectual capital is composed of human capital, structural capital and relationship capital [2, 5, 7, 8]. An illustration of the structure of this model is given in fig.1.

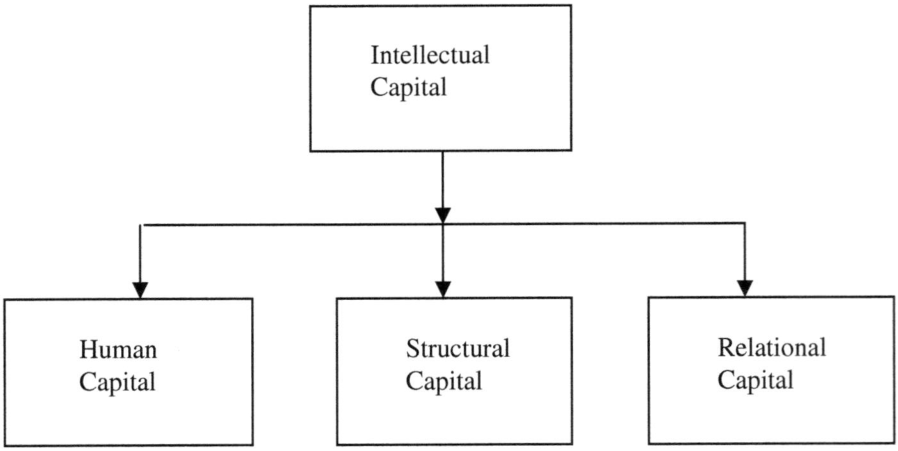

Fig. 1 Operational structure of the organizational IC

The human capital contains all the knowledge, intelligences and values employees may have together. It is a kind of summation of all individual contributions coming from employees. The most important contribution is knowledge. Both explicit knowledge and tacit knowledge are considered [13-15], and intelligences are considered in the perspective of the multiple intelligence model developed by Gardner [16]. Values are considered here in the cultural framework of a given society, and they are associated with the organizational behavior and corporate social responsibility [17-19]. Structural capital contains internal functional organization, all internal regulations, formal and informal communication channels, intellectual properties and organizational culture. It constitutes the intangible infrastructure of the organization and the functional support of the operational management. Relationship capital represents the result of all relations the organization has with suppliers, customers and other important stakeholders.

The canonical model or the standard paradigm of the intellectual capital is based on several assumptions that are summarized by Viedma [5] as follows: (1) the accounting view; (2) the strategy implementation view; (3) breakdown of intellectual capital; (4) cause-and-effect relationship; (5) relatively static approach to value-creation processes; (6) limitation of concept of intellectual capital; (7) attempts to treat intangible assets as if they were tangible. *The accounting view* is concerned with the practical possibility of measuring the value of the company intangible assets. However, that means to extend a linear and materialistic approach to a nonlinear and non-materialistic entity. The result could be a gross approximation with large errors of interpreting the resulting values. *The strategy implementation view* is related to the strategic management at the implementation level. However, the static nature of the canonical model cannot yield the best approaches for a successful implementation. It could be a mistake to consider the same approach for intangible like for tangibles. *The breaking down of intellectual capital* is actually a common denominator of all models based on a linear thinking. In linear thinking, a problem can be decomposed into smaller problems, and each smaller problem has to be solved for a solution. By aggregating these smaller problems' solutions one may get the solution of the initial problem. This approach works well only in a linear environment and for linear problems. However, intellectual capital is strongly nonlinear and breaking it down into human capital, structural capital and relationship capital is arbitrary and based on a wrong assumption that these entities are independent. *Cause-and-effect relationship* logic applied to this canonical structure cannot yield adequate results since the division between human capital, structural capital and relationship capital is arbitrary. This model reflects *a relatively static approach* to value-creation processes. According to Viedma [5, p.247], "The artificial categorization of intellectual capital lacked consideration of how firms actually deploy their resources through their organizational core activities. Because of this, the above-mentioned models fall short in explaining how firms effectively compete and how they recreate the sustainable competitive advantage that gives rise to value creation". There is an obvious *limitation of using the concept of intellectual capital.* The stress is put on the intellectual attribute, which cannot cover the whole spectrum of the intangible aspects. Usually, *intellectual* means knowledge and rational decision making. When we discuss about intangible assets we mean also values, symbols, talent, motivation, employee commitment and organizational culture. That means that it is necessary to extend the semantic interpretation of the concept such that we can include all of these intangible assets. This above limitation can be seen also from the *tendency of using the same models and methodologies to manage intangibles and produce external reports*. This tendency is supported by the general approach of transferring models used for tangibles to the intangibles, although they have a different nature. It is a problem coming from education and our way of thinking in a Newtonian vision that is characterized by determinism, linearity and materiality.

2.4 Toward the Dynamic Dimension of the Intellectual Capital

In a thoughtful analysis of the intellectual capital models, Kianto [20] finds that most of these models are based on the static view of the intellectual capital like a stock. "When intellectual capital is viewed as a stock, it is assumed that it is something that can be relatively easily identified, located, moved and traded, much like some sort of a package, albeit an intangible one. When intellectual capital is framed in this manner, it is typically understood to be a possession or owned properly of the organization, manifested for example as patents, trademarks, business applications and brands" [20, p.344]. This perspective is linked also with the knowledge metaphor of "stuff", as demonstrated by Andriessen [21, 22]. In a static perspective, time does not exist. Intellectual capital is considered as a potential for company and as an operational stock of resources, which is far from the operational management. The static intellectual capital approach is suitable for producing snapshots in time of the current intangible assets of the company.

The dynamic perspective of the intellectual capital is based on the idea that knowledge can be considered as *flow*, or *stock and flows*. According to Nissen, "To the extent that organizational knowledge does not exist in the form needed for application or at the place and time required to enable work performance, then it must flow from how it exists and where it is located to how and where it is needed. This is the concept of knowledge flows [23]. Also, intangible assets may have interactions and can be transformed through production process such that they are in a continuous dynamics. As Kianto explains [20], dynamism can be injected into the static resource architectures by examining interrations and interdependencies between various resources. However, this interpretation is incomplete since flow implies a field of pressure which is not evident in this situation. Moreover, the dynamics of intellectual capital comes not only from the resources transformation, but also from the capacity of the organization to transform intellectual capital itself from the state of a potential into the state of an operational capital. According to Joshi and Ubha [24, p.576], "Managing a knowledge organization necessitates a focus on the critical issues of organizational adaption, survival, and competence in the face of ever-increasing, discontinuous environmental change. The profitability of a knowledge firm depends on its ability to leverage the learnability of its professionals and to enhance the reusability of their knowledge and expertise". This goal can be attained only when intellectual capital is understood as a dynamic system underpinned by a leadership process.

3 The Theory of Nonlinear Integrators

The concept of *integrator* for a generic organizational system and its intellectual capital has been first defined and then developed in a series of published papers by Bratianu [25-27]. In the next sections we present the definition, interpretation and use of this new concept in the intellectual capital theory.

3.1 Definition

By definition, *an integrator is a powerful field of forces capable of combining two or more elements into a new entity, based on interdependence and synergy. These elements may have a physical or virtual nature, and they must possess the capacity of interacting in a controlled way.*

The *interdependence* property is necessary for combining all elements into a system. The *synergy* property makes it possible to generate an extra energy or power from the working system. In a knowledge system, synergy may be used with some extended semantics to reflect the nonlinear increase in understanding as a result of knowledge processing. Also, it may be introduced the concept of *synowledge* to reflect the same final result. It makes the difference between a linear system and a nonlinear one. In the case of the linear system the output is obtained through a summation process of the individual outputs. For instance, a mechanical system made of rigid frames works in a linear regime, while an electrical system works in a strongly nonlinear regime. In the first example there is only interdependence and no synergy. In the second example there is both interdependence and synergy. In organizational behavior, we can talk about linear work in groups and nonlinear work in teams. In the first case, sharing the same goal but not the same responsibility leads to interdependence and a linear behavior. In the second case, sharing the same goal and the same responsibility leads to interdependence and synergy, which means a nonlinear behavior. However, synergy is not a guaranteed effect. It must be obtained by an intelligent team management.

An integrator is a powerful field of forces. In order to understand the meaning of such a field of forces we may think of the gravity field. We live in this field and apparently we cannot feel it. However, the moment we jump we feel the powerful force of attraction toward the earth surface. If we extend the semantic of the physical field to a non-physical field we may identify many fields of forces within an organization. These are organizational fields of forces. The main fields of forces in a given company are: processes, work legislation, cognitive knowledge, emotional knowledge and organizational values. There are integrators that act within a specific field of forces, and integrators that act across several fields of forces. There are linear integrators and nonlinear integrators. The *linear integrators* are specific to mechanical and sequential processes characterized by a summative aggregation. The *nonlinear integrators* are specific to complex informational, social, cultural, organizational and psychological processes. These differences come from the difference between linearity and nonlinearity concepts [28-30].

Integrators manifest themselves in different ways within organizations having different consequences. On one side, different integrators may have different actions within the same company, and on the other side the same integrator can manifest in different ways in different companies. To be able to evaluate these different manifestations we introduce the concept of *operational intensity*. By definition, *operational intensity* represents the capacity of an integrator to act in a given organizational context. This means that the operational power of an integrator depends both on the quality of the integrator as well as on the organizational context in which it acts. Integrators act on the *potential* intellectual capital of any organization and transform it into the *operational* intellectual capital. Thus, an organization with powerful integrators will have a higher level of operational intellectual capital than any other organization with less powerful integrators. For instance, a university with excellent professors, researchers and students may have modest outputs if the university management is weak or noncompetitive. In time, excellent professors will leave for other universities, and excellent students will go to attend better universities. In this example the university management is not a powerful organizational integrator.

3.2 Technology and Processes

In any organization we can make a distinction between the production process and the managerial process which are in fact interconnected. The production process consists of a certain technology and the work processes associated with it. Together they constitute an integrator, in the sense of generating a force field capable of concentrating and structuring the necessary work force for the production process. When the production process is equipment structured, being a spatial distributed process, the operational intensity of the integrator is relatively small. When the production process is continuous and the workers are aggregated on technological lines, the operational intensity of the integrator sensibly grows. Because the workers participate on mechanical sequences, well determined and quantified, according to predetermined programs and not with their entire knowledge capacity, the integrator is linear. However, the integrator acts only upon knowledge because the solutions are predetermined and learned by the workers. They do not solve problems and do not make decisions that imply their value systems. When the technological process is not conceived as a string of linear sequences but rather as a complex of relationships and interactive sequences that intertwined and interrelated, having as support a good IT system, the technological process becomes nonlinear. Its integration capacity of individual knowledge and intelligence grows, thus generating interdependency and synergy. In the case of creative technological processes, like different artistic productions the integrator acts both on knowledge and intelligence, as well as cultural values. In these situations, the integration process is much more powerful and the synergy effect is much more obvious.

3.3 Management

Management is the process through which an organization achieves its objectives efficiently and effectively. This means that management is by nature an *integrator*, more powerful than technology and associated processes. Unlike technology that is a very specific and somewhat stiff integrator, management is a generic and flexible integrator. At the same time, the integration capacity of management consists in the manager abilities to create intelligent solutions to the problems they face, avoiding routine and standardization. Management is both science and art, combining power of principles with the value of the experience. Achieving performance in management requires going beyond the mechanical model of thinking that comes from the industrial era. In this new context, management works on individual knowledge by turning it into organizational knowledge and on individual intelligence by transforming it into organizational intelligence. Technology as an integrator is able to act mainly on explicit knowledge, which is coded. The management integrator acts both on explicit and tacit knowledge [2, 13, 15]. The essence of management is the decision making process. That means that management as an integrator acts upon one's individual value system, generating organizational values, as a part of the organizational culture [17-19]. Management is strongly linked to the production process. If there is an industrial type of a linear production process, then management will act as a linear integrator. On the other hand, specific companies of the new economies, where technology is a highly nonlinear integrator, management should adapt to the new context and should become a creative process. The final result in this last case involves a strong synergy and it is an intellectual capital with a high operational level.

3.4 Leadership

There is a whole debate in the literature on the semantic distinction between management and leadership, managers and leaders [31, 32]. In essence, management ensures the objectives undertaken by an organization in terms of efficiency, effectiveness, and control. In other words, management is an operational process that ensures the organization's status quo. Managers are those who have been invested with institutional authority to perform the functions of planning, organizing, leading and control. Although management is not a standardized process, it requires compliance with the requirements stated above. Unlike management, leadership is the process by which the organization is proposing a series of changes, either for the need to adapt to today's dynamic external business environment, to achieve a competitive advantage, either as a result of the business vision. In this perspective, leadership must define the vision for change, set direction for change and motivate people to achieve the objectives of change. "Leadership is thus the process by which a person can influence a group of others in order to achieve a common goal" [32, p.3]. Leadership is thus a process based on the reasoning and emotional power to influence people to make a change, to achieve a particular purpose. Change is not an end in itself, but only the process that can implement a specific strategy and achieve a particular purpose. Leaders may have managerial functions or not, and they hold the art and power to influence those around them. Leaders have vision and a series of personal attributes that in a given managerial context can trigger and implement a process of organizational change. Leaders have the ability to resonate with emotional states of the people around them and with their requirements. Leadership is much more powerful than the new type of management, acting on individual intelligence and on the individual values of the employees. While management supports the process of integrating individual knowledge and intelligence, leadership puts particular emphasis on the integration of individual intelligence and values. That makes leadership a very powerful integrator, with a greater impact on generating intellectual capital. Companies led by leaders succeed in generating intellectual capital much better than companies led by managers [33, 34].

3.5 Organizational Culture

Peters and Waterman have been among the first authors who emphasized the importance of the organizational culture, in achieving excellence. As concluded in their study about the best managed companies, "excellent companies are marked by very strong cultures, so strong that you adapt or leave. There are no half measures for most people in these companies" [35, p.77]. A strong culture is a fundamental value system, traditions, symbols, rituals and informal rules that indicate how people should behave in most of the time. Organizational culture is a very powerful nonlinear integrator since it acts mainly on individual intelligence and the fundamental values of each individual, thus creating a spirit of excellence. However, organizational culture can cause side effects if fundamental values are based on fear and punishment, and there is a disagreement between the company interests and the values of the individuals.

4 The New Dynamic and Integrative Structure of the Intellectual Capital

There are two main interventions into the structure of the canonical intellectual capital. The first one consists in considering the standard components of the intellectual capital as intermediate entities and not as the final and basic

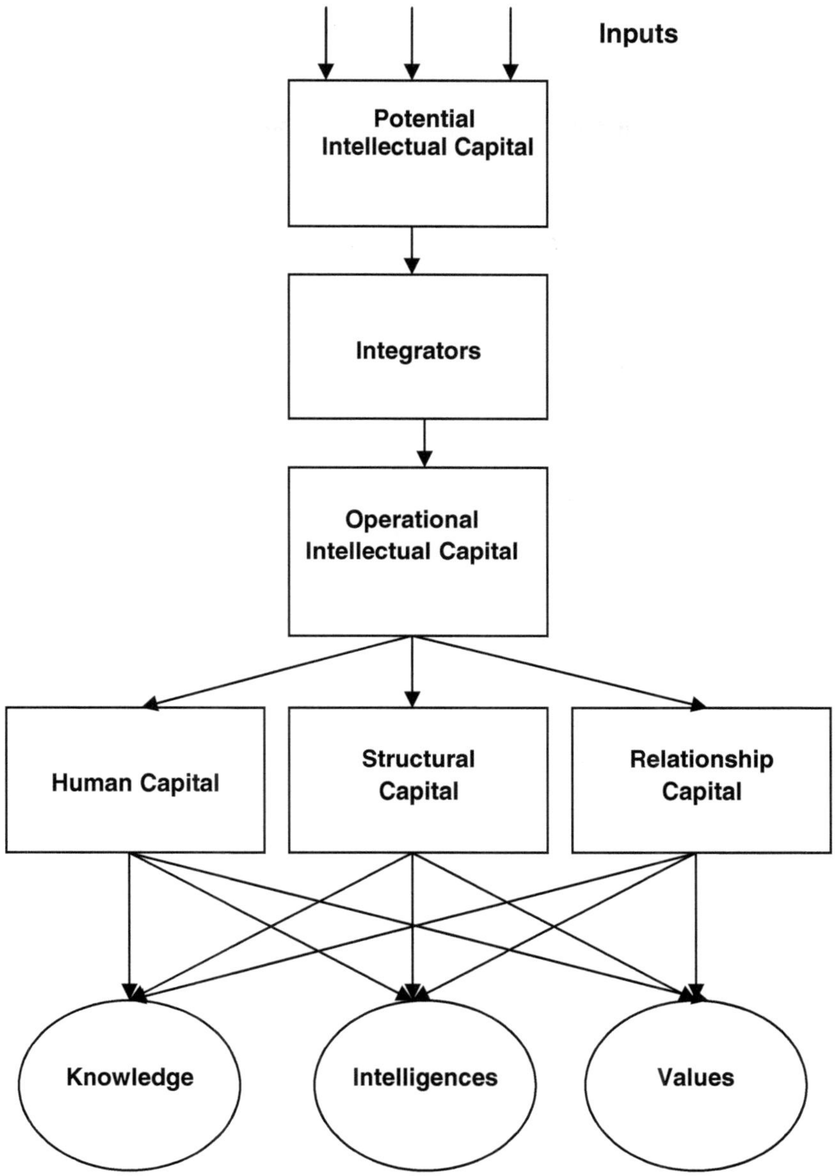

Fig. 2 The new structure of the dynamic integrative intellectual capital

entities because they are not independent entities. Each of them is composed of knowledge, intelligence and values, that represent independent entities. The second intervention is to introduce the integrators as a dynamic component of the whole structure, such that we can understand the transformation of potential intellectual capital into the operational one. The final result of these changes is illustrated in fig.2.

5 Conclusions

The purpose of this paper is to introduce in the framework of intellectual capital the concept of integrators. By definition, an integrator is a powerful field of forces capable of combining two or more elements into a new entity, based on interdependence and synergy. The role of integrators is to transform the potential of the organizational intellectual capital into an operational intellectual capital. The more powerful the integrator is, the higher the level of the operational intellectual capital that can be achieved. By introducing this new concept we also get a dynamic structure of the intellectual capital, that is much more adequate to the strategic framework than the canonical static structure. Human capital, structural capital and relationship capital are not independent entities and thus, the canonical structure should be changed to avoid measuring twice same components. In this perspective, we considered that knowledge, intelligence and values could be considered the new basic entities of the intellectual capital. In conclusion, employees enter into any organization with their individual knowledge, intelligence, and values, and the nonlinear integrators transforms these individual inputs of the intellectual capital potential into the organizational knowledge, organizational intelligence and organizational values.

References

1. Al-Ali, N.: Comprehensive intellectual capital management. Step-by-step. John Wiley & Sons, Hoboken (2002)
2. Andriessen, D.: Making sense of intellectual capital. Designing a method for the valuation of intangibles. Elsevier, Amsterdam (2004)
3. Edvinsson, L.: What you need to know to navigate the knowledge economy. Prentice Hall, London (2002)
4. Roos, G., Pike, S., Fernstrom, L.: Managing intellectual capital in practice. Elsevier, Amsterdam (2005)
5. Viedma Marti, J.M.: In serach of an intellectual capital comprehensive theory. Electronic Journal of Knowledge Management 5(2), 245–256 (2007)
6. Schiuma, G.: The managerial foundations of knowledge assets dynamics. Knowledge Management Research & Practice 7, 290–299 (2009)
7. Edvinsson, L., Malone, M.: Intellectual capital: realizing your company's true value by finding its hidden brainpower. Harper Business, New York
8. Stewart, T.: Intellectual capital. The new wealth of organizations. Nicholas Brealey Publishing House, London (1999)
9. Sveiby, K.E.: The new organizational wealth. Managing & measuring knowledge-based assets. Berret-Koehler Publishers, San Francisco
10. Sullivan, P.H.: Profiting from intellectual capital. Extracting value from innovation. John Wiley & Sons, New York
11. Sveiby, K.E.: Intellectual capital and knowledge management, `http://sveiby.com/articles/IntellectualCapital/html` (retrieved June 2, 2010)
12. Sveiby, K.E.: The new organizational wealth. Managing & measuring knowledge-based assets. Berret-Koehler Publisher, San Francisco

13. Davenport, T., Prusak, L.: Working knowledge. Harvard Business School, Boston (2000)
14. Polanyi, M.: The tacit dimension. Peter Smith, Gloucester (1983)
15. Becerra-Fernandez, I., Sabherwal, R.: Knowledge management. Systems and processes. M.E.Sharpe, New York (2010)
16. Gardner, H.: Multiple intelligences. New horizons. Basic Books, New York (2006)
17. Schein, E.H.: Organizational culture and leadership, 3rd edn. Jossey-Bass, San Francisco (2004)
18. Weick, K.E.: Making sense of the organization. Blacwell Publishing, Oxford (2001)
19. Zohar, D., Marshall, I.: Spiritual capital. Wealth we can live by. Berret-Koehler Publisher, San Francisco (2004)
20. Kianto, A.: What do we really mean by the dynamic dimension of intellectual capital? International Journal of Learning and Intellectual Capital 4(4), 342–356 (2007)
21. Andriessen, D.: On the metaphorical nature of intellectual capital: a textual analysis. Journal of Intellectual Capital 7(1), 93–110 (2006)
22. Andriessen, D.: Knowledge as love. How metaphors direct our efforts to manage knowledge in organizations. Knowledge Management Research & Practice 6, 5–12 (2008)
23. Nissen, M.E.: Harnessing knowledge dynamics. Principled organizational knowing & learning. IRM Press, London (2006)
24. Joshi, M., Ubha, D.S.: Intellectual capital disclosure: the search for a new paradigm in financial reporting by the knowledge sector of Indian economy. Electronic Journal of Knowledge Management 7(5), 575–582 (2009)
25. Bratianu, C.: An integrative perspective on the organizational intellectual capital. Review of Management and Economic Engineering 6(5), 107–113 (2007)
26. Bratianu, C.: A dynamic structure of the organizational intellectual capital. In: Naaranoja, M. (ed.) Knowledge Management in Organizations, Vaasan Yliopisto, Vaasa, pp. 233–243 (2008)
27. Bratianu, C.: A new perspective of the intellectual capital dynamics in organizations. In: Vallejo-Alonso, B., Rodriguez-Castellanos, A., Arregui-Ayastuy, G. (eds.) Identifying, Measuring, and Valuing Knowledge-based Intangible Assets: a New Perspectives, pp. 1–21 (2011)
28. Bratianu, C.: The frontier of linearity in the intellectual capital metaphor. Electronic Journal of Knowledge Management 7(4), 415–424 (2009)
29. Bratianu, C.: Thinking patterns and knowledge dynamics. In: Proceedings of the 8th European Conference on Knowledge Management, Barcelona, September 6-7, pp. 152–157 (2007)
30. Bratianu, C., Vasilache, S.: A factorial analysis of the managerial linear thinking model. International Journal of Innovation and Learning 8(4), 393–407 (2010)
31. Adair, J.: The inspirational leader. How to motivate, encourage and achieve success. Kogan Page, London (2003)
32. Daft, R.L.: The leadership experience, 4th edn. Thomson South-Western, New York (2008)
33. Collins, J., Porras, J.: Built to last. Successful habits of visionary companies. Harper Business, New York (2002)
34. Welch, J., Welch, S.: Winning. Harper Business, New York (2005)
35. Peters, T., Waterman Jr., R.H.: In search of excellence. Lessons from America's best-run companies. Harper Colins Business, London (1982)

Semantic Technologies in Business Process Management

András Gábor[1] and Zoltán Szabó[2]

[1] Corvinno Technology Transfer Center Ltd,
Közraktár utca 12/a, 1093, Budapest, Hungary
agabor@corvinno.hu
[2] Corvinus University of Budapest, Fővám tér 13.,
1093 Budapest, Hungary
szabo@informatika.uni-corvinus.hu

Abstract. Business process management (BPM) is a key managerial approach to improve competitiveness and organizational performance. The paper addresses the knowledge management aspects of BPM. Starting from several outcomes of process modeling, job role and position related competences, IT requirement specification, organizational learning and knowledge transfer, the common ground of knowledge management, semantic technologies and business process modeling will be discussed. The questions to be answered: how can process models be utilized and integrated with Knowledge Management Systems (knowledge representation, semantic technologies)? How can the knowledge transfer activities be supported, that are central issues in BPM initiatives? How can Knowledge Management Systems (KMS) underpin the long term sustainability and institutionalization of BPM based innovations? Is there a role of process-oriented KMS in BPM-related system development projects? The paper will give an overview of the "big picture" and also outlines a few applications as proof of concept. The final conclusion leads to a high level model and approach that can be used to harmonize BPM initiatives with KM concerns.

Keywords: Business Process Management, process modeling, knowledge representation, semantic technologies.

1 Introduction

Organizations today try to survive in turbulent economic environments. The typical modern enterprise grows in complexity and scope that makes process orientation and process thinking more and more important. Global business trends of increasing competition, globalization, deregulation and operational challenges like shorter product life-cycles, personalized customer needs enforce firms to find new ways of operation, competition and co-operation [1]. Business Process Reengineering (BPR) was one of the first options to answer these challenges.

Reengineering as a radical change program and Business Process Management (BPM) as a set of continuous efforts towards process excellence are knowledge-intensive activities. Knowledge related issues can be barriers of both implementation and long-term sustainability of process management. On the other hand, knowledge management methods and systems can be major facilitators of BPM initiatives.

BPM provides several benefits: a strategy-driven, transparent, traceable, flexible and responsive organization can be designed, based on the improved alignment between business and IT. BPM initiatives have traditionally three focal points: cost, quality and time – the magic triangle of BPR.

The semantic technologies have been developed very fast in the last decade, partly due to the overall success and coverage of the Internet, partly because of the demand for added value IT services. The e-commerce, the e-government also brought the users' community closer with the technology and while technology is always smart, the usability already depends on the right positioning of the business process, the allocation of tasks, responsibilities, etc. This way, the advanced technology and advanced business process modeling go hand-in-hand.

In this paper, the authors try to point out where and what the connections are between process modeling and semantic technologies. Later, several examples will be given. The examples are cases from different projects, with one or two particular aspects in the focus. By putting the particular cases together we get a wider picture.

2 Reengineering and Knowledge Management

In the next sessions, the basic concepts of BPM and its knowledge related aspects will be discussed and problems that have knowledge management (KM) origins are addressed. Innovative approaches that integrate KM and BPM are discussed in the last subsection.

2.1 Characteristics of Reengineering and the Knowledge Related Issues

Since the seminal work of Michael Hammer [2], Business Process Reengineering or Redesign (BPR) has become one of the most popular and successful business movements. Organizations were strongly criticized for inefficient and ineffective business structure and outdated processes. Firms usually left the existing processes intact and utilized computers simply to speed them up, without addressing their fundamental performance deficiencies. It became widely accepted, that job designs, workflows, control mechanisms and organizational structures have come from a different competitive era and that unarticulated operating rules must be redesigned. BPR can be defined as the critical analysis and radical redesign of existing business processes to achieve breakthrough improvements in performance measures. [3] In the last two decades, the know-how of BPR was developed in huge steps. It was recognized in the early years that redesign must penetrate the company's core organizational elements: roles and responsibilities, measurements and incentives, organizational structure, information technology, shared values and skills [4]. Reengineering is a major change program, it is rather a strategic change project and it requires strong leadership from the top management. The change management process is a critical part of this kind of efforts. The pace of the deployment of the new, innovative solutions is very much depending on the level of the applied process management. Process management means in this context the exhaustive description of the processes and the definition of the underlying and interrelating connections. Recently, the Business Process Management as a complex management toolset integrates BPR concepts, quality management approaches, change and project management methods and, as a critical component, IT related tools and methods. There are several BPM methodologies [5] available in the market, a generalized life-cycle has several phases or sub-systems:

- Business process strategy
- Process documentation
- Process analysis and design
- Implementation and change management
- Process operation
- Process controlling/monitoring

IT and BPM have a recursive relationship. IT capabilities should support business processes, and business processes should be in terms of the capabilities IT can provide. The technology centered development strategy moves toward the knowledge intensive solutions. BPM is tightly linked with Service Oriented Architecture (SOA); their synergistic relationship enables extensive integration of the new applications with traditional ones.

According to the traditional approach, BPM technologies can be classified into two main groups: business process description/modeling and automation, including workflow, groupware, Business Activity Monitoring (BAM) applications. In addition to these categories, there are a wide range of technologies focusing on the knowledge aspect and supporting the two main tasks (modeling and automation): process mining, business rule mining, enterprise content management, semantic technologies.

BPM defines activities, the input-output information, also the organizational context, decision points, competencies and infrastructural preconditions, which are needed for the execution of the given activity.

The harmonization of competencies, responsibilities and also the information flow and organization, is a necessary but not sufficient precondition of the process formulating. The optimization of the process-structure can be performed along the above mentioned conditions and if it is successful, then the specification of the most up-to-date and innovative info-communication technologies can be set up, through which the savings will be realized on the expenditure side and other non-pecuniary objectives will be achieved

Efficient management of information and knowledge should move toward the efficient management of changes in the information and knowledge. However, existing approaches are inadequate for highly dynamic and volatile processes whose steps cannot be planned in advance and during which new, unanticipated "knowledge needs" frequently arise. The fast development in the field of economics, technology and informatics requires the ability of fast adaptation of individuals and organization, being either public administration institutions, corporations or even citizens themselves.

There are several knowledge management challenges in a BPM initiative:

- Knowledge discovery and codification is a key enabler of the initial phases – process documentation/modeling, analysis and (re-) design. In the implementation and operation phase dissemination of knowledge required by the new/redesigned process is a central issue too. New employees should be trained too, which is a key issue of sustainability.
- Knowledge transfer: process models and documentation have a central role in this aspect, integrating the business with technology domain (process to automation), cross-functional collaboration requires also efficient transfer. Traditional models, especially in the first aspect, have several limitations.
- Knowledge sharing: facilitates the cooperation and teamwork, especially in the design phase, and it provides a solid base for the operation and monitoring phases.
- Knowledge utilization: in the operation phase, knowledge that underpins process execution should be explicitly available and internalized by the participants.
- Knowledge renewal: as BPM is a cyclic approach, it must be able to integrate new knowledge and adapt the whole BPM system and the organizational process know-how accordingly.

Due to the complexity and the dynamic nature of modern multinational organizations, a focused management of process knowledge is necessary, that is sophisticated, professional and facilitates knowledge engineering. Semantic technologies as facilitators of transforming process models to executable applications are recently discussed in the literature [6]. In this article we will extend this to the operational phase of the BPM life-cycle. Borrowed from the ITIL methodologies, a kind of Knowledge Configuration Management might be applied: management of process and job related knowledge elements, enabling customized training programs and the efficient maintenance of knowledge. This approach can be implemented using semantic technologies.

2.2 The Business Process Modeling

Although BPM has a very large literature, still there are different views, concepts and misconceptions in this area. Normally, BPM has four basic pillars: modeling activities, which will form a set of steps in the process, defining the task(s) to be executed, responsibilities, reporting, the parallelism or strict order of execution, exception rules, resource allocation and connectivity. In association with the task (activity) / process description, all the competences can be / should be given on task level. By competences we will understand knowledge, skill and attitude that is necessary to sufficient execution. From the triad, only the knowledge element can be modeled sufficiently, the hands-on skills and psychological/sociological attitudes are difficult; the latter can not even be interpreted on task level.

The second pillar is the information flow, input/output information is linked to the activity, hence the information processing and the transfer can also be rightly modeled.

The third pillar is the organizational view. If the processes are mapped against the organization, the first level link is the role. The job roles are associated with process steps and activities. One or more job roles are assigned to a position, the positions fill up the organizational unit and the full organization.

Finally, the fourth pillar is the output and outcome of the BPM. The most general utilization of BPM is regulation. The regulatory actions may lead to a more advanced use of BPM: the optimization of processes, first of all, the radical simplification of the process connectivity, the simplification of decision, reporting routes and cutting through the useless iterations. Since the information flow and information processing is modeled in a very detailed level, the information systems requirement specification is naturally given. There are two more outcomes of the BPM which are hardly investigated: how to use BPM for intra- and inter-organizational knowledge transfer and how to make use of matching the activity related and job role related competences. These two definitely lead to knowledge management – looking for semantic solutions. As an interesting outcome, the semantic technology based approaches also fertilized the information systems requirement specification.

3 Knowledge Engineering and BPM

3.1 Semantic Business Process Management

BPR and BPM are traditionally tools that transform business requirements to system specifications. BPM as a continuous effort toward process excellence has the lead role in the integration of Business and IT architecture by systematically aligning organization structure, process and technology. This alignment is biased by the knowledge related issues.

In a typical project, the conceptual model facilitates the exploration, documentation and validation of requirements. This modeling approach, due to its implicit semantics, has limitations in the automation of further processing (e.g. generation of applications). BPM has the potential to transform the traditional solution development lifecycle.

The ultimate purpose of BPM is the description, automation, monitoring and improvement as a part of a cycle of continuous innovation.

A recent promising tendency in application development is business process design based software development. The main challenge in BPM is the continuous and seamless translation between the business requirements view and the IT systems and resources. Semantic Business Process Management (SBPM) is a new approach that can increase the level of automation in the translation between these two domains. According to the core paradigm of SPBM, the two levels can be represented using ontology languages and automated translation. [7]. As Kramler and Murzek assert [8], ontologies provide the semantics and they describe both the modeling language constructs and the model instances, facilitating also the automatic creation of workflow processes.

Due to the organic development of business processes and IT solutions, there is a significant margin between them. If technically executable processes and workflows were derived from business process structures and a standardized repository of business and IT logic was used, the gap between the business and the IT would be closed by automatic generation of workflow systems – based on BPM defined ontologies.

The extended use of BPEL (Business Process Execution Language) as modeling language, the BPMN (Business Process Modeling Notation) as modeling notion brought closer the business line and the IT staff. The enterprise architecture and, even more, the enterprise architect emphasize the importance of understanding both sides, the application of the seamless modeling methodology. The "missing link" is the ontology: this knowledge representation form helps the architect to map business processes against the information processing technology. The models, represented in the modeling tool, are exported and annotated into ontology and the ontology already gives ample munitions to behave as input to the workflow modeling. This is how the loop is closed.

There are big expectations and of course the methods are not sensitive enough to cope with every kind of business processes, also standardization, interoperability, or ASP solutions may mean constraints.

3.2 *Knowledge Transfer in BPM*

Knowledge management and BPM do not necessarily meet. It is a matter of definition of knowledge management, however, it is widely accepted that knowledge management has a human resource management view, an artificial intelligence view and an organizational, systems development view. From organizational point of view, the most important issue is the proper management of the intellectual capital, saving the intellectual asset of an organization. As we emphasized in the previous section, the process modeling can not be complete without the exact definition of the activities related competencies – the knowledge management option is there. The articulation of the organizational knowledge, the representation of the related knowledge makes the knowledge transfer relatively easy. The maintenance of the articulated and represented knowledge is still a delicate issue; it can be managed only with very careful manual work.

3.3 *Competence Matching*

Competences listed on the activity side and those being on the job role side can be and should be matched. Matching is a must, because the differences show how adequate the job role definitions are (right people on the right places doing the right thing…). The discrepancies are direct indications of in-house or other, formal, informal and non-formal training needs. The job role requirements can hardly be understood or interpreted without the process context and this context is provided by BPM. (From an HRM point of view, the difference is that recruitment/selection is done to fill an open position. As we saw earlier, the position might contain more than one job role. Therefore, the competence requirement rather characterizes the position and the person who will be in the position than the role. Therefore, it is very important to distinguish between role and position.)

It is also important to emphasize that competence as a phenomenon is rather used in the sense of a meta concept. The instance of the competence is the piece of concrete knowledge – a procedure, a rule, etc. The abstraction enables the users to compare two structures, two ontologies – and it is already a matter of properly chosen algorithm [9].

4 Proof of Evidence

In this section, several examples will be outlined shortly as a proof of evidence. We provide an integrated concept, following a model-building approach by combining the results and experiences of our previous projects (see Figure 1.). Detailed documentation of evidences and results of these R&D projects are available. Using the outcomes of the individual projects, a new conceptual model has emerged. During the last decade, several applications were developed by our team – participating in several national and EU projects. The projects are different and they depend very much on the ideas and the underlying objectives of the funding agencies, but on the long term it was possible to follow a strategic direction aimed to build up the "big picture" piece by piece. This "lego" type research strategy resulted in the cases to be introduced shortly. The summary follows the history of the last decade.

Having dealt with systems development earlier, our group started to deal with intelligent systems, applied artificial intelligence and expert systems from the mid 80s. Later, knowledge management provided a good intellectual-theoretical umbrella to connect intelligent systems with business oriented approach. The expert system development resulted in the use of a rule based system as the everyday practice, and due to the success and convincing power of CommonKADS, the **ontology** became the primary knowledge representation tool. [10, 11]

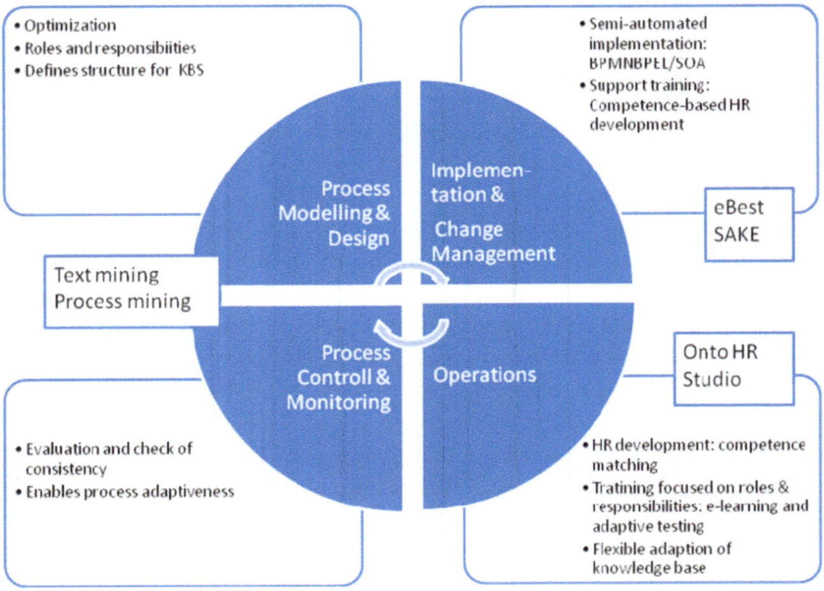

Fig. 1 Conceptual model

One of the application areas was an e-learning platform – **STUDIO**. The main idea was to represent the basic concept hierarchy of a given domain in ontology and structure around the learning content and to create an engine that first tests where the user's knowledge gaps are and then accordingly customizes the learning material. The main benefits of this approach were twofold: it directly helps the learner to learn according to his/her learning style, pace, in an informal or non-formal learning style. Indirectly, the system is domain specific; in the corporate environment the construction of the system is equal to knowledge articulation, knowledge elicitation and knowledge transfer. The approach, the platform, became widely used in training and education, the statistics based evaluation gives a very detailed and fine picture about the learners' activity, their learning habits and the relevance of the learning material. Up to now, there were no corporate knowledge management applications. The truth is that the maintenance of the ontology is still a problem and can be done only manually and very carefully [12, 13].

Under the **Structural and Cohesion Fund** framework a large, nationwide project was launched in 2004-2007 in Hungary. [14] The objective of the project was to create the reference process model of the HEI with the contribution of 13 universities (representing 42% of the total number of the students). The reference model serves two purposes: the first one is the starting point for the individual institutions to develop their own localized and customized solution, the second one is a common platform of understanding between the institutions and the ministry to control the feasibility and effect of the planned measures. The model contains over 400 processes grouped into 14 main categories. It covers every aspect of university management from strategic planning to the detailed bookkeeping action, from student administration to facility management. Apart from the professional experiences, the best and a little bit surprising experience of the project was the involvement of over 200 academics into process modeling. They understood the logic of an administrative environment, which was quite different from their strict academic interests. The common platform provided a solid base to the further developments. Later, high level executive information systems, decision support systems were introduced using data warehouse technology and several levels and solutions of business intelligence. The concept of using the reference process model fits to the most advanced enterprise architecture principles (e.g. TOGAF 9) and in 5 cases served directly as information systems requirement specification. Debrecen University used it for SAP implementation.

A part of the reference process model addresses the student mobility questions – mainly aroused from the transition to the Bologna system. Another project, also under the **Structural and Cohesion Fund** framework in the cooperation of 12 universities, tried to find a solution. The problem the consortium tried to solve was how to homogenize the input competencies on master level if the students may come with any kind of bachelor degree. As a test laboratory, the business informatics master was chosen. Assumed the biggest mobility, the incoming students should have had very different knowledge levels, while the minimum requirement on the input side was known. The solution developed to handle the case was the discovery of individual knowledge gaps and giving an individually customized learning material to the learner. The iterative use of the system resulted in a sufficient and homogenous knowledge level. [15]

Lessons learned from the project were twofold: first, the use of the competence based learning outcome is a very good basis for designing academic processes and second, in a conservative environment like academia, it still sounds frivolous and more refused than accepted. Vas, Kő, Kismihók [16, 17, 18] continued research on this line. The main focus had been given to the competence matching. First the **SAKE** project gave a unique opportunity to model output and outcome competencies in ontology and to embed it into an agile knowledge intensive groupware and decision support system. The most delicate part of the research project was the competence matching – on one hand the educational sector's output (supply) and on the other hand the required competencies (demand) by the labor market. The test scenario we tried to follow was offering and looking for systems analysts. There were some interesting results. A well selected, designed and customized intelligent groupware system can manage very well if there is an output from the competence matching. In combination with other techniques the time horizon gaps can be bridged over: the gap is between the demand, which is in the present, and the output, which is in 3 or 5 years. The SAKE system could provide a good environment to integrate other forecasting methods with economic growth assumptions, forecast of investment climate, etc. Concerning the competence matching, the most difficult part was (and still is): how to translate the job seekers' communications (ads) into job role competences.

The connection between the business process modeling and software development became an important and hot topic in the mid 90s. The integrated information systems already became accepted as minimum requirements in the business community, the 3A layer architecture, but most of all, the introduction of the SOA architecture opened a new horizon. Instead of showing to each other, the business side and IT jointly looked for the opportunity of the "wall-to-wall" design, starting the design procedure from high level, what the business line can understand (and they also understand business logic), almost to the automatic code generation. BPEL, BPMN became very popular. Ternai [19] conducted several experiments with ARIS, later with ADONIS to connect process model with ERP systems. The trial through Websphere was supported by IBM, but also Microsoft and SAP products were tested from the aspect of connectivity. The results of these tests proved the strength of the connectivity; it verified and validated the priority of the BPM modeling.

The success of the first attempts of the wall-to-wall modeling lead us to make a step forward to the Semantic Business Process Modeling. Mapping the processes seems to be logical; the main challenge in BPM is the *continuous* translation between the business requirements view and the IT systems and resources. Semantic Business Process Management (SBPM) is to represent the two levels using ontology languages and to employ automated translation. [7] Business processes have to perform well within the dynamic organizational environments. Conceptual modeling captures the semantics of an application through the use of a formal notation, but the descriptions are intended to be used by humans and not machines. The semantics contained in these models are especially implicit and cannot be processed. With the semantic schema the creation and the use of the conceptual models can be improved, furthermore, the implicit semantics having been contained in the models can be – partly – articulated and used for further processing. Ternai and Török [19] developed, in the **eBest** project [21], a solution that starts from the high level process model and the workflow is automatically generated through an ontology interpretation. The real benefit of the solution is not the first application but the fact that it can be "cashed" after the first modification. It is important to notice that workflow management and business process management are very often mixed up, especially by IT people. In the mentioned approach, the end result is a work flow system, but since it is generated from a business process model, there is no contradiction at all. [22]

The comparison of output and outcome competences having been earned in the education with the job role competences stayed longer in the focus of the research interests. In the **OntoHR** project [23], the problem was analyzed and tried to be solved from a Human Resource Management point of view. First, from the main three activities of a typical HR consulting firm: recruitment, selection and executive search; the third was taken out from the scope. After thorough analysis, pre-selection has been proved to be the proper scope. At this point, the comparison between competence based job role requirements and the individual competencies have become a crucial point. The job seekers' competencies are defined by the knowledge gap discovery, the solution that has been introduced earlier, but is now marketed under the brand name of STUDIO. The comparison between the demand

and supply is paired with the investigation of the mental ability of the job seekers. It gives more clues to conclude with the pre-selection.

5 Conclusions

There are many interesting opportunities for further development; the most integrative direction is to connect job role modeling and pre-selection with the process modeling. As it was stated earlier, in process modeling the activities are associated with the description of flow of information and the organizational view: what activity is done by which role, position or organizational unit. It is assumed that the link between the activities (process steps) and the job role is the competence. In our understanding, the competence is a composition of knowledge, skills and attitude. In the semantic approach, the only thing we can handle operationally is the piece of knowledge which is necessary to complete the given process step. Hence, the competences (part of them) are connected to the job role competences. This way, the HR pre-selection, selection work is embedded into an organizational and, what is even more important, process context. We have to differentiate between job role competence and competences associated with the position, which gives an interesting organizational behavioral insight as well.

References

1. Stewart, R.: Managing Today & Tomorrow, pp. 147–190. Macmillan (1991)
2. Hammer, M.: Reengineering Work: Don't Automate. Obliterate. Harvard Business Review 68(4), 104–112 (1990)
3. Davenport, T.H., Short, J.E.: The New Industrial Engineering: Information Technology and Business Process Redesign. Sloan Management Review, 11–27 (Summer 1990)
4. Hall, G., Rosenthal, J., Wade, J.: How to Make Re-engineering Really Work. Harvard Business Review, 119–131 (November-December 1993)
5. Anonymous: ARIS Value Engineering-Concept. Whitepaper. IDS Scheer AG (June 2005), http://www.sdn.sap.com/irj/scn/go/portal/prtroot/docs/library/uuid/eae8e311-0b01-0010-0f9c-8d26e2714a91?QuickLink=index&overridelayout=true&5003637725232
6. Hepp, M., Hinkelmann, K., Karagiannis, D., Klein, R., Stojanovic, N.: Proceedings of the Workshop on Semantic Business Process and Product Lifecycle Management (SBPM 2007) 3rd European Semantic Web Conference (ESWC 2007), Innsbruck, Austria (June 7, 2007), http://www.heppnetz.de/files/SBPM2007_Proceedings_A4.pdf
7. Cardoso, J., Hepp, M., Lytras, M.D.: The Semantic Web: Real-World Applications from Industry. Springer (2007) ISBN: 0387485309
8. Kramler, G., Murzek, M.: Business Process Model Transformation Issues (2006)
9. Ildikó, S.: Comparing The Competence Contents of Demand and Supply Sides on the Labour Market. In: 33rd International Conference on Information Technology Interfaces, Cavtat, Croatia (2011)

10. Bálint, M., Andrea, K., Péter, F., András, G.: Advisor - How can we support the employee and the employer in fringe benefit construction? In: Second European Conference on Knowledge Management, UK, Dublin (2001)
11. Andrea, K., Zoltán, S.: The Value to Knowledge Management Using IT Standards - Assessment and Audit Issues of the Knowledge Management Applications Development. In: The 7th European Conference on Knowledge Management, Budapest, September 4-6 (2006), `http://www.academic-conferences.org/pdfs/eckm06-proceedings_booklet.pdf`
12. Vas, R.F., Kovács, B., Kismihók, G.: Ontology-based Mobile Learning and Knowledge Testing. International Journal of Mobile Learning and Organization 2, 128–147 (2009)
13. Kismihók, G., Kovács, B., Vas, R.F.: Integrating Ontology-based Content Management into a Mobilized Learning Environment. In: Caballé, S., Xhafa, F., Daradoumis, T., Juan, A.A. (eds.) Architectures for Distributed and Complex M-Learning Systems: Applying Intelligent Technologies, pp. 192–210. IGI Global, Hershey (2009)
14. Gábor, A., Szabó, Z.: Noisy Reform or Silent Revolution in the Higher Education? A Hungarian Overview. In: Cunningham, P., Cunningham, M. (eds.) eAdoption and the Knowledge Economy: Issues, Applications, Case Studies, pp. 1774–1780. IOS Press, Amsterdam (2004)
15. Gábor, A., Szabó, Z.: Knowledge based institutional capacity building in the Hungarian Higher Education. In: Berdai, A., Sekhari, A. (eds.) The Proceeding of the International Conference on Software, Knowledge and Information Management and Applications (SKIMA 2009), Fez, Morocco, October 21-23, pp. 78–85 (2009) ISBN: 9781851432516
16. Kő, A., Gábor, A., Kovács, B.: Agile Knowledge-Based E-Government Supported By Sake System. Journal of Cases on Information Technology (JCIT) 3, 1–20 (2011)
17. Kovács, B.: Improving Content Management - a Semantic Approach. Acta Cybernetica 4, 579–593 (2008)
18. Vas, R., Kovács, B.: Ontology-based Content Management Systems in Public Administration. In: Schweighofer, E. (ed.) Legal Informatics. The LEFIS Series, vol. 2, pp. 45–62. Prensas Universitarias de Zaragoza, Zaragoza (2008)
19. Ternai, K.: A New Approach in the Development of Ontology Based Workflow Architectures. In: 17th International Conference on Concurrent Enterprising, Aachen, Germany, June 20-22 (2011)
20. Ternai, K., Török, M.: Semantic modeling for automated workflow software generation – An open model. In: 5th International Conference on Software, Knowledge Information, Industrial Management and Applications (SKIMA 2011), Benevento, Italy, September 8-11 (2011)
21. eBEST: Empowering Business Ecosystems of Small Service Enterprises to Face the Economic Crisis. The project co-funded by the European Commission, FP7-SME-2008-2 No. 243554. WWW page, `http://www.ebest.eu/` (accessed November 2, 2011)
22. Kő, A., Dr. Ternai, K.: A Development Method for Ontology Based Business Processes. In: eChallenges e-2011 Conference, October 26-28 (2011)
23. Kismihók, G., Mol, S.: The OntoHR project: Bridging the gap between vocational education and the workplace. In: Book of Abstracts, EAWOP 2011, pp. 194–195. University of Maastricht (2011)

Integrating Knowledge Management in the Context of Evidence Based Learning: Two Concept Models Aimed at Facilitating the Assessment and Acquisition of Job Knowledge

Stefan T. Mol[1], Gábor Kismihók[2], Fazel Ansari[3], and Mareike Dornhöfer[3]

[1] HRM-OB Group Rm. M.2-36, Amsterdam Business School,
University of Amsterdam, Plantage Muidergracht 12,
1018 TV Amsterdam,
The Netherlands
s.t.mol@uva.nl

[2] Department of Information Systems,
Corvinus University of Budapest, 1093 Budapest,
Fővám tér 8, Hungary
kismihok@informatika.uni-corvinus.hu

[3] Institute of Knowledge Based Systems & Knowledge Management,
University of Siegen, Hölderlinstr. 3,
57068 Siegen, Germany
{fazel.ansari,m.dornhoefer}@uni-siegen.de

Abstract. Within the field of Human Resource Management (HRM), the role of individual knowledge has received limited research attention despite offering the promise of superior job performance and improved managerial decision-making. In part, this lack of research may be attributed to the difficulty and laboriousness inherent to the adequate and accurate modeling of job relevant knowledge, particularly since such knowledge by definition varies from job to job. Despite this caveat, there is much to be gained from a knowledge based approach to (managing) human resources. The current paper presents two ontology based concepts for modeling job relevant knowledge, namely Meta-Practitioner and Med-Assess. The former focuses on availing to a practitioner audience the evidence that has accumulated in the academic literature, whereas the latter focuses on the facilitation of personnel selection and training in the medical field through a detailed assessment of individual job knowledge and general mental ability. Ultimately both concepts are aimed at knowledge provision to job applicants and incumbents alike. Having discussed the concepts, the paper summarizes the gains that may be expected from their implementation by presenting an integrated framework. The framework focuses on integrating aspects of Knowledge Management (KM) in the context of Evidence Based Learning (EBL) for business organizations. The paper concludes by addressing the challenges that lie ahead, highlighting some of the limitations of this approach and offering suggestions for further research.

Keywords: Knowledge Management, Evidence Based Management, Evidence Based Learning, Meta-Analysis, Ontology, Personnel selection.

1 Introduction

1.1 Knowledge Management

Today the inadequate use of documented knowledge has resulted in a *know-do* gap in enterprises, particularly for improving the quality of business processes and managing human resources. The reason for this inadequate use of knowledge is not only the result of a lack of knowledge resources, but also the lack of comprehensive frameworks in which accumulated data/information may be stored and updated so as to facilitate their continued (re)usability within organizations [1, 2]. Therefore actionable information (knowledge) is not available at the right time and place for decision makers. Particularly undocumented knowledge (e.g. tacit knowledge or experiences [3] of domain experts) constitutes an implicit resource which is not easily modeled and extracted [1, 2]. Therefore many enterprises are continually missing opportunities to capitalize on intellectual capital (knowledge) for decision-making. In this era, the major question is how Knowledge Management (KM) could improve the proper provision and (re)usability of organizational knowledge (classified either as explicit, tacit or latent).

KM encompasses those processes, technologies and resources used by enterprises to inspire, acquire, gather, manage, share and distribute knowledge and information [1]. Maier defines KM as *"the management function responsible for the regular selection, implementation and evaluation of goal-oriented knowledge strategies that aim at improving an organization in order to improve organizational performance"* [2]. Moreover, KM is aimed at integrating various knowledge intensive business processes, systems and disciplines (e.g. Human Resource Management (HRM), performance management, competence management, and innovation management), which yields knowledge for decision-making. The objective of KM is to empower managers (decision makers) to develop customer-oriented approaches and to convert knowledge into added value and profits in the long-term to *"turn information into actionable knowledge, foster innovation, enable learning from mistakes and best practices, and promote effective knowledge sharing"* [4]. Knowledge in this context is information that has been processed, combined, presented and argued in a meaningful and useful way so as to form a concrete basis for decision-making [1, 2].

Knowledge enables people to make decisions, take action and solve problems, to allow the implementation of strategies and achievement of objectives [5]. In the last two decades, companies are increasingly being transformed into intelligent enterprises, in which knowledge is being produced, absorbed and commercialized

[6]. Enterprises, which process higher quality and more current knowledge than their competitors, are able to develop customer-oriented approaches to convert knowledge into added value and profit [6].

Four perspectives for KM research are envisaged in the extant literature. These result from combining the perspectives on knowledge (epistemology) and social reality (ontology) [1]. These four types are: (1) Cybernetic perspectives, (2) Scientific Management, (3) Soft Systems, and (4) Organizational Development [1]. These perspectives differ on: i) the basic definition of KM about process and purpose, ii) the basic requirements for KM such as data and views, iii) the definition of knowledge actors as a group or an individual, and iv) the definition of the knowledge that changes under the influence of learning [1]. In an integrated framework, KM encompasses a hybrid approach to using all four perspectives. In this context KM fosters the utilization of organizational knowledge and HRM. KM is associated with information and communication technology (ICT) usage (e.g. websites, software solutions, or social networks), where it is more than a tool or a database [1, 2]. Remarkable aspects of KM are knowledge discovery, -sharing and -transfer with great potential for innovation, time and cost savings, and other benefits in terms of new forms of Knowledge Management Systems (KMS) [1, 2]. Thus access to a variety of information sources is enabled, by paying attention to the personalization of data views and data confidentiality issues with respect to user privacy and organizational ethics. The variety of KM tools are distinguishable in three types: collaborative, content management, and business intelligence tools [7]. As stated by Moffett and McAdam "*Collaborative tools include groupware technology, meeting support systems, knowledge directories, and intranets/extranets*" [7]. And "*content management tools include the Internet, agents and filters, electronic publishing systems, document management systems, and office automation systems*" [7]. Finally, "*business intelligence tools include data warehousing, Decision Support Systems (DSS), Knowledge Based Systems (KBS) and workflow systems*" [7]. Having presented a general overview of the field of KM, the next section focuses on the role of knowledge in HRM.

1.2 The Role of Knowledge in HRM

Within the HRM domain, the importance and centrality of organizationally relevant knowledge is increasingly being recognized. The Evidence Based Management (EBM) movement for instance, is premised on the idea that managerial decision-making is improved when it is based on the best available (scientific) evidence. In part such evidence, or decision relevant knowledge, may be derived from systematic reviews of the extant academic literature, which contains a vast body of knowledge that is seldom consulted by managers. Indeed, of 1140 respondents (including but not limited to national, regional, and local governments, research funding organizations, international organizations, and scientific and professional associations) who participated in a survey [8] conducted under the auspices of the EU 7th framework program, a staggering 84% disagreed or disagreed strongly with the statement that there is *NO access problem to scientific publications in Europe*. By facilitating the practitioner conduct of systematic reviews of the academic literature, and therewith access to the scientific knowledge that may form the basis of managerial decision-making, it is proposed here that the concomitant Evidence Based Learning (EBL) can result in a competitive advantage for organizations.

A different HRM domain in which the role of (job specific) knowledge is also increasingly being recognized is the field of personnel selection. Most personnel psychologists nowadays seem to agree, that empirically at least, general mental ability (or intelligence) is the single best predictor of job performance, regardless of job type [8, 10, 11, 12]. That is, of all predictors of job performance available today, general mental ability appears to correlate most strongly with job performance. Strangely, however, we know very little about why general mental ability is related to job performance, although leading authors in this field consistently claim that people who score higher on general mental ability acquire more job knowledge more quickly and are therefore able to demonstrate superior job performance (see for example [13]). Job knowledge therewith appears to be a more proximal predictor of job performance than general mental ability.

In sum, although job knowledge appears to play a central role in both EBL and personnel selection alike, researchers and practitioners in the field of HRM seem to have encountered great difficulties in capturing knowledge and using it to facilitate superior managerial decision-making. It would seem that KM could provide at least some of the answers that these parties are looking for by facilitating the acquisition and sharing of job knowledge and by setting out to transform tacit knowledge into explicit knowledge. Based on our research, we are not aware of many initiatives that have sought to exploit the potential synergies between HRM and KM. Indeed, Carter and Scarbrough in [14] concluded that "*there is a pressing need for a second generation of KM...*" [that puts] "*people-issues at the centre stage of discussion, theorizing and practice*".

The aim of the current paper is to present two preliminary concepts in which we take a multidisciplinary approach to the mapping and assessment of, and learning from, organizationally relevant knowledge. These cases are built on the state of the art in the fields of KM, HRM and ICT. It is our contention that some of the challenges that were raised in the above may only be met through cross-disciplinary collaboration. This paper will proceed by first presenting the Meta-Practitioner concept, aiming at facilitating EBL. Next the Med-Assess concept is introduced. Med-Assess is aimed at assessing knowledge, abilities, and competences of both applicants and employees in the medical field. Subsequently, an integrated framework will be presented in which we will further elaborate on the unique gains, which may be realized through this multidisciplinary approach to knowledge, and we will discuss the challenges that lie ahead in the full implementation of these concepts. This paper will conclude with a summary, discussion of the limitations, and suggestions for future research.

2 The Meta-Practitioner Concept

The Meta-Practitioner endeavor aims to investigate whether EBL and EBM can be promoted through a dedicated ICT interface that allows HRM practitioners (i.e. HR managers) to generate meta-analytically derived summaries of the academic literature pertaining to an HRM-OB (Human Resource Management Organizational Behavior) related problem.

Currently, few HR managers consult the academic literature, and when they do, they are likely to be overwhelmed by the sheer volume of the extant research body. Indeed, staggering numbers (i.e. more than 50%) of HR practitioners disagree with, or lack knowledge of, key findings in the research literature, such as the finding that intelligence predicts job performance better than conscientiousness [15, 16] . Since research findings of individual empirical investigations are oftentimes idiosyncratic to the specific context in which the study was conducted, and might indeed even contradict the findings of other investigations, a promising way in which practitioners may be able to benefit from research findings in academia is through systematic reviews of the literature. Of the different kinds of systematic reviews of the academic literature that could be conducted, namely quantitative, qualitative and theoretical [17], the concept of Meta-Practitioner focuses on facilitating the practitioner conduct of quantitative systematic reviews, also referred to as meta-analyses.

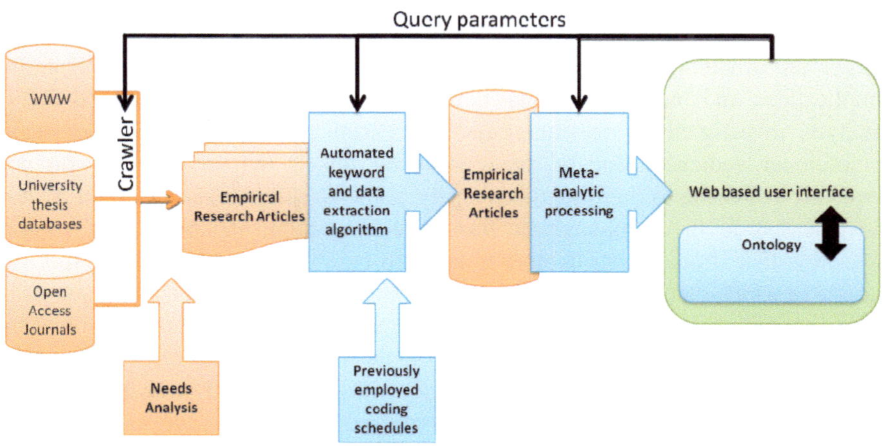

Fig. 1 Workflow of Meta-Practitioner

As depicted in Fig. 1, meta-analysis is a technique of summarizing and aggregating the findings of large numbers of previous investigations into the strength of the relationship between two or more variables (for instance the correlation between general mental ability and job performance). It is through meta-analysis that we can make sense of voluminous and often contradictory research findings. The manual conduct of meta-analysis is a labor intensive affair, in which tens if not hundreds of both published and unpublished primary studies investigating a particular phenomenon must be painstakingly identified, located, and retrieved. Subsequently the correlation matrices and other study characteristics reported in these primary studies must be coded, after which statistical processing and reporting can commence. Within meta-analysis, the oftentimes contradictory findings from individual academic studies, are meticulously collected, coded, and essentially averaged to yield insight into the "true" relationships between the variables of interest to the meta-analyst. Furthermore, meta-analysis can grant insight into the extent to which such "true" relationships generalize from one context to the next. For instance, it was through the conduct of meta-analysis that general mental ability (or intelligence) was identified as one of the strongest predictors of job performance [10], regardless of the type of job, organization, or culture in which it is performed. For these reasons meta-analyses give us insight into the state of the art within a particular management domain, and it is this cutting edge knowledge that has the potential to provide practitioners with the evidence they need to support their decision-making.

Although we do not deny that practitioners may benefit from qualitative and theoretical systematic reviews, the decision to focus on quantitative systematic reviews is premised on the idea that such reviews are more amenable to the automated processing that is envisaged in the current concept than qualitative or theoretical reviews. Up until now, the conduct of meta-analysis has mainly been limited to academics who have been trained in the underlying statistical techniques, and who publish their meta-analyses in academic journals that mainly target an academic audience. Unfortunately, this means that the managerial implications and interventions that could be drawn from these published meta-analytic investigations are oftentimes inaccessible to the non-academic target practitioner audience. One of the reasons for this is that practitioners do not typically subscribe to the oftentimes costly academic journals and as such do not have access to the meta-analyses that have been published therein. A second reason is that it takes considerable training to be able to understand the specific jargon in which meta-analytic investigations are reported. This being the case, it is not surprising that managers oftentimes see themselves forced to rely on their intuition as opposed to actively learning from the evidence that has accumulated in the extant academic literature. A final issue with published meta-analyses is that they become outdated as of the time they are published. Although meta-analyses typically report estimates of the number of primary studies needed to overthrow the conclusions that are drawn, it would obviously be much more desirable to include new studies in the meta-analytic estimates as they become available. This is especially the case when one considers the fact that journal editors may be reluctant to publish a new meta-analysis on a topic on which a meta-analysis has already been published. In sum, managers have little or no access to a vast body of knowledge that could be expected to facilitate learning and result in improved decision-making. This state of affairs may be redressed by utilizing cutting edge information communication technologies to bridge the divide between oftentimes highly technical academic articles and real-life managerial problems that derive from a highly contextualized organizational environment. The aim of the Meta-Practitioner concept is to i) facilitate the practitioner conduct of meta-analytic investigations and ii) to compile and present meta-analytic findings in a practitioner-friendly and accessible interface that will promote the conduct of EBM and EBL. The first aim will be accomplished by means of dedicated web-crawlers that will scour the internet for potentially relevant primary studies. Sources that will be specifically targeted are university thesis databases and peer-reviewed (open access) journals. Based on an automated analysis of bibliographic details, variables investigated, study context, sample characteristics, and correlation matrices (or other measures of effect sizes between variables), studies will be coded, classified, and stored in a database, ready for meta-analytic processing. The interface will present the identified studies into a particular relationship or relationships to the practitioner meta-analyst, who may then proceed to identify those studies most relevant to his or her management problem by fine-tuning several query parameters, such as publication status, context, and measures included. Since this process is a technical affair, in which individual studies need to be evaluated (particularly concerning their relevance, quality, and various statistical parameters), the interface will be pre-configured with defaults on key parameters. The meta-analytic processing itself is fully automated. Since the current endeavor is limited in scope, we envisage this project as a proof of principle and limit ourselves to an investigation of those primary studies that have been published in the HRM-OB discipline.

In order to ensure that the outcomes of Meta-Practitioner will be relevant to and usable by HR practitioners and students alike, the first step in the realization of this concept will be a needs analysis. The aim of this analysis is to i) identify particular examples of managerial decisions that could benefit from the academically accrued evidence and ii) to examine how such evidence is best presented to a practitioner audience. Subsequently the development of a literature search algorithm that will be equipped to locate, identify, screen, and download empirical journal articles, theses, and open access reports, that have focused on investigating a particular HRM-OB related phenomenon, will follow. The so identified articles will be stored in a database for further processing. A detailed search will also be performed for the coding schedules that have been assembled and used in previously published meta-analyses in the extant HRM-OB literature. On the basis of these coding schedules a master coding list will be assembled that will form the input of the keyword and data extraction algorithm that will code the downloaded primary studies on those characteristics needed for meta-analytic processing. In addition to keyword detection methods, for increasing the effectiveness of search queries to potential users, a second method – the so called Association Measures [18, 19, 20] – will also be applied, as this will add semantic information about words and their interrelationships. The resultant database will form the input of a third algorithm that will meta-analytically process and automatically statistically summarize the information obtained from the primary studies. Relationships revealed through meta-analytic processing are then represented in an ontology based system, with which users will interact via a web-based user interface that is capable of navigating through the ontology concepts and that is able to present the related meta-analytic output, background information, and (bibliographic details of) the primary studies involved.

Web-services will be provided to both practitioners and academics/students, who (on the basis of a query of concepts) may retrieve evidence relevant to their specific managerial problem or research question, in the form of customized learning content. The interface will be equipped to facilitate queries of the academic research base with regard to specific variables and their interrelationships. The interface will allow users to indicate whether they are "novice" or "expert". The "novice" interface will be equipped with defaults on key technical parameters, whereas the "expert" interface will allow user to specify these parameters. Depending on whether the user is novice or expert, the meta-analytic output will either be presented in non-technical or technical language. In order to assess the impact of Meta-Practitioner on the learning of practitioners and students alike, two pilot studies will be organized. Since one of the key objectives of the current investigation is to bridge the science-practitioner divide by facilitating practitioner learning, one pilot study will assess practitioner experiences. Here the focus will be on whether practitioners i) found and applied to their decision-making the evidence pertaining to their management dilemma, ii) found the evidence useful in resolving the issue at hand, iii) have recommended/would recommend the system to others, and iv) felt they learned something valuable through their interactions with the system. A second student pilot study will investigate whether students i) mastered relevant content as required by the particular HRM-OB course in which they are participating, ii) have acquired technical know-how with regards to conducting meta-analysis, and iii) would consider using or actually use the system in writing their master theses. With these aims and objectives the accessibility of and learning from the academic literature among both practitioner and student audiences should be increased. In doing so, Meta-Practitioner will promote and further disseminate the conduct of EBM and EBL.

In sum, the new and innovative aspects of Meta-Practitioner are that i) it facilitates the conduct of meta-analysis by practitioners and academics alike and ii) that it aims to bridge the science-practitioner divide by giving practitioners the opportunity to inform their managerial decisions with evidence that may be obtained through an interface that summarizes research findings in a user friendly and accessible format.

3 The Med-Assess Concept

Having introduced the concept of Meta-Practitioner, this section focuses on a second concept called Med-Assess. The abbreviation is a short term for *Adaptive Medical Profession Assessor*. Med-Assess focuses on the measurement of the job knowledge and general mental ability of job applicants and employees in the medical field. The focus here is on the selection of employees on the basis of an assessment of their work related knowledge (e.g. treatment of patients suffering neurological diseases), and the provision of recommendations for additional training courses, qualification measures, or required learning material. Current employee selection practices concentrate on broad educational qualifications and personality tests. These, however, are either too unrefined or unsuited for employee development and the adequate provision of feedback to the candidate. Therefore this approach utilizes job knowledge as a predictor of future job performance. Job knowledge, as opposed to personality, has the major advantage of being malleable, thus allowing for a person centered approach to personnel selection. With the help of novel semantic technology, applicants' real previous experiences (both professional and educational) are evaluated against specific job requirements (general mental ability and job knowledge based technical competencies) for a particular position.

Moreover, the Med-Assess concept supports Vocational Education and Training (VET) on the job and furthers competencies in a certain context (i.e. human health services and medical profession). The concept is founded on an ontology based knowledge representation method as well as KM, which bridges HRM and the labor market in the human health services and medical profession. In the personnel selection and job matching process, the system is set up to measure the knowledge required to fulfill particular tasks (i.e. skills, competency, and actionable information). Based on the Organization for Economic Co-operation and Development (OECD): *"a competency is more than just knowledge and skills. It involves the ability to meet complex demands, by drawing on and mobilizing psychosocial resources (including skills and attitudes) in a particular context"* [21]. However, for the purposes of Med-Assess competency is essentially defined as the behavioral outcome of job knowledge. Med-Assess provides an adaptive solution for clinics that may be expected to facilitate and improve on-the-job training. The adaptive characteristic of Med-Assess is to assess the knowledge of an employee and to offer relevant training materials or courses. The target group consists of medical professionals such as nurses or other care givers (e.g. ward nurses, medical imaging nurses, and physiotherapists). Furthermore, Med-Assess also supports hospital management in selecting new employees. The solution is adaptable to different medical qualification areas. This way each clinic will be able to create individual Med-Assess measuring knowledge bases (i.e. neuroscience, internal medicine, sports medicine, etc.).With regard to an international context, the concept of Med-Assess allows for the determination of whether a foreign job applicant holds the prerequisite knowledge and qualifications for the target job, or whether he or she will need additional training to fulfill prerequisite job related tasks. The proposed Med-Assess system allows employers to assess their job candidates and give them concrete feedback as to where their qualifications are still lacking or need special improvement. In this way the system is set up to propose specific VET courses and/or programs that may assist candidates in improving their qualifications and therewith to increase their chances of finding employment. At the same time, the system may also be deployed for purposes of training needs analysis of existing medical staff by identifying missing knowledge and offering individualized learning pathways.

The foundation of Med-Assess is to model job descriptions based on a combination of requirements postulated by recruitment agencies, the experiences of medical partners, internet resources and other national e.g. [22]/ European [23] and international guidelines [24]. This way the assessment of knowledge and competencies of a migrant job applicant would be more formalized and comparable to the local applicants for a medical institution. With hindsight to the recommendation of qualification measures, a knowledge base containing measures such as local job training centers, vocational or medical schools, training on the job, or commercial courses will be established. Additionally bachelor or master nursing studies at a university or a distance learning university could be recommended to professionals on the basis of their individually tailored job knowledge profile. The Med-Assess outcome is a kind of certificate, which reveals and documents the knowledge, ability and competence of the candidate (examinee) in a specific medical field.

As already mentioned above, the basis for Med-Assess are models of job descriptions and requirements for these occupations as well as possible qualification measures stored in the knowledge base. In this context Med-Assess applies the already existing solution of OntoHR [25, 26] developed as part of a favorably evaluated (8.5/10) EU project under the same title. OntoHR supports the creation of ontology based job knowledge models, which allows HRM managers to generate highly personalized jobs, job descriptions and associated training material for job applicants or existing employees. These jobs are described in terms of a certain set of technical competencies, each of which consists of a large number of knowledge elements. Since general mental ability is likely to facilitate the rapid acquisition of job knowledge, the job modeling and evaluation module of Med-Assess acknowledges this issue by lowering/raising the threshold needed to receive a passing score for a particular knowledge domain. The combination of these two automatically adds up to an employee knowledge and training profile with the relevant learning materials. For the knowledge base and ontology development, job descriptions and evaluation criteria/requirements for the specific jobs have to be identified and collected. The ontologies for the job descriptions have to be modeled after which test questions may be generated. During the development of OntoHR the target group consisted of IT professionals. This area and the associated requirements are of course entirely different to the medical field. Therefore it is not possible to simply offer the same tests to IT and medical employees. As a consequence the main focus is on the job models, which have to be assembled and adapted for this specific area. A valid knowledge assessment is only possible by representing highly specific and contextualized job content in the system.

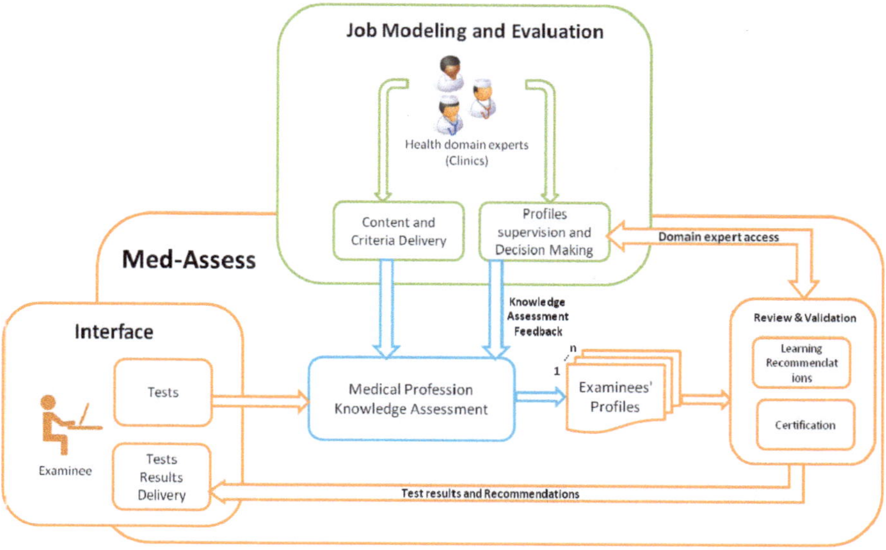

Fig. 2 Concept image of Med-Assess

Fig. 2 shows the concept image of the Med-Assess system. On the left side, the examinees, which may be new job applicants or job incumbents, will take part in the knowledge assessment for certain job related areas. The assessment of job knowledge is supported by the help of the internal logic of Med-Assess, the medical profession knowledge assessment module. The modeling of target jobs takes place with the help of ontologies. For these models health domain experts have to provide content and job criteria to a knowledge engineer. These job specific criteria and this content will be tailored for medical professions. Besides the support in creating criteria and requirements, health domain experts are responsible for the profile supervision and decision-making process (i.e. by deciding which knowledge domains are (ir)relevant and (un)essential for which jobs and by deciding on pass/fail scores to apply to each of the knowledge domains). Med-Assess will not perform as a static system; instead it will provide measures to adapt the content, criteria, and requirements flexibly. The medical profession knowledge measuring module is closely connected to the profile of the examinee, in this way the system is enabled to personalize recommendations for individual learning. Based on the test results, recommendation logic provides different learning opportunities such as courses, additional readings or on the job measures. The health domain experts have a possibility and an access to modify or update learning recommendations. The same applies for the certifications, which are given to the examinees, based on their test results.

The realization of Med-Assess will be an ICT system. The examinee will experience a web based frontend, where he or she will not only complete the assessment test, but also receive the evaluation, learning recommendations and related certificates. In sum, the Med-Assess concept is customizable to other job areas, and indeed such a capability has arisen from designing an adaptive and robust ICT solution.

4 Applicability of Merging Meta-Practitioner and Med-Assess

As explained earlier, Meta-Practitioner is aimed at facilitating practitioner access to academically accrued evidence through the automated conduct of meta-analysis. Med-Assess is a personnel selection and training platform that takes an individualized approach to the assessment and development of job specific knowledge. Therefore both concepts can be used within the process of job knowledge provision and delivery for improving individual job performance especially vis-à-vis decision-making. In addition, both concepts are bridging education, job knowledge and job performance as three pillars of success in business organization. While Med-Assess is concerned with general mental ability and knowledge assessment, Meta-Practitioner is mainly designed to provide academically driven evidences or know-how for supporting effective decision-making.

Med-Assess is basically designed and proposed for the medical profession market and Meta-Practitioner for HRM. Despite the divergence of application areas, both concepts are adaptable and customizable to be used either in tandem or in other domains of application. For instance, once domain experts provide decision making evidences in the frame of Med-Assess; the question is how far a candidate is aware of the relevant literature? And vice versa, once Meta-Practitioner provides academic evidences, how far are these evidences used to make better decision in personnel selection? Such applicability is rooted in the general approach used in both concepts' architectures, which encompass adaptive ICT based methodologies and techniques. Therefore an integrated framework is required to utilize both proposed concepts for efficient job knowledge provision towards improving job performance, which is discussed in section 5.

5 Integrated Framework - Applying Knowledge Management and Evidence Based Learning in Business Applications in the Context of Meta-Practitioner and Med-Assess

Meta-Practitioner and Med-Assess use techniques from both KM and EBL and combine them into a single application domain. During the introduction of the paper the definition of KM of Maier [2] was presented. This definition focuses on the improvement of *"organizational performance"*. The widespread implementation of Meta-Practitioner may be expected to improve organizational performance by availing practitioners with managerially relevant evidence that is based on sound academic research. It goes without saying that decision-making that is based on the best evidence available will be superior to decision-making that is based on hunches or intuition. The main idea of Med-Assess is to measure the concrete knowledge of hospital employees or job applicants. Applying Med-Assess in hospitals and furthering the knowledge as well as technical competencies of employees and applicants alike, may also be expected to improve organizational performance and organizational development overall. The potential courses and training measures further the knowledge of the employees, allow for the flexible deployment of staff to different job related tasks, and therewith increases their job performance. Meta-Practitioner on the other hand applies Evidence Based Learning and Management through meta-analysis and therewith supports practitioner decision-making. Decision-making, a field in the context of Knowledge Management, bridges the gap between KM and EBL.

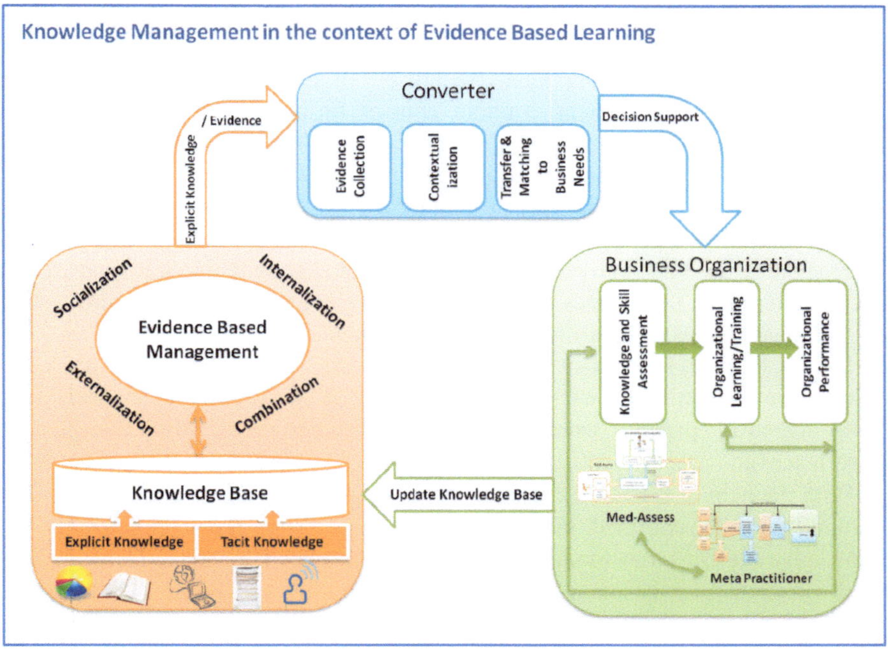

Fig. 3 Knowledge Management in the context of Evidence Based Learning

Knowledge Management and Knowledge Based Methods, in these cases ontologies, are applied to the creation of the Meta-Practitioner and Med-Assess systems. A knowledge engineer builds up the system in close contact with the knowledge domain experts (in these cases the HRM practitioners and health domain experts). The knowledge engineer models the requirements based on the input from the experts and external resources. He or she is also responsible for the adaptiveness and flexibility of the solutions. The systems have to be flexible enough so that their content can be modified later on, and allow experts to define variations of settings. As already mentioned before, Meta-Practitioner is even planned in two modes, one for “experts” and one for “novices”.

Having presented the two systems the figure above conceptualizes the integrated framework. Fig. 3 shows the context of KM in EBL as well as the integration of the aforementioned concepts of Meta-Practitioner and Med-Assess into a business organization. The depicted process flow (see Fig. 3) consists of the evidence management module, the evidence converter and the business organization. On the right hand side of the figure, the business organization is sketched, with the key indicators of knowledge and skill assessment, organizational learning and performance. All three aspects are closely interlinked and supported by the concepts of Med-Assess and Meta-Practitioner.

Both of the aforementioned concepts (Med-Assess and Meta Practitioner) are application scenarios to further the aspects of organizational and individual learning within a business organization. Med-Assess supports assessing the knowledge and skills of individuals within the organization and offers relevant personalized learning content, training courses or other additional knowledge sources based on the assessment results. Meta-Practitioner will be used to support decision-making by supplying additional knowledge from a scientific knowledge base to the organization and thus promoting the organizational performance. On the left side of the figure, the knowledge base is situated inside the EBM module. The sources of the knowledge base are explicit and implicit knowledge accumulated from the business organization as well as from external sources such as web pages, books, scientific papers or documented practical experiences such as lessons learned, success stories or good practices. Remarkably the knowledge sources are distinctive for each of the concepts (e.g. Meta-Practitioner deals only with scientific papers, whereas Med-Assess deals also with documented experiences).

With the help of the EBM cycle, the knowledge, which may have been implicit previously, is transformed into an explicit form. After the externalization of knowledge the converter module translates and matches knowledge elements into valid contexts and business scenarios. Through this converter, based on the extracted and contextualized knowledge, an active decision support is possible.

6 Conclusion and Outlook

Within this paper, two concepts were presented in which knowledge plays a central role. While the Meta-Practitioner concept is aimed at improving practitioner access to and use of academically accrued knowledge, the Med-Assess concept is aimed at facilitating personnel selection in the medical field and/or conducting training needs analyses. The current paper presents two concepts that tackle the challenges associated with building value from knowledge. The concrete outcome of the implementation of Meta-Practitioner would be improved access to state of the art scientific knowledge, and therewith improved decision-making on the part of practitioners employed in the field of

HRM. Concrete outcomes of the implementation of Med-Assess would be the accurate assessment of individual job knowledge, the facilitation of selecting context decision-making, the provision of remedial and customized training, and (therewith) the development of a knowledgeable workforce. Notably it should be feasible to integrate both systems into a single unified system. Such an approach (i.e. Equipping the Med-Assess concept with a Meta-Practitioner module) would facilitate job incumbents in filling knowledge gaps that are detected by the Med-Assess knowledge assessment.

The fact that organizations and academics alike have seldom addressed job knowledge of applicants and incumbents raises a number of important questions and concerns. First and foremost is the question of whether organizations intend to be helped in the first place. Certainly, in leveraging science into practice one needs to be concrete. It is our contention that the intimate collaboration between academics and practitioners that would be crucial for the implementation of either concept is likely to kindle and evoke enthusiasm for these concepts in practice. One foreseeable challenge is the delineation of the business case for developing and implementing the laborious system which requires considerable resources. In defining this business case it will be essential to demonstrate the practical utility of the concepts that are outlined above. Assuming that the correlation between job knowledge, knowledge acquisition and subsequent job performance can be ascertained, it should be possible to compute the increase in Euro payoff of the selected/trained group by applying a utility model e.g. Cronbach and Gleser [27].

Further challenges arise when we approach Evidence Based Management and Learning from a theoretical perspective. Since these research efforts need a multidisciplinary approach, an important aspect of this endeavor is the establishment of a common vocabulary. Further research should be conducted on grasping tacit knowledge and converting that into explicit evidence for decision-making. This also implies that the evidence base should be maintained and updated regularly. This work also encompasses some technical challenges, including suitable user interfaces (in case of Meta-Practitioner the interface should integrate web-mining, data-mining, text-mining, ontology engineering, and data query services). Besides integration, usability is also an issue, as the interfaces should be designed to be comprehensive enough for non-technical users.

In order to overcome these challenges we suggest following a four step protocol, when deploying similar systems. Step 1 is a needs analysis and problem description. Step 2 is evidence extraction from knowledge databases, based on step 1. The 3rd step is to apply those evidences to the particular business case / decision-making situation, and finally the 4th step is to evaluate and feed back into any of the following steps.

The importance and centrality of job knowledge in today's knowledge based economy is undeniable. It is only through a sustained and multidisciplinary effort that we can start picking the fruits of the approaches that are outlined in this paper. Although we believe both concepts to be viable, certain research challenges remain. For instance, the Meta-Practitioner concept is currently mainly concerned with explicit knowledge that has in some way shape or form been documented.

Med-Assess absorbs knowledge (know-how) of domain experts as well. One challenge that awaits here is facilitating the automated processing and mapping into the ontologies of such knowledge. It should also be recognized that job incumbents are likely to have significant know-how, and incorporating this latent knowledge into either system will be another significant challenge.

References

1. Wijnhoven, F.: Knowledge management: more than a buzzword. In: Knowledge Integration, pp. 1–16. Physica-Verlag, Germany (2006)
2. Maier, R.: Knowledge Management Systems, Information and Communication Technologies for Knowledge Management, 3rd edn. Springer, Germany (2007)
3. Nonaka, I., Von Krogh, G.: Tacit Knowledge and Knowledge Conversion: Controversy and Advancement in Organizational Knowledge Creation Theory. Organization Science 20(3), 635–652 (2009)
4. Eppler, M.J.: Managing Information Quality: Increasing the Value of Information in Knowledge-intensive Products and Processes. Springer, Berlin (2006)
5. Pfeifer, T.: Quality Management: Strategies, Methods and Techniques. Carl Hanser Verlag, Germany (2002)
6. Rampersad, H.K.: Total Quality Management, an Executive Guide to Continuous Improvement. Springer, Germany (2001)
7. Moffett, S., McAdam, R.: Contributing and enabling technologies for knowledge management. International Journal of Information Technology and Management 2(1/2), 31–49 (2003)
8. European Commission, Online survey on scientific information in the digital age (2012), http://ec.europa.eu/research/science-society/document_library/pdf_06/survey-on-scientific-information-digital-age_en.pdf
9. Salgado, J.F., Anderson, N., Moscoso, S., Bertua, C., de Fruyt, F., Rolland, J.P.: A Meta-Analytic Study of General Mental Ability Validity for Different Occupations in the European Community. Journal of Applied Psychology 88(6), 1068–1081 (2003)
10. Schmidt, F.L., Hunter, J.E.: Select on Intelligence. In: Locke, E.A., Malden, M.A. (eds.) The Blackwell Handbook of Principles of Organizational Behavior, pp. 3–14. Blackwell Publishing (2000)
11. Schmidt, F.L., Hunter, J.: General Mental Ability in the World of Work: Occupational Attainment and Job Performance. Journal of Personality and Social Psychology 86(1), 162–173 (2004)
12. Schmidt, F.L., Hunter, J.E., Raju, N.S.: Validity Generalization and Situational Specificity: A Second Look at the 75% Rule and Fisher's Z Transformation. Journal of Applied Psychology 73, 665–672 (1988)
13. Schmidt, F.L.: Cognitive Tests Used in Selection Can Have Content Validity as Well as Criterion Validity: A Broader Research Review and Implications for Practice. International Journal of Selection and Assessment 20(1), 1–13 (2012)
14. Carter, C., Scarbrough, H.: Towards a second generation of KM? The people management challenge. Education +Training 43(4), 215–224 (2001) (Permanent link to this document: http://dx.doi.org/10.1108/EUM0000000005483)

15. Rynes, S.L., Colbert, A.E., Brown, K.G.: HR professionals' beliefs about effective human resource practices: Correspondence between research and practice. Human Resource Management 41, 149–174 (2002)
16. Rynes, S.L., Giluk, T.L., Brown, K.G.: The very separate worlds of academic and practitioner periodicals in human resource management: Implications for evidence-based management. Academy of Management Journal 50, 987–1008 (2007)
17. Briner, R.B., Denyer, D., Rousseau, D.M.: Evidence-Based Management: Concept Cleanup Time? The Academy of Management Perspectives, Archive 23(4), 19–32 (2009)
18. Manning, C., Schütze, H.: On Foundations of Statistical Natural Language Processing. The MIT Press, Cambridge (1999)
19. Pearce, D.: A comparative evaluation of collocation extraction techniques. In: Third International Conference on Language Resources and Evaluation, Las Palmas, Spain (2002)
20. Evert, S.: The Statistics of Word Co-occurrences: Word Pairs and Collocations. Ph.D. thesis, University of Stuttgart (2004)
21. Organization for Economic Co-operation and Development (OECD): The definition and selection of key competencies - Executive Summary (2005), http://www.oecd.org/dataoecd/47/61/35070367.pdf
22. Arbeitsagentur (Federal Employment Agency in Germany), General job description of a nurse (2012), http://berufenet.arbeitsagentur.de/berufe/docroot/r2/blobs/pdf/archiv/8791.pdf
23. European Association for Neurosciences Nurses, EANN (2012), http://www.eann.net
24. World Federation of Neuroscience Nurses (2012), http://www.wfnn.nu
25. Kismihók, G., Mol, S.T.: The OntoHR Project: Bridging the Gap between Vocational Education and the Workplace. In: Book of Abstracts, EAWOP 2011, pp. 194–195. University of Maastricht (2011)
26. OntoHR project homepage (2012), http://www.ontohr.eu
27. Cronbach, L.J., Gleser, G.C.: Psychological tests and personnel decisions, 2nd edn. University of Illinois Press, Urbana (1965)

Towards an Integrated Platform for Big Data Analysis

Mahdi Bohlouli[1], Frank Schulz[2], Lefteris Angelis[3], David Pahor[4], Ivona Brandic[5], David Atlan[6], and Rosemary Tate[7]

[1] Institute of Knowledge Based Systems & Knowledge Management, University of Siegen, Hölderlinstr. 3, 57068 Siegen, Germany
mbohlouli@informatik.uni-siegen.de
[2] SAP Research, Karlsruhe, Germany
frank.schulz@sap.com
[3] Aristotle University, Thessaloniki, Greece
lef@csd.auth.gr
[4] Arctur d.o.o, Slovenia
david.pahor@arctur.si
[5] Technical University of Vienna, Austria
ivona@infosys.tuwien.ac.at
[6] Phenosystems SA, Belgium
atlan_d@web.de
[7] University of Sussex, United Kingdom
rosemary@sussex.ac.uk

Abstract. The amount of data in the world is expanding rapidly. Every day, huge amounts of data are created by scientific experiments, companies, and end users' activities. These large data sets have been labeled as "Big Data", and their storage, processing and analysis presents a plethora of new challenges to computer science researchers and IT professionals. In addition to efficient data management, additional complexity arises from dealing with semi-structured or unstructured data, and from time critical processing requirements. In order to understand these massive amounts of data, advanced visualization and data exploration techniques are required.

Innovative approaches to these challenges have been developed during recent years, and continue to be a hot topic for research and industry in the future. An investigation of current approaches reveals that usually only one or two aspects are addressed, either in the data management, processing, analysis or visualization. This paper presents the vision of an integrated platform for big data analysis that combines all these aspects. Main benefits of this approach are an enhanced scalability of the whole platform, a better parameterization of algorithms, a more efficient usage of system resources, and an improved usability during the end-to-end data analysis process.

Keywords: Scalable Decision Support, Complex Event Processing, Big Data, Cloud Computing.

1 Introduction

The amount of data produced and processed is expanding at an extreme pace. Two sources of data can be distinguished: human-generated data and machine-generated data, and both present huge challenges for data processing. The big data phenomenon cannot be defined by data volume alone. Additional layers of complexity arise from the speed of data production and the need for short-time or real-time data storage and processing, from heterogeneous data sources, from semi-structured or unstructured data items, and from dealing with incomplete or noisy data due to external factors. All these aspects render the analysis and interpretation of data a highly non-trivial task. It becomes even more challenging when data analysis and decision making needs to be carried out in real time. The information processing capacity of humans is highly limited. One highly cited study showed that only about seven pieces of information can be held in short-term memory [20]. Therefore a suitable technological support is strongly needed in order to present the information in a more accessible form.

Taking these points into consideration, the following definitions have been proposed to capture the notion of big data: "big data is when the size of the data itself becomes part of the problem" (Loukides in [22]) or "data that becomes large enough that it cannot be processed using conventional methods" (Dumbill in [22]). Other authors define big data by the three dimensions of volume, velocity and variety [26].

1.1 Big Data Examples

There are numerous domain examples, including web applications, recommender systems for online advertising, financial decision making, medical diagnostics, or the operation of social networks or large IT infrastructures. For instance, Google was processing 20 petabytes (10^{15} bytes) per day in 2008. In 2011, and was able to sort one petabyte of 100-byte-strings in 33 minutes on an 8000 machine cluster. Amazon.com reported peak sales of 158 sold items per second on November 29, 2010, and Walmart retail markets handle more than 1 million of transactions per hour. Nowadays, the amount of data from automated sensors, RFID tags or mobile devices surpasses the human-generated data by far. According to Teradata, a single six-hour flight of a Boeing 737 airplane produces 240 terabyte of sensor data. It was estimated by IBM that currently 2.5 x 10^{18} bytes of raw data are created every day by humans or machines [17].

1.2 Business Impact

In 2011 for the first time, Gartner Market Research added the term "'Big Data' and Extreme Information Processing and Management" to their annually published hype cycle for emerging technologies. The business value of advanced analytics of huge amounts of data has been widely recognized as a key driver for future business growth. The analyst firm Wikibon published a report that estimates an annual growth of the Big Data market of over 50 % for the next five years, resulting in a market volume of 53 billion US-$ in 2017. Key drivers are the competitive advantage and the increased operational efficiency gained by advanced analytic capabilities.

This paper presents the first ideas and goals of an initiative that was originated as a collaboration of researchers from European universities and companies, aiming to develop a generic, sophisticated, and customizable platform able to extract information from extra-large data sources and streams from the Cloud environment or physically situated resources. Using pattern recognition, statistical and visual analytic techniques, the goal of the integrated platform is to present the information in a helpful form, enhancing decision-making across many domains.

The initiative aims to combine advanced techniques to enable: (a) applications across a wide range of domains; (b) integration of large-scale data from disparate resources and streams; (c) scalability and elasticity on cloud infrastructures; (d) statistical identification and discovery of complex events that would be imperceptible for standard analyses; (e) effective and meaningful decision support, and (f) continuous quality control of results.

In section 2, existing technologies for big data are reviewed. Section 3 discusses some specific use cases and derives requirements for an integrated platform. Section 4 outlines the proposed architecture and key aspects of such a platform. In section 5, related work on end-to-end data analysis platforms is discussed.

2 Existing Technologies in Data Analysis and Machine Learning

Decision support systems are nowadays ubiquitous in industrial and research applications, and a large variety of commercial and open source tools and libraries exist. Furthermore, there is a rich theoretical background from various disciplines such as statistics and operations research that lays a solid foundation for decision making systems. The use of statistical and data mining methods has been limited to specific data from specific sources, depending on the application domain.

Notable open source tools include the R project [25] for statistical analysis, the WEKA project [15] for data mining, the KNIME platform [6] for data analytics, and the Apache Mahout [4] project for machine learning on top of the map reduce framework Hadoop. The R statistical language and the Predictive Model Markup Language (PMML) offer the opportunity to combine a wide range of statistical methodologies and models that are able to cooperate for processing massive data from diverse sources and producing output for feeding the decision support systems.

2.1 R Project and PMML

R is an open source statistical language, in fact a comprehensive suite of tools providing to the users a vast variety of statistical and graphical techniques for data analysis, and most importantly, the facilities to expand the language by programming new routines, functions and to add new packages. Furthermore, R can be linked with other languages (C, C++) and can be used for advanced massive data analysis.

The Predictive Model Markup Language (PMML) is an XML-based language developed by the Data Mining Group (http://www.dmg.org/) providing ways to represent models related to predictive analytics and data mining. PMML enables the sharing of models between different applications which are otherwise incompatible. The primary advantage of PMML is that the knowledge discovered can be separated from the tool that was used to discover this knowledge, so it provides independence of the knowledge extraction from application, implementation platform and operating system.

2.2 WEKA

The Weka workbench [15] is a collection of state-of-the-art machine learning algorithms and data pre-processing tools. It is very flexible for users who can easily apply a large variety of machine learning methods on large datasets. It can support the whole process of data mining, starting from the preparation of data to the statistical evaluation of the models. The workbench includes a wide variety of methods such as regression, classification, clustering, association rule mining, and attribute selection. Furthermore, it supports streamed data processing. The system is open-source software, written in Java and freely available.

2.3 KNIME

According to [6], the Konstanz Information Miner (KNIME) is a modular environment, developed as an open source project (http://www.knime.org) which enables easy visual assembly and interactive execution of data pipelines. It is designed as a teaching, research and collaboration platform and provides integration of new algorithms and tools as well as data manipulation or visualization methods in the form of new modules or nodes. Its great advantage is the powerful user interface, offering easy integration of new modules and allowing interactive exploration of analysis results or models. Combined with the other powerful libraries such as the WEKA data mining toolkit and the R statistical language, it provides a platform for complex and massive data analysis tasks. KNIME is continuously maintained and improved through the efforts of a group of scientists and is offered freely for non-profit and academic use.

2.4 Apache Mahout

Mahout is an open source software project hosted by the Apache foundation [4]. It provides a machine learning library on top of Hadoop, with the goal to provide machine learning algorithms that are scalable for large amounts of data. The development has been initiated with the paper [8]. Up to now, several dozens of algorithms have been implemented for data clustering, data classification, pattern mining, dimension reduction and some others. All algorithms are written in Java and make use of the Hadoop platform.

3 Requirements

This chapter states some requirements on the envisioned integrated platform for big data analysis. Based on the analysis of several use cases from different domains, the following areas have been identified as crucial building blocks.

- Functional requirements
- Data integration: For addressing problems from real-world application domains, the platform must be capable to access multiple different data sources and to deal with inconsistent or noisy data.
- Statistical analysis: The analysis of data can be simple (like counting) or complex (like the calculation of a Bayesian network for prediction). The platform must support different types of data analysis, including the calculation of statistical key figures like quantiles or correlation coefficients.
- Interactive exploration: The platform has to support intuitive visualization for visual analytics and easy interaction with the data.
- Decision support: In addition to the analysis of data, the platform should also provide mechanisms for domain-specific data interpretations that are valuable for decision making. Manual analysis, semi-automated decision support or fully automated systems should be provided depending on the application area.
- Non-functional requirements
- Scalability: The platform and its various constituents have to be able to handle huge amounts of data. All components must be designed in such a way that they can be deployed in a distributed computing environment.
- Near real time: Fast processing is the main requirement of some use cases. The core platform must be able to support near real time processing in combination with selected components.
- Resource efficiency: While keeping the objectives of throughput and speed, the system resources should be utilized in an efficient way. This has to be achieved by a system management component which is part of the platform.

These requirements can also be categorized with respect to research area:

- Database Technologies: Distributed databases, parallelism, NoSQL approaches.
- Information Systems: Design of an integrated platform with scalability of all components and efficient usage of IT resources, making use of current system architectures (multi-core) and increased availability of main memory.
- Algorithmics: Design of efficient algorithms for external memory, algorithms fitting into the MapReduce paradigm or other parallelization patterns. Streaming algorithms for efficient processing of amounts of data that are so huge that scanning it more than once or a few times only is infeasible, or for processing data that naturally arrives as an event stream.

The main challenge exists in the combination and simultaneous realization of these requirements.

4 Solution Approach

The proposed platform for data analysis aims to address the requirements given in section 3. The following figure 1 describes its main components.

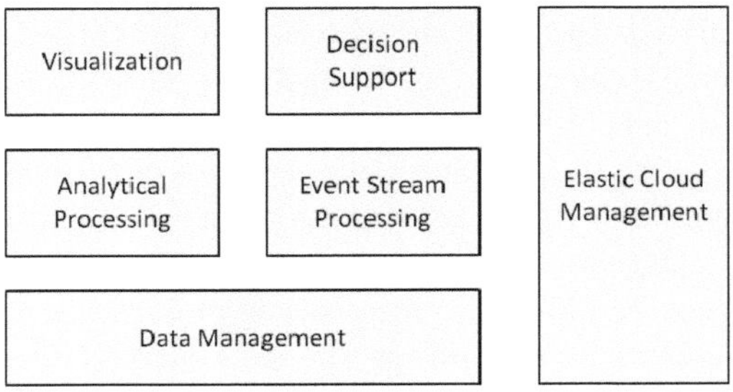

Fig. 1 Building blocks

The data management component is responsible for providing access to heterogeneous data sources, data integration and pre-processing. It contains a storage component and interfaces for efficient data retrieval and aggregation. Depending on the requirements, either a batch-oriented analytical processing or a near real-time event stream processing can be performed on the data. In both components, efficient algorithms will provide parallel processing for statistical analysis and complex evaluations. The results can be visualized for human consumption or used as input for a decision support component that provides semi or fully automated solutions in decision problems. While all components have to be scalable in order to cope with large amounts of data, a dedicated management component for elastic clouds will control the infrastructure resources provided to each component for an efficient and balanced operation of the whole system.

In addition, this component takes service level objectives into account for ensuring the requested end user experience.

For Big Data manipulation and processing we have concluded that we must use NoSQL databases, which add affordable *horizontal scalability* (scale-out) of storage spreading over nodes, over clusters and eventually over data-centers to *vertical scalability* (scale-up) and enable large data-throughputs - especially write-to-storage. With this decision, we are consciously sacrificing the RDBMS capabilities of orthogonalized data schemes consisting of tables and complex relationships (like joins) and a powerful query language (SQL).

At the same time it is to be stressed that the fundamental differences between today's leading NoSQL solutions are much greater that the differences between different "strains" or products of RDMSs. The NoSQL landscape is filled with disparate and - sometimes - diverging solutions of optimization for Big Data handling that can be complementary only if a unified platform with a common systems' API is implemented. NoSQL databases scale in very different ways, having greatly differing data models and specific mechanisms for data querying. The latter are - on the main part - much more primitive than SQL although attempts are being made recently to bring more structure to querying in certain NoSQL databases - for example by developing SQL-like interfaces, such as Pig, Hive and unQL top of the MapReduce mechanism. Furthermore, there are also significant differences in the type of scaling NoSQL products support. Some of them enable good scaling of the *data-set size*, some grow well in the *volume of concurrent transactions*, some excel only in the brute *speed of data storage read or write,* while others have a hybrid set of the before mentioned scalability capabilities but with significant compromises stemming from this.

The danger of using the wrong NoSQL tool for a specific large data-set processing problem is thus much more pronounced than choosing the "wrong" RDBMS for classic relational processing. Also the implementation, systems integration and programming of some of the NoSQL databases is much more challenging that the incorporation of relational database technologies in applications and middleware due to the young age and documentation scarcity of some of the NoSQL products. Another fact is that not all NoSQL databases are good at (horizontal) distribution over nodes and not all NoSQL databases support effective replication (especially master-to-master) between server clusters. Usually, good scalability paired with excellent node-distribution means the underlying data model is primitive, and vice-verse. A good case in point are graph databases which are very single-node scalable and transaction-throughput efficient but are not optimized for efficient horizontal distribution of processing.

How can we, then, propose our *Platform,* if the NoSQL landscape is so divergent and - partially - exclusive? For the purpose of establishing our *Integrated Platform for Big Data Analysis* we propose the realization of a practical solution for Big Data storage and management that is standardized and formalized as much as possible, while at the same time supports different aspects of strengths of separate NoSQL store-type solutions. On the other hand our combined NoSQL system should be manageable and controllable, so we must limit the number of databases used. Our solution is to use a hybrid middleware NoSQL system for Big

Data storage, composed of 3 different databases, each optimized for a specific data model/performance and scalability case. Large data-sets will be thus stored in several different NoSQL (non-relational) databases in the back-end of the *Platform* infrastructure, depending on the type, amount and stream bandwidth of the input data, so that the most optimal database managing system will be used for the appropriate type. It is expected that the different use-case scenarios will provide data and data-streams of events that will demand specialized database processing. Complex querying of sufficiently small sub-sets of data will still be possible with a RDMS.

According to **Brewer's Theorem** - also known as the **CAP Theorem**, distributed computer systems cannot guarantee all of the following:

- **Consistency**, with all network nodes seeing the same data at the same time;
- **Availability**, with a guarantee that every request to the system resources receives a response about whether it was successful or failed;
- **Partition Tolerance**, with the system continuing to operate despite arbitrary loss of messages between nodes.

Since the CAP theorem states that it is impossible to have both ACID (*Atomic, Consistent, Isolated, Durable*) database consistency and high data availability, we shall use the above described hybrid NoSQL infrastructure that will enable consistency (ACID) for certain usages and high availability for others - depending on the use case. Dissimilar data sources in use cases for the *Platform* will be, therefore, handled by the hybrid storage back-end of the elastic cloud of the *Platform* infrastructure, consisting of an SQL database for smaller data-sets and three specialized types of distributed, multi-nodal NoSQL databases for large data storage, each of them optimized for a certain use scenario (only one database product per store-type):

- **Document Store**, like, for example Apache CouchDB (BigCouch) or MongoDB;
- **Wide Column Store**, such as, for example HBase or Cassandra;
- **Key Value Store**, like, for example MEMBASE, Riak or Redis;

There is a fourth type of NoSQL store - the **Graph Database**, for example InfoGrid, Neo4J or Infinite Graph, which implements flexible graph data models. For our *Platform* we have decided to minimize complexity and have determined that we can cover all major large-data processing with the combination of some or all of the above mentioned store-types, without introducing the added management and programming overhead of Graph Databases.

The infrastructure will have an open services interface based upon a RESTful open data protocol handler positioned between the back-end distributed resources (applications and data) and the service consumers (clients of the use case scenarios) in the elastic cloud. Quality of service is undoubtedly an important factor in today's distributed IT systems with loosely coupled client applications connecting in large numbers to services interfaces of back-end distributed applications in server grids. The *Platform* infrastructure will address this through its specialized framework. The above mentioned NoSQL database systems do not require high-specification servers [29]. In this regard, the systems will not displace current working machines with new resources, but the development of the elastic infrastructure will be accomplished by adding new machines to the current working cluster.

5 Conclusion and Outlook

This paper presents the vision of an integrated platform for big data analysis. The vision encompasses the applicability to wide range of domains, the integration of heterogeneous data resources and streams, the efficient usage of computing resources, statistical analysis and machine learning, and an effective and meaningful decision support.

The paper provides a thorough overview of relevant technologies and related work on big data platforms. The key requirements and a high level solution are described. Main benefits of the intended platform are an enhanced scalability of the whole system, a better parameterization of algorithms, a more efficient usage of system resources, and a better usability during the end-to-end data analysis process.

References

1. Alexandrov, A., Ewen, S., Heimel, M., Hueske, F., Kao, O., Markl, V., Nijkamp, E., Warneke, D.: MapReduce and PACT - Comparing Data Parallel Programming Models. In: Proceedings of the 14th Conference on Database Systems for Business, Technology, and Web (BTW), pp. 25–44 (2011)
2. Agrawal, D., Das, S., El Abbadi, A.: Big Data and Cloud Computing: Current State and Future Opportunities. In: 14th International Conference on Extending Database Technology, EDBT (2011)
3. Apache Cassandra, `http://cassandra.apache.org/`
4. Apache Mahout, `http://mahout.apache.org/`
5. Banker, K.: MongoDB in Action. Manning Publications Co. (2012)
6. Berthold, M.R., Cebron, N., Dill, F., Gabriel, T.R., Kötter, T., Meinl, T., Ohl, P., Sieb, C., Thiel, K., Wiswedel, B.: KNIME: The Konstanz Information Miner. In: Studies in Classification, Data Analysis, and Knowledge Organization, GfKL (2007)
7. Chang, F., Dean, J., Ghemawat, S., Hsieh, W.C., Wallahch, D.A., Burrows, M., Chandra, T., Fikes, A., Gruber, R.E.: Bigtable: A Distributed Storage System for Structured Data. In: Seventh Symposium on Operating System Design and Implementation, OSDL (2006)
8. Chu, C.T., Kim, S.K., Lin, Y.A., Yu, Y., Bradski, G.R., Ng, A.Y., Olukotun, K.: Map-Reduce for Machine Learning on Multicore. In: Twentieth Annual Conference on Neural Information Processing Systems (NIPS), pp. 281–288 (2006)

9. Condie, T., Conway, N., Alvaro, P., Hellerstein, J.M., Elmeleegy, K., Sears, R.: MapReduce Online. In: Proceedings of the 7th USENIX Conference on Networked Systems Design and Implementation (NSDI), p. 21 (2010)
10. Czajkowski, G., Dvorsky, M., Zhao, J., Conley, M.: Sorting Petabytes with MapReduce (September 2011)
11. Das, S., Sismanis, Y., Beyer, K.S., Gemulla, R., Haas, P.J., McPherson, J.: Ricardo: Integrating R and Hadoop. In: SIGMOD, pp. 987–998 (2010)
12. Dean, J., Ghemawat, S.: MapReduce – Simplified data processing on large clusters. In: Proceedings of the Sixth Symposium on Operating System Design and Implementation (2004); Journal Version: Communications of the ACM 51(1), 107–113 (2008)
13. Gartner Research: Hype Cycle for Emerging Technologies (July 2011), `http://www.gartner.com/it/page.jsp?id=1763814`
14. Ghemawat, S., Gobioff, H., Leung, S.T.: The Google File System. ACM SIGOPS Operating Systems Review 37(5), 29–43 (2003)
15. Hall, M., Frank, E., Holmes, G., Pfahringer, B., Reutemann, P., Witten, I.H.: The WEKA Data Mining Software: An Update. SIGKDD Explorations 11(1) (2009)
16. Hameurlain, A., Küng, J., Wagner, R., Böhm, C., Eder, J., Plant, C. (eds.): Transactions on Large-Scale Data- and Knowledge-Centered Systems IV. LNCS, vol. 6990. Springer, Heidelberg (2011)
17. IBM: Bringing big data to the enterprise, `http://www-01.ibm.com/software/data/bigdata/`
18. Kelly, J.: Big Data Market Size and Vendor Revenues, Wikibon Report (March 2012), `http://wikibon.org/wiki/v/Big_Data_Market_Size_and_Vendor_Revenues`
19. Manyika, J., Chui, M., Brown, B., Bughin, J., Dobbs, R., Roxburgh, C., Byers, A.H.: Big data: The next frontier for innovation, competition, and productivity. McKinsey Report (May 2011)
20. Miller, G.A.: The Magical Number Seven, Plus or Minus Two: Some Limits on Our Capacity for Processing Information. The Psychological Review 63, 81–97 (1956)
21. Neumeyer, L., Robbins, B., Nair, A., Kesari, A.: S4: Distributed Stream Computing Platform. In: The 10th IEEE International Conference on Data Mining (ICDM) Workshops, pp. 170–177 (2010)
22. O'Reilly Media: Big Data Now (September 2011)
23. Olston, C., Reed, B., Srivastava, U., Kumar, R., Tomkins, A.: Pig latin: a not-so-foreign language for data processing. In: SIGMOD, pp. 1099–1110 (2008)
24. Pavlo, A., Paulson, E., Rasin, A., Abadi, D.J., DeWitt, D.J., Madden, S., Stonebraker, M.: A Comparison of Approaches to Large-Scale Data Analysis. In: SIGMOD, pp. 165–178 (2009)
25. R Development Core Team. R: A language and environment for statistical computing. R Foundation for Statistical Computing, Vienna, Austria (2012) ISBN 3-900051-07-0
26. Russom, P.: Big Data Analytics. TDWI Report (Q4 2011)
27. Stratosphere Research Initiative, `http://www.stratosphere.eu/`
28. Thusoo, A., Sarma, J.S., Jain, N., Shao, Z., Chakka, P., Anthony, S., Liu, H., Wyckoff, P., Murthy, R.: Hive – a warehousing solution over a Map-Reduce framework. PVLDB 2(2), 1626–1629 (2009)
29. Warden, P.: Big Data Glossary. O'Reilly Media Publications, USA (2011)
30. White, T.: Hadoop: The Definitive Guide. O'Reilly Media Publications, USA (2009)

Towards a Smooth E-Justice: Semantic Models and Machine Learning

Elisabetta Fersini[1], Francesco Archetti[1,2], and Enza Messina[1]

[1] University of Milano-Bicocca, Viale Sarca 336,
20126, Milan, Italy
fersini@disco.unimib.it
[2] Consorzio Milano Ricerche, Viale Cozzi 53,
20126, Milan, Italy
{messini,archetti}@disco.unimib.it

Abstract. The dynamic deployment of Information and Communication Technologies in the judicial field, together with the dematerialization of proceedings pushed by e-justice plans, is encouraging the introduction of novel litigation support systems. In this paper we present two innovative systems, JUMAS and eJRM, which take up the challenge of exploiting semantics and machine learning techniques for managing in-court and out-of-court proceedings respectively. JUMAS stems from the homonymous EU research project ended in 2011. It provides not only a streamlined content creation and management support for acquiring and sharing the knowledge embedded into judicial folders, but also a semantic enrichment of multimedia data towards a better usability of judicial folders. eJRM arises from the related ongoing research project funded in the framework PON "Ricerca e Competitività 2007-2013". It exploits semantic representation and machine learning reasoning mechanisms towards a support system for online mediation to encourage the resolution of out-of-court disputes and consequently to increase access to justice.

Keywords: machine learning, semantics, e-justice, integrated systems.

1 Introduction

The judicial sector represents one of the largest information bound professional communities, where cooperation among parties, easy access to information and time/costs savings represent critical issues. The use of Information and Communication Technologies (ICT) is considered one of the key elements, to support both citizens and legal professionals, for overcoming some of the deficiencies of the "justice system". Most of the current ICT development efforts have been focused on the deployment of case management systems and ICT equipments offered at different organizational levels (courts and districts). Although these developments have been pursued at different levels in various EU countries, the trend toward a full e-justice is clearly still in progress. Judicial cooperation as well as accessibility and usability of judicial folders are still affected by a traditional support toolset. However, the growing amount of digital judicial information calls for the development of novel knowledge management mechanisms and their integration into justice management systems. Different commercial solutions have been proposed on the market for addressing data, knowledge and e-discovery issues related to the judicial field (see [1] as survey). Most of the available products provide tools to identify, collect, process, review and produce information related to cases. The functionalities related to those tools are mainly concerned with the automation of manual processes such as archiving trial details, acquiring scanned paper records and OCR, retrieval and consultation of digital material (as for instance minutes and normative), management of courts (event scheduling, court personnel assignment) and drafting dispositions. The keyword that could describe these toolsets is automation, where the role of semantics and the potential support of machine learning are completely disregarded. This opens up new opportunities, both for in-court and out-of-court issues, to develop advanced ICT systems to speed up judicial proceeding, improve efficiency and effectiveness of law disputes and promote confidence in the justice system. In this paper we present two initiatives aimed at introducing semantics and machine learning in ICT judicial systems: JUMAS and eJRM. JUMAS, mainly related to in-court proceedings, faces the issue of a better usability of multimedia judicial folders by addressing three main issues: automatic transcription, text/audio/video annotation and semantic search. eJRM, focused on visualizing the relationship between "Citizen" and "Justice System" for out-of-court proceedings, deals with semantic representation and reasoning mechanisms for improving the awareness of citizens to personally evaluate the outcome of a potential litigation, to be guided to a non-conflict settlement and to be assisted in selecting the eventual legal support.

The paper is structured as follows. In section 2 the JUMAS integrated system is presented by focusing on its relevance for in-court trials and by detailing the main components. In section 3 the eJRM system is outlined by pointing out the dimension of out-of-court law disputes and by describing its relevant functionalities. In section 4 advantages and opportunities related to the proposed system are briefly outlined.

2 The JUMAS System: Support to In-Court Litigations

The progressive deployment of ICT technologies in the courtroom (audio and video recording, document scanning, courtroom management systems), jointly with the requirement for paperless judicial folders pushed by e-justice plans [2], are quickly transforming the traditional judicial folder into an integrated multimedia folder, where documents, audio recordings and video recordings can be accessed usually via a web-based platform. This trend is leading to a continuous increase in the number and the volume of case-related digital judicial libraries, where the full content of each single hearing is available for online consultation. A typical trial folder contains: (1) audio hearing recordings; (2) video hearing recordings; (3) transcriptions of hearings; (4) hearing reports; (5) attached documents (scanned text documents, photos, evidences, etc..). The usability of electronic judicial folders is still affected by traditional support toolsets: information search is limited to text search, transcription of audio recordings (mandatory for text search) is still a "slow" and fully manual process and information extraction is still a manual activity. Moreover, information embedded in audio and video recordings, describing not only what was said in the courtroom, but also the way and the specific trial context in which it was said, still needs to be exploited. Although "information is there", information extraction and semantically empowered judicial information retrieval still waits for proper exploitation tools.

The JUMAS project (*JUdicial MAnagement by digital libraries Semantics*), funded by the EU-7FP and ended in 2011, was aimed at defining multimedia judicial folders and powerful toolsets able to fully address the knowledge embedded in trial recordings. In order to explain the relevance of the JUMAS objectives we report some volume data related to the judicial domain context. Consider for instance the Italian context, where there are 167 courts, grouped in 29 districts, with about 1400 courtrooms. In a law court of medium size (10 court rooms), during a single legal year about 150 hearings per court held with an average duration of 4 hours. Considering that approximately in 40% of them only audio is recorded, in 20% both audio and video while the remaining 40% has no recording, the multimedia recording volume we are talking about is 2400 hours of audio and 1200 hours of audio/video per year. According to this data we can figure out a hypothesis of storage space of about 8.7 MB/min for audio and 39 MB/min for audio/video. During the definition of the space dimension required on a single site, the estimation need also take into account that a trial includes some additional data: (1) textual source as for example minutes in .doc and .pdf format; (2) images; (3) other digital material. Under these hypotheses, the overall size hypothesis (related to criminal trials) for the Italian in-court system in one year is about 800 terabyte.

JUMAS, in order to manage such quantity of complex data, was aimed to: (1) optimize the workflow of information through search, consultation and archiving procedures; (2) introduce a higher degree of knowledge through the aggregation of different heterogeneous sources; (3) speed up and improve decision processes discovering and exploiting knowledge embedded into multimedia documents in order to consequently reduce unnecessary costs. These goals have been achieved by developing functionalities to collect, enrich and share multimedia judicial documents annotated with embedded semantics and automatically generated speech transcriptions.

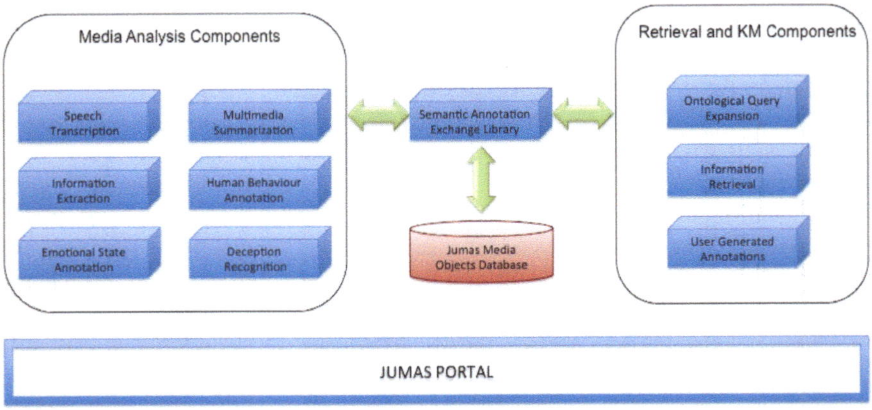

Fig. 1 JUMAS Components

The JUMAS system, as reported in figure 1, is based on a collection of key building blocks: a central database of media objects, a user interface on a web portal, a set of media analysis components, a set of information retrieval and knowledge management components and finally a semantic annotations exchange library. According to this architecture, when a judicial trial is recorded in a courtroom, it is submitted to the Media Analysis Components for extracting relevant semantic tags for the subsequent retrieval activities. The main semantic tags automatically extracted are related to: (1) speech transcriptions, (2) video annotation as human behaviors, (3) audio annotations as emotional states and (4) text annotations as deceptive states and trial relevant details. After this analysis, the media stream together with the extracted semantic tags are indexed into the Jumas Media Object Database through the Semantic Annotation Exchange Library. The Exchange Library supports also the Retrieval and Knowledge Management Components for advanced searching functionalities. Once the judicial trials have been processed and stored, the end user can search for and within trials in an enriched space by using traditional text-based retrieval functionalities as well ontologically augmented query services.

2.1 Media Analysis Components

Speech Transcription. A first fundamental information source, for a digital library related to the courtroom debate context, is represented by the audio recordings of actors involved into hearings/proceedings. The automatic transcription is provided by an Automatic Speech Recognition (ASR) system [3] trained on real judicial data coming from courtrooms. Currently two languages, Italian and Polish, have been considered for inducing the models able infer the transcription given the utterance. Since it is impossible to derive a deterministic formula able to create a link between the acoustic signal of an utterance and the related sequence of associated words, the ASR system exploits a statistical-probabilistic formulation based on Hidden Markov Models [4]. In particular, a combination of two probabilistic models is used: an acoustic model, which is able to represent phonetics, pronounce variability and time dynamics (co-utterance), and a language model able to represent the knowledge about word sequences. The audio acquisition chain in the courtroom has been designed specifically to improve the Word Error Rate (WER) using lossless compression such as FLAC and cross-channels analysis. This allows a good trade-off between the conflicting needs of a manageable dimension of the audio file and good quality recording.

Emotional State Annotation. Emotional states represent a bit of knowledge embedded into courtroom media streams. This kind of information represents hidden knowledge that may be used to enrich the contents available in multimedia digital libraries. The possibility for the end users to consult the transcriptions, also by considering the associated semantics, represents an important achievement that allows them to retrieve an enriched sentence instead of a flat one. This achievement radically changes the consultation process: sentences can assume different meanings according to the affective state of the speaker. In order to address the problem of identifying emotional states embedded into courtroom events, an emotion recognition component based on Multi-layer Support Vector Machines (SVMs) [5] is comprised into the JUMAS system.

Human Behavior Annotation. A further fundamental information source, for a semantic digital library into to the trial management context, is concerned with the video stream. Recognizing relevant events that characterize judicial debates have great impact as well as emotional state identification. Relevant events happening during debates “trigger” meaningful gestures, which emphasize and anchor the words of witnesses, highlighting that a relevant concept has been explained. The human behavior recognition modules enclosed in the JUMAS system capture relevant events that occur during a trial in order to create semantic annotations that can be retrieved by the end users. The annotations are mainly concerned with the events related to the witness: change of posture, change of witness, hand and body gestures, fighting. The modules are based on motion analysis and are able to combine localization and tracking of significant features with supervised classification approaches [6,7]. The set of annotations produced by the human behavior recognition modules provide useful information for the information retrieval process and for the creation of a meaningful summary of the debates.

Deception Detection. The discrimination between truthful and deceptive assertion is one of the most important activity performed by judges, lawyers and prosecutors. In order to support their reasoning activities, aimed at corroborating/contradicting declarations (lawyers and prosecutors) and judging the accused (judges), a deception recognition module has been developed as a knowledge extraction component. The deception detection module, which stands at the end of the data processing chain combining the output of the ASR, Video Analysis, and Emotion Recognition modules, is based on Naïve Bayes and Support Vector Machines classifiers [8]. The distinction between true and deceptive statements requires a labeled corpus with truth value associated with each statement. The underlying models are concerned with lies, contradictory statements, quotations and expressions of vagueness. The deception indications are provided by highlighting relevant statements (derived from verbal expression of witnesses, lawyers and prosecutors) in the text transcription. The identified statements may support the reasoning activities of the judicial actors involved in a trial by triggering relevant portion of the debate representing cues of vagueness and contradiction.

Information Extraction. The current amount of unstructured textual data available into the judicial domain, especially related to transcriptions of debates, highlights the necessity to automatically extract structured data from the unstructured ones for an efficient consultation processes. In order to address the problem of structuring data coming from the automatic speech transcription system, we defined an environment that combines regular expression, probabilistic models and relational information. A probabilistic framework based on Conditional Random Fields (CRFs) [9] for labeling a set of trial transcriptions, has been developed for JUMAS. CRFs, which are discriminative graphical models, have been trained by using both transcriptions as training examples and domain knowledge as additional information. The structured information are exploited in two different ways: (1) to provide additional information to the Retrieval component for an efficient storage and search of trials; (2) to provide a structured sketch of a trial contents for consequently speeding up the consultation process.

Multimedia Summarization. Digital videos represent a fundamental informative source of those events that occur during a trial: they can be stored, organized and retrieved in short time and with low cost. However, considering the dimension that a video source can assume during a trial recording, several requirements have been pointed out by judicial actors: fast navigation of the stream, efficient access to data inside and effective representation of relevant contents. One of the possible solutions to these requirements is represented by multimedia summarization aimed at deriving a synthetic representation of audio/video contents, characterized by a limited loss of meaningful information. In order to address this problem, a summarization environment based on an unsupervised learning approach has been developed to provide both offline [10] and online summaries [11]. Concerning the offline approach, a storyboard of a trial is derived, with no user involvement, by exploiting automatic transcriptions, emotional states and human behaviors. As far is concerned with the online summarization, user query and legal domain ontology are exploited for creating a user-centered storyboard. Both approaches are enclosed and extended a clustering algorithm known as Induced Bisecting K-means [12].

2.2 Retrieval and Knowledge Management Components

Ontological Query Expansion. Textual-based retrieval functionalities are not sufficient for finding and consulting transcriptions (and other documents) related to a given trial. A first contribution of the ontology component developed in the JUMAS system is concerned with its query expansion functionality. Query expansion aims at extending the original query specified by the end users with additional related terms (with a given confidence). The main objective is to narrow the focus (AND query) or to increase recall (OR query). A further functionality offered to the end user is related to the possibility of knowledge acquisition. The ontology component offers to the judicial users the possibility of acquiring specific domain knowledge, i.e. the opportunity of specifying semantic relationships among concepts embedded in trial transcriptions.

Information Retrieval. Currently the retrieval process of audio/video materials acquired during a trial needs the manual consultation of the entire multimedia tracks. The identification of a particular position on multimedia stream, with the aim at looking/listening at/to specific declarations, participations and testimonies, is possible either by remembering the time stamp in which the events were occurred or by watching the whole recording. The conjunction of automatic transcriptions, semantic annotations and ontology representations allow us to build a flexible retrieval environment based not only on simple textual queries, but on wide and complex concepts. In order to define an integrated platform for cross-modal access, a retrieval model able to perform semantic multimedia indexing and retrieval has been developed [13]. In particular, a linear combination of the following information has been developed: (1) similarity of representative frames of shots, (2) face detector output for topics involving people, (3) high level feature considered relevant by text based similarity, (4) motion information extracted from videos, and finally (5) text similarity based on ASR lattices. To this purpose three main retrieval functionalities have been provided to the end user: (a) Basic Search: specification of list of keywords; (b) Advanced Search: specification of linguistic query weights associated with single terms, specification of linguistic quantifiers (most, all, at least n) to aggregate the terms and query translation based on bilingual dictionaries; (c) Semantic Search: specification of wide and complex concepts based on multimedia content or ontological information.

User-Generated Annotations. Judicial users usually tag manually some papers for highlighting (and then remembering) significant portion of a debate. An important functionality offered by the JUMAS system relates to the possibility of digitally annotating relevant arguments discussed during a debate. In this context, the user-generated annotations may help judicial users for future retrieval and reasoning processes. The user-generated annotation module gives three main contribution: (1) it enables judges, prosecutors, lawyers and court clerks to work collaboratively on a trial, e.g. a prosecutor who is taking over a trial can build on the notes of his/her predecessor, (2) it allows end-users to assign free tags to multimedia contents in order to organize the trials according to their personal preferences, (3) it recommends tags for annotating a given trial. The recommendation of tags has been developed as a meta-recommender that integrates collaborative filtering and occurrence-based recommendations.

3 eJRM: electronic Justice Relationship Management

Several studies conducted by the European Commission about out-of-court law disputes have stressed the relevance of ICT to facilitate the resolution of litigations arising both in domestic and cross-border environments. The important contribution that integrated ICT systems could provide in this context can be grasped by pointing out the dimension of litigations addressed through Alternative Dispute Resolution (ADR), i.e. proceedings with no formal court hearing or litigation. According to the 2011 report[1] disclosed by the European Parliament (Economic and Scientific Policy Department) the increasing trend in the use of ADR counts about 410.000 cases in 2006, 473.000 in 2007 and more than 500.000 in 2008. More recent and impressive statistics are related to the Italian context[2], with a particular focus on mediation (one of the available schema for ADR). Since the legislation about mandatory mediation was in force (April-September 2001), about 39.000 cases started. According to the forecast provided by the Italian Ministry of Justice, based on historical data and seasonal trend and taking into account a progressive adoption of mediation by citizens, about 280.000 cases are planned to be addressed through ADR in 2012.

This numbers have envisaged ICT to be the key action in this area, encouraging therefore shifting from Alternative Dispute Resolution to Online Dispute Resolution (ODR). ODR, born from the synergy between ADR and ICT, is a type of dispute resolution involving technology and Internet to facilitate and speed up the resolution of out-of-court disputes. Several initiatives have been investigated for supporting ODR. While commercials products offer Internet-based support toolsets as video conferencing, chat room and templates-based case definition, research initiatives are mainly focused on developing advanced intelligent technologies for helping the resolution of the disputes. A first attempt to apply computational intelligence approaches to ODR is represented by a template-based system known as DEUS [14] that, by requiring the specification of goals and beliefs of litigants, calculates the agreement level in family law property negotiation. In case a settlement is not achieved, the collected information help a mediator to understand what issues are in dispute. More sophisticated systems are represented by Split-Up [15] and Family-Winner [16]. Split-Up is a hybrid framework that combines rule-based systems and neural networks to assist disputes about properties distribution, which is able to provide a “best alternative to a negotiation agreement” to litigants. Family-Winner, which is a game theory based approach for Australian family negotiations, asks to disputants to provide a list of items involved in the litigation and to assign a corresponding “relevance” value to each item. Given this information the system, by reformulating influence diagrams, uses game theory and heuristics to determine a suitable trade-off between claims. A more recent approach is represented by the BEST-project [17], whose main goal is to investigate semantic web technologies as support to law cases retrieval. Ontology-based search, together with ranking functionalities, provides to the parties the opportunity to evaluate claims and liabilities.

---

[1] `http://www.europarl.europa.eu/activities/committees/studies.do?language=EN`

[2] `https://www.giustizia.it/`

All these systems highlight several limitations: (1) claims are collected by a fixed-structure template to be filled in by parties, with no possibility for litigants to provide claims and argumentations by using natural language; (2) the litigations are managed as negotiation, with no direct involvement of a mediator; (3) the outcome of a litigation, when provided to parties, is mainly determined by domain-dependent heuristics; (4) the mechanisms for determining the outcome of a litigation does not take into account potential dependencies among claims and requirements. The call for overcoming these limitations can be therefore easily explained. eJRM, acronym of *electronic Justice Relationship Management*, represents an Italian ongoing initiative aimed at dealing with semantic representation and machine learning reasoning mechanisms for improving the awareness of citizens to personally evaluate the outcome of a potential litigation, to be guided to a non-conflict settlement and to be assisted in selecting the eventual legal support. The main goal of eJRM consists in the development of a platform for managing the relationships between citizen and justice system in order to radically improve two main processes:

***Online Trial*:** Online management of activities related to the mediation process (eFolder management, virtual room meetings, online template filling, etc...)

***Self-Litigation*:** Capability of a citizen to autonomously classify (Case Discovery), formalize (Case Definition) and solve (Case Resolution) a dispute with a third party, with no involvement of legal actors (judges, clerks and lawyers).

These goals are achieved by developing a system infrastructure as depicted in Fig. 2. The bottom part of the architectural overview represents the basic ICT and Information Management infrastructure underlying the eJRM platform.

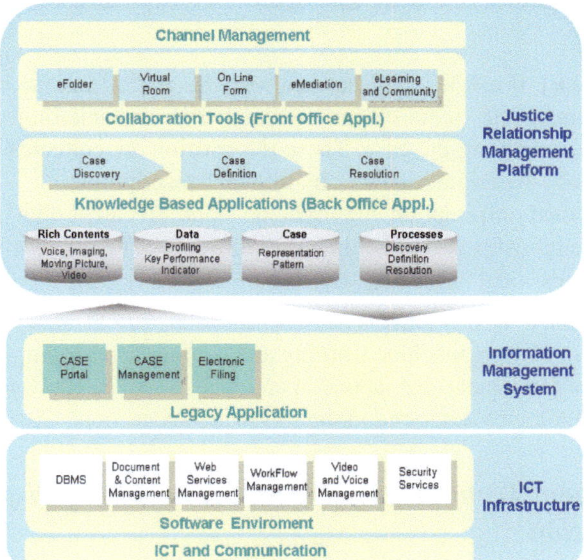

Fig. 2 eJRM functional architecture

The current solutions are mainly related to electronic filing (OCR, picture scanning, etc…), document archiving (DBMS), security services (authentication protocols), video/voice management (audio video acquisition) and case management tools (case archiving and retrieval). The upper part describes the Justice Relationship Management platform, providing both collaboration tools and knowledge management functionalities. The basic applications are mainly related to the set of collaboration tools enabling the dematerialization of the Trial Folder (eFolder), the management of online face-to-face meetings (Virtual Room), the assisted fill in of templates (Online Template) and the support for the online mediation process (eMediation). More advanced functionalities are provided as knowledge management tools, which allow a litigant to determine the type of judicial dispute related to his/her claims (Case Discovery), to collect and analyze relevant argumentation able to describe the case (Case Definition) and finally to provide possible outcome of the dispute together with potential legal support for finalizing the litigation process (Case Resolution). eJRM is based on a steering process for addressing a potential dispute through a nested step-by-step procedure based on automatic reasoning mechanisms. Given a set a formal representation defined by experienced mediators, the functional workflow of the system can be summarized as follows:

[*Case Discovery*]. When a citizen wants to start a mediation process, he/she is firstly guided to identify the nature of the dispute he/she is involved in (family rights, condominium, heritage, etc...).

If eJRM is not able to deal with the given case (no similar cases are available), the system provides to the user a short list of "resolution professional" selected by matching the user needs with the skills and experience of the resolution professionals.

[*Case Definition*]. If eJRM is able to deal with the given case, the system collects pertinent claims and requirements needed for instantiating the concepts representation underlying the case. These information are collected either through a guided ontology-based interview or by free text descriptions.

If the collected information highlight an "outlier case" with respect to the repository of analogous disputes, the system provides to the user a short list of "resolution professional" as in step 1.1.

[*Case Resolution*]. If the case is well formulated, eJRM starts the settlement simulation by exploiting machine learning and automatic reasoning mechanisms.

In order to implement the above mentioned steps, eJRM encloses several intelligent technologies described in the following subsection.

3.1 Intelligent Technologies as Support to Online Dispute Resolution

The eJRM system provides two main building blocks based on semantics and machine learning as support to ODR.

Knowledge Management: Retrieval and Extraction

***Knowledge retrieval*:** Indexing and retrieval of legal documents help both mediators and intelligent mediation support tools. Mediators might need to consult similar cases and norms for proposing suitable agreement to disputants, while intelligent representation and reasoning mechanisms behind "Case resolution" require judicial cases to train instance-based models and semantic-based reasoning mechanisms for subsequently providing possible outcomes of a dispute. In this context eJRM extends keyword-based retrieval functionalities for dealing with semantic and uncertain data related to judicial cases.

***Knowledge extraction*:** A fundamental requirement for instantiating a mediation process is to have a set of key information about the dispute, to be used both by a mediator and by automatic reasoning mechanism. Although traditional ODR systems are based on fixed archetypes (templates) to collect relevant information about a dispute, novel litigation systems should support the need of litigants to express claims and motivations by natural language. Extracting the core of a dispute as structured information from a non-structured text is a key challenge for speeding up both the intervention of judicial operators and for supporting "automatic mediation" mechanisms. Concerning this issue Natural Language Processing techniques, and in particular Conditional Random Fields [9], are investigated for defining a suitable solution able to deal with the legal mediation domain, that is huge, ambiguous and strongly heterogeneous. In order to structure past cases, distinctive claims of previous disputes and key sentences of corresponding verdicts are extracted as knowledge to be exploited by representation and reasoning mechanisms.

Dispute Modeling: mediation representation and semantic-based reasoning

Mediation representation: A semantic representation of a dispute is a first step for allowing mediators to guide the negotiation process as well for supporting reasoning mechanisms to simulate the potential outcome of a dispute. eJRM defines a core ontological representation related to mediation documents (e.g. norms, settlement documents, contracts) and relevant terms (e.g. mediation, parties, family law), for then modeling specific concepts related to a dispute and the relationships among them (e.g. mediation topic, mediation parties involved in the litigation).

Semantic-based Reasoning: The awareness of a citizen about the potential outcomes of a litigation represents a challenging prospective goal. eJRM provides a set of semantic-based reasoning mechanisms to help a disputant to understand, on the basis of past similar disputes, the possible scenarios related to his/her concerning. In particular, by exploiting ontological instances of former similar litigations and the applied dispute case, the system derives a set of characteristics suitable to train machine learning models (Neural Network, Support Vector Machines and Dynamic Bayesian Networks) and to infer the dispute outcomes.

It's easy to guess that eJRM builds upon components and methodological competencies accrued during JUMAS. From a technological point of view

Information Management System and eFolders are adapted to handle mediation data. Concerning computational intelligence issues related to Case Discovery, Definition and Resolution, eJRM encloses and extends mechanisms developed in JUMAS (CRFs, semantic retrieval and ontological representation) to enable automatic mediation processes.

4 Discussion

The wider ranging implications of e-justice over the judicial system as a whole are mainly related to three factors: access to law and information at EU and national level, electronic communication between a judicial authority and the citizen and secure communication between judicial authorities in domestic and cross-border context. The specific contributions given by the proposed systems can be pointed out individually. Concerning JUMAS, the following impacts have been highlighted:

Time savings: it highlights a potential reduction of time spent for the transcription close to 70% and consultation around 90%.

Costs savings: the services needed for providing hearing transcription, per legal year along the national territory, can be reduced of 50%. For the Italian context this implies an approximate cost reduction from 20K€ to 10K€.

Overcoming of geographical limits: no geographical limits are settled for consulting judicial folders. Judicial actors involved in a trial can consult and enrich the multimedia judicial folder according to their roles and rights.

Enhancing consultation: it enhances the quality of the judicial decisions by providing semantic annotations of trial events. The JUMAS portal has been perceived as a straightforward tool for supporting common daily activity such as retrieval of transcriptions, consultation of multimedia streaming and annotation of relevant contents.

Enhancing traceability of trials: JUMAS, thanks to a cross-retrieval functionality, can improve the awareness of linked trials enabling then connections and similarities between different cases.

The expected impacts of eJRM are:

Time savings: while litigation may take place months after the event, mediation through eJRM can takes place immediately after the dispute arises. Moreover, virtual meetings among parties could be arranged at close intervals, which are typically shorter than the in-court litigations.

Costs savings: the ODR services provided by eJRM limit the traveling expenses, increasing therefore the access to justice for many disputants.

Overcoming of geographical limits: no geographical limits are settled for approaching a dispute through eJRM. Parties can either negotiate autonomously a dispute or start a settlement simulation by taking advantage of Internet.

Enabling asynchronous proceedings: e-mediation is flexible thanks to asynchronous communication services provided to the parties. Disputants can participate in a mediation proceeding at convenient times.

Enhancing e-mediation: litigants can be guided to automatically identify the nature of the dispute, to provide pertinent claims and requirements and to finally reach an agreement.

Face-to-face contact: eJRM allows parties to participate in mediation as well as Alternative Dispute Resolution, enabling oral discussion and consequently overcoming speech and body language limitations.

Enabling natural language argumentation: eJRM provides to disputants the possibility to argument their claims and requirements by using free text given in natural language statements.

The most valuable contribution is related to the intelligent support the JUMAS and eJRM can provide not only as tools for a better usability of "trial" folders, but also input for a more future oriented specification of e-justice systems.

Acknowledgments. This work has been supported by the European Community under the JUMAS project (ref.: 214306), by the framework PON "Ricerca e Competitività 2007-2013" under the eJRM projects (re.: PON01_01286), and by "Dote ricercatori"- FSE, Regione Lombardia.

References

1. Kershaw, E., Howie, J.: eDiscovery institute survey on predictive coding. TR-Electronic Discovery Institute (2010)
2. Council of the European Union. Multi-annual european e-justice action plan 2009-2013. The Official Journal of the European Union, C75/1 (March 31, 2009)
3. Falavigna, D., Giuliani, D., Gretter, R., Loof, J., Gollan, C., Schlueter, R., Ney, H.: Automatic transcription of courtroom recordings in the jumas project. In: Proc. of ICT Solutions for Justice (2009)
4. Rabiner, L., Juang, B.H.: Fundamentals of Speech Recognition. Prentice-Hall Inc. (1993)
5. Fersini, E., Messina, E., Archetti, F.: Emotional state in judicial courtrooms: an experimental investigation. Speech Communication 54(1), 11–22 (2012)
6. Briassouli, A., Tsiminaki, V., Kompatsiaris, I.: Human motion analysis via statistical motion processing and sequential change detection. EURASIP Journal on Image and Video Processing (2009)
7. Kovács, L., Utasi, Á., Szirányi, T.: VISRET – A Content Based Annotation, Retrieval and Visualization Toolchain. In: Blanc-Talon, J., Philips, W., Popescu, D., Scheunders, P. (eds.) ACIVS 2009. LNCS, vol. 5807, pp. 265–276. Springer, Heidelberg (2009)
8. Ganter, V., Strube, M.: Finding hedges by chasing weasels: hedge detection using wikipedia tags and shallow linguistic features. In: Proc. of the ACL-IJCNLP, pp. 173–176 (2009)
9. Lafferty, J.D., McCallum, A., Pereira, F.: Conditional random fields: Probabilistic models for segmenting and labeling sequence data. In: Proc. of the 18th International Conference on Machine Learning, pp. 282–289 (2001)

10. Fersini, E., Messina, E., Archetti, F.: Multimedia Summarization in Law Courts: A Clustering-based Environment for Browsing and Consulting Judicial Folders. In: Proc. of the 10th Industrial Conference on Data Mining (2010)
11. Fersini, E., Sartori, F.: Semantic storyboard of judicial debates: a novel multimedia summarization environment. Program: Electronic Library and Information Systems 42(2) (2012)
12. Archetti, F., Campanelli, P., Fersini, E., Messina, E.: A Hierarchical Document Clustering Environment Based on the Induced Bisecting k-Means. In: Larsen, H.L., Pasi, G., Ortiz-Arroyo, D., Andreasen, T., Christiansen, H. (eds.) FQAS 2006. LNCS (LNAI), vol. 4027, pp. 257–269. Springer, Heidelberg (2006)
13. Darczy, B., Nemeskey, D., Petrs, I., Benczr, A.A., Kiss, T.: Sztaki@trecvid 2009 (2009)
14. Zeleznikow, J., Meersman, R., Hunter, D., van Helvoort, E.: Computer tools for aiding legal negotiation. In: Proc. of the 6th Australasian Conference on Information Systems (1995)
15. Stranieri, A., Zeleznikow, J.: The Split_Up system: Integrating neural networks and rule-based reasoning in the legal domain. In: Proc. of the Fifth International Conference on Artificial Intelligence and Law (1995)
16. Bellucci, E., Zeleznikow, J.: Family winner: A computerised negotiation support system which advises upon australian family law. In: ISDSS 2001, pp. 74–85 (2001)
17. Uijttenbroek, E.M., Lodder, A.R., Klein, M.C.A., Wildeboer, G.R., Van Steenbergen, W., Sie, R.L.L., Huygen, P.E.M., van Harmelen, F.: Retrieval of Case Law to Provide Layman with Information about Liability: Preliminary Results of the BEST-Project. In: Casanovas, P., Sartor, G., Casellas, N., Rubino, R. (eds.) Computable Models of the Law. LNCS (LNAI), vol. 4884, pp. 291–311. Springer, Heidelberg (2008)

Weaving Personal Knowledge Spaces into Office Applications

Heiko Maus, Sven Schwarz, and Andreas Dengel

Knowledge Management Department,
German Research Center for AI, Trippstadterstr. 122, Germany
andreas.dengel@dfki.de

Abstract. The paper presents recent developments in our research on Semantic Desktop for personal knowledge management supported by an ecosystem of applications and plug-ins using the knowledge worker's Personal Information Model (PIMO) – a formal representation of his mental model for knowledge work – in everyday applications. We explain how the infrastructure enables the availability of the PIMO as one vocabulary throughout different applications as well as mobile access, the importance of the mental model in the PIMO, and how to get direct benefits from the PIMO in daily activities. We also address steps towards building a Group Information Model from individual PIMOs within the ecosystem.

Keywords: Semantic Desktop, Personal Information Management, PIMO, Knowledge Management.

1 Introduction

The modern working environment places high requirements on knowledge workers: they are confronted with various applications, are involved in several projects and processes, work in changing teams, are on the road with a mobile office, and finally, face an ever increasing flow of information. The resulting knowledge spaces are complex, dynamic, distributed over several applications, and use different vocabulary. It is hard to keep the overview in the resulting personal knowledge space.

This challenge is addressed with the concept of the Semantic Desktop [1,2]. It follows the strategy to embed the mental model of the knowledge worker in daily work by means of a Personal Information Model ("PIMO"). The user's mental model in the PIMO consists of concepts (called "Things" such as specific topics, projects, persons, tasks, …), associations between them (persons are *member of* projects, a task *has topic* "Semantic Desktop", ...), and finally, associated resources (documents, e-mails, web pages, pictures, …) (see [3] for a detailed motivation of the PIMO).

The PIMO serves as an easy to understand conceptualization of the knowledge worker's mental model, which can be used as a common vocabulary across different applications. Therefore, the PIMO provides the means required for a multi-criterial document classification considering the user's subjective view [4]. Fig. 1 shows an example of a PIMO graph with resources from the file system, the web, a task tool. The resources are associated with topics, events, tasks, etc. This graph then serves in different applications to find and access resources or things, to annotate and to relate them (this vocabulary is used in the upcoming figures as well).

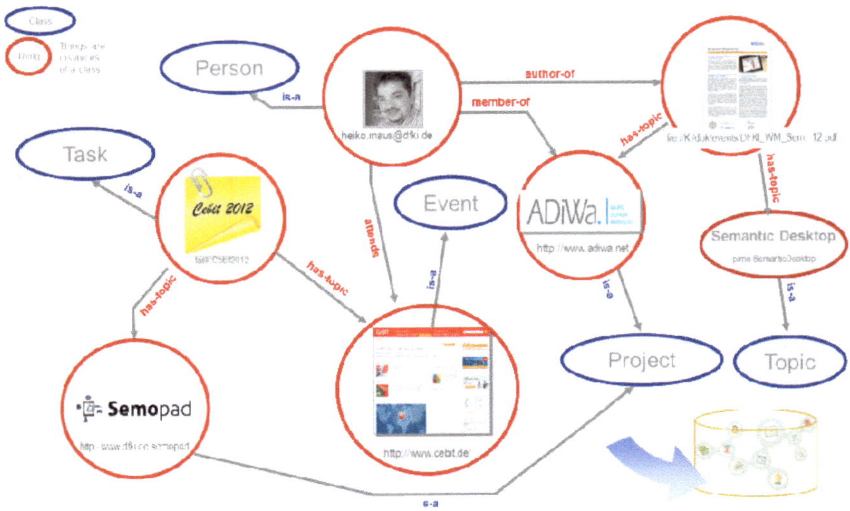

Fig. 1 A schematic excerpt of a Personal Information Model: Things, classes, resources, and associations

The PIMO uses the semantic power of the formal representation of the PIMO ontology[1] [3], thus, introducing a knowledge representation layer on the user's computer. Besides enabling to annotate and interconnect resources over application borders, further semantic services are possible which make use of the semantic representation of the user's mental model in the PIMO.

Challenges of the Semantic Desktop are the initial bootstrapping of a user's PIMO to cover a relevant part of his current mental model of his knowledge work and the ubiquitous availability of the things and resources in the user's daily work activities. To support this, the Nepomuk[2]-project developed a personal knowledge workbench for the Semantic Desktop [5] providing a comprehensive user interface to create, access, and maintain a PIMO as well as introducing new resources or things into it, e.g., by annotating files, e-mails, and web-pages or writing notes in

[1] http://www.semanticdesktop.org/ontologies/2007/11/01/pimo/

[2] http://nepomuk.semanticdesktop.org/

the semantic wiki. Although the workbench provides valuable means to work with the PIMO for Personal Information Management [6], a comprehensive integration into the user's daily applications was minimal and, hence, the users lacked in-situ support in applications such as email client, web browser or file explorer.

This paper presents an advanced infrastructure for the PIMO ecosystem (section 2) enabling a plug-in architecture for in-situ PIMO access throughout different applications (section 3), thus, providing the knowledge worker one vocabulary for his work, regardless of the application or location. In contrast to the Nepomuk PIMO, which was stored on the user's desktop, this new architecture allows ubiquitous access by storing PIMO data in the cloud, thus, allowing to apply sharing to group members.

2 PIMO Infrastructure as Semantic Middleware on the Desktop

As mentioned above, the PIMO is introduced as a knowledge representation layer on the user's desktop. Now, the PIMO is a cloud-based service and provides a service API based on JSON RPC (see Fig. 2). The PIMO Service API uses the PIMO schema with its classes and properties and intended semantics, relies on URIs[3] to identify things and resources, and most importantly, defines a set of methods to access and manipulate the PIMO. In contrast to typical semantic web approaches, the service API does not allow direct access of the core data. So, for instance, the data cannot be read or modified using SPARQL[4] or alike. Instead, a designated set of methods guarantees a consistent and privacy-safe access to the PIMO. It also provides specialized services such as proposals of relevant things for a given text (see information extraction in [8]), a history of used/modified resources and things, as well as, a feed for recently shared concepts, etc. With this approach, we also connected the concept map based knowledge base used in the agile knowledge workflow-tool TaskNavigator [14].

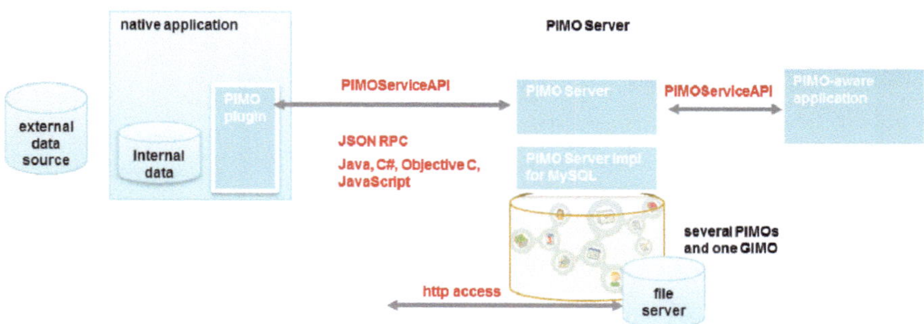

Fig. 2 PIMO architecture

[3] Universal Resource Identifier.

[4] `http://en.wikipedia.org/wiki/SPARQL`

On the client side, we have implementations in different languages such as Java, C#, Objective C, Mozilla's XUL, and JavaScript, allowing to embed in-situ access to the PIMO as a plug-in (such as a sidebar) in different applications (see Fig. 3 to Fig. 5), thus, making it available in standard office applications such as e-mail clients, web browsers, and even in the Windows File Explorer. With the same plug-in mechanism, we also implemented various observation components in office applications to observe the user's actions and information items used, e.g., with our DragonTalk system for the Mozilla suite[5].

This PIMO infrastructure serves as a *semantic middleware* on the user's desktop realizing the knowledge representation layer and interconnecting various types of applications.

Annotating a new resource will create a new "thing" for that resource in the PIMO. For example, when the PDF file containing the flyer in Fig. 1 (top right) is annotated with the project "ADiWa", a new instance of type *pimo:Document* is created for the file resource. The new thing gets a new, unique URI and is added to the PIMO. This procedure is called "rebirth" as the resource already exists on the file system but is now also represented in the PIMO. The resource's originating location (file path, URL, etc.) is stored as "grounding occurrence" in the PIMO, that way, the provenance is captured and the PIMO GUI can easily "open" the native file when double-clicking on the resource, for example. Metadata such as title/subject, author, recipient, sender, etc. are also present in the PIMO.

Once reborn, annotating a thing with some concept in the PIMO is done by adding an association using the PIMO property *pimo:hasTopic*. If the user is willing to disclose a thing, which has an associated grounding occurrence, the user

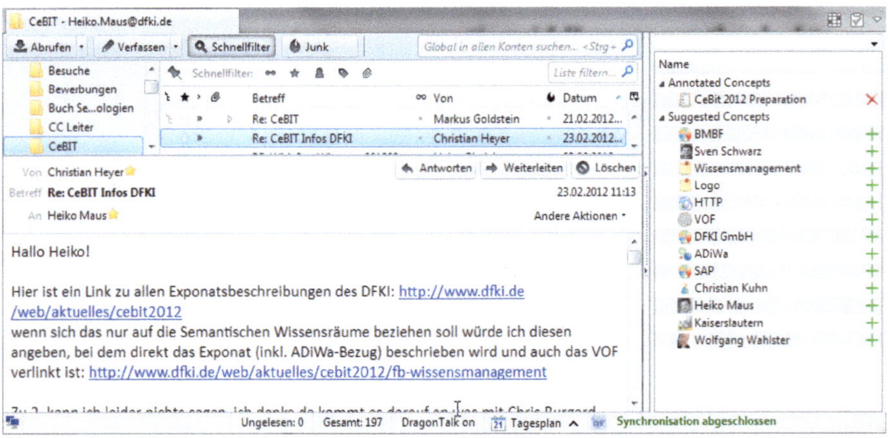

Fig. 3 PIMO sidebar in Mozilla Thunderbird with one annotated thing (a task for a trade fair) and suggesting further things based on analysis of the full text. Annotating a thing allows then to directly open the email from whatever application the thing is accessed.

---

5 `http://dragontalk.opendfki.de/`

is asked whether the resource should also be shared. Sharing a native file resource then means that the file gets uploaded to the PIMO server and from then on it is available for other users as well as from different devices.

For the research prototype, currently, only a very simple sharing mechanism is implemented, neglecting the task to detect local or remote version changes and updating the versions automatically. Thus, this feature currently only supports *publishing* content like papers, slides, or web pages for a certain topic. However, we understand the importance of an advanced approach for a full-fledged working environment; so, a more sophisticated mechanism is on the agenda.

3 PIMO-Enabled Applications for Knowledge Work

From our longstanding research in this area – with questions tackling the PIMO, value-added semantic services, and several implementations of a Semantic Desktop (EPOS, gnowsis, Nepomuk, Refinder[6] from our spin-off gnowsis.com, as well as the PIMO ecosystem presented here) – we see that the main challenge is to face one of the main hindrances in (personal) knowledge management: the individual effort to be invested for getting benefits out of the system. This divides into the individual effort for the ramp-up (how fast can the user start to work beneficially with such a system), into the effort for actually constantly using it, and into effort required for maintaining the knowledge base. And finally, the cognitive effort required for the user to understand the PIMO and to work with it. In the following, we want to address several foci of our research to reduce this effort.

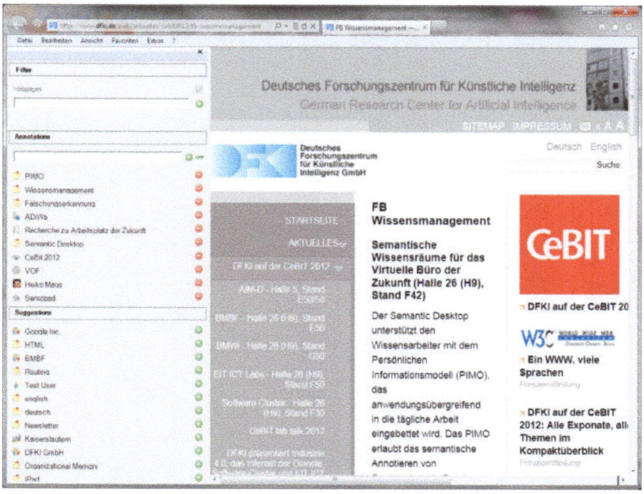

Fig. 4 PIMO sidebar embedded in MS Internet Explorer

[6] `http://getrefinder.com`

Availability in daily activities: First of all, for today's knowledge workers, the PIMO has to be ubiquitous for them and must embrace the information items of their knowledge work. A single Semantic Desktop application alone would require bringing information items to this central place, requiring effort and risking to create yet another knowledge base isolated from the information items in the applications. Now, the previously introduced semantic middleware allows to use the PIMO also in standard applications as long as they can be extended with code using the PIMO Service API. Our current focus is on office applications such as e-mail clients (Mozilla Thunderbird as in Fig. 3 and MS Outlook SmartOffice-Plugin), web-browsers (MS Internet Explorer (see Fig. 4) and Mozilla Firefox), personal information management tools (Nepomuk Personal Semantic Workbench [5]), and task management tools (ConTask [9]). Basic functionality of all plug-ins is to search and access things and to annotate resources with things directly from within the application (see Fig. 3 for an annotated e-mail). That means, users now have the benefit of interconnecting resources from and within different (native) applications using the same vocabulary.

Also of importance for the user is the ease of file handling on his desktop, because the standard office applications rely on files to store documents. There is a lot of structuring done by the user reflecting his mental model from choosing names for files and folders to clustering files into folders up to the folder hierarchy. Handling files on the desktop is still an essential activity in today's knowledge work. Therefore, we also embedded PIMO access into the Windows File Explorer (see Fig. 5, realized as a sidebar plug-in and a namespace extension) with the possibility to easily annotate files and folders with things (and thus, also rebirth files in the PIMO, or to create things out of folders). The sidebar also shows the annotated concepts for selected files and folders and allows a concept-based search/filter for files or folders using PIMO things. That way, the PIMO provides a means for navigation directly in the file explorer itself.

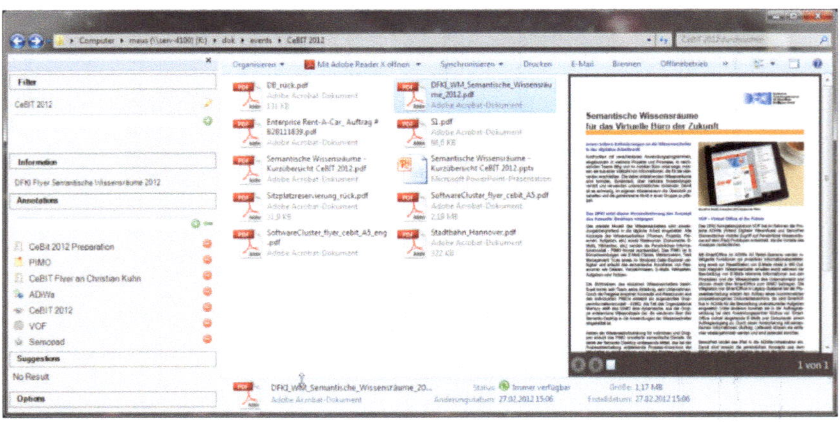

Fig. 5 Semantic File Explorer: PIMO vocabulary embedded in MS File Explorer. Filter Event "CeBIT 2012" is set, a flyer is selected and annotated things are listed.

Mental model for knowledge work: To get the most benefit of the PIMO for knowledge work, the relevant part of the user's mental model has to be reflected in the PIMO. We addressed this in [17] by populating the PIMO with the user's native structures (folders-hierarchies in file system, e-mail folders, etc.) found on his desktop as digital footprints of his mental model [17]. A recent approach is bootstrapping a PIMO from a user's e-mail by crawling his e-mail account and identifying PIMO-relevant things [12]. More things, resources and relationships between them are then gathered by the PIMO enhanced daily usage of office applications. After a few days, the PIMO will already provide the network of concepts and resources centered on a user's tasks. Although the formal representation of the PIMO allows for a rich semantic modeling of concepts and their relationships, the study of several PIMOs and their usage over time in [6,16] showed that for most users it is sufficient to see that things are connected and there is no need for a more specific semantic relationship between things for their purposes. Again, to reduce the cognitive effort for users, we apply tagging as an easy-to-use functionality that helps users to quickly weave their personal knowledge space without forcing decisions about the actual semantic relationship. We successfully applied this also in other situations, e.g., tagging tasks in agile knowledge workflows with concepts in the TaskNavigator [14]. However, we still stick to semantic relations, but focus on automatic predictions of relationship types. Dedicated applications can decide automatically on the right (domain specific) semantic relationships. This works in the task management domain (subtasks, executors, used resources), or in a tool for meeting protocols (attendees, agenda). Furthermore, some metadata can be sensed or observed and stored automatically, e.g., in the mobile scenario we can use the sensors of a mobile device (time, location), or, user observation software can observe user behavior and resource usage automatically (opened files, visited folders or web pages; see [15]).

Direct benefits from the PIMO: Again, derived from the challenge of a satisfactory return on invested effort for the user, direct benefits for the user need to be established. We accomplish this by providing an ecosystem of tools and services using the PIMO and supporting the user's daily requirements for personal knowledge management. These tools and plug-ins are designed to provide direct benefits for the user:

- Tools for personal knowledge management allowing to acquire, create, classify, and organize information items using the PIMO such as the SemanticFileExplorer in Fig. 5.
- Information retrieval components providing fast access to information items via common keyword search, associative search, semantic search, and combinations of them based on the semantic annotations using the PIMO and the full text of the information items.
- Ubiquitous PIMO: One vocabulary throughout all applications, to be accessed via plug-ins to organize and interconnect information items. The PIMO is

available everywhere for ease of access and for annotating information items on the fly in order to constantly evolve the PIMO. Further assistance and coverage is achieved by an automatic interpretation and analysis of information items in focus and proactively suggesting annotations and related things (such as the suggesting relevant things for an opened email in Fig. 3).

- Allowing to use information from the group and share information with the group easily (see next section).

Furthermore, the availability of a machine-readable vocabulary of the user's mental model allows easy development of new added-value services supporting the user in his personal knowledge management.. Examples are the semantic search on the user's personal knowledge space [7], ontology-based information extraction using the PIMO as background knowledge [8], personal image collections using the PIMO [10], semantic annotation embedded in office documents [11], and personal trend recognition in the PIMO [18]. From a research perspective, the PIMO allows to get a more precise understanding of the user's current activities and topics. Therefore, in our user observation for context-aware services in [15,17], the PIMO serves as a vocabulary for identifying and formally representing the user's interests, current activities, and the context he is in. With this infrastructure as a basis, we developed context-aware services providing proactive information delivery based on the user's context [15]. Moreover, an agile personal task management embedded in the Semantic Desktop (ConTask, see [9]) uses the PIMO for organizing a user's tasks and their contents, applying task and context oriented proactive information delivery. It observes the user's actions to connect them to the respective tasks and identifies task switches of the user to keep the proactive information delivery always aligned to the user's current work. All these plug-ins and tools are part of the presented PIMO ecosystem providing direct support of activities in the user's knowledge work.

Extending the Personal Knowledge Space to the Group: The views of individual knowledge workers also influence their team, their department, their company. By allowing to share things and resources from individual PIMOs a Group Knowledge Space is evolving. As this is based on the PIMO, we refer to this as the *GIMO* – the *Group Information Model.* We apply a bottom-up approach, allowing users to easily share things and resources to the group as well as being able to directly use things and resources from the group for the individual work. Again, to infuse the GIMO to the group members and keeping a momentum of sharing and using of information, several value added services are required. In our current ecosystem we use the GIMO also for concept proposals for annotating (personal) information elements, provide RSS feeds for activities on the GIMO (shared concepts and resources, changes, annotations), and being a source for proactive information delivery, e.g., in agile knowledge workflows as in TaskNavigator [14]. Therefore, the GIMO is a dynamic knowledge base evolving from individuals interacting in the group and resembling a part of the Organizational Memory. By means of the Semantic Desktop, it is in turn still available within the knowledge worker's applications, and thus, group knowledge is always available during the individual knowledge work.

With the evolving GIMO from individual PIMOs, we see several research questions arising such as maintenance issues, community building, as well as cooperative domain ontology creation from individual PIMOs (see also earlier work in [13]).

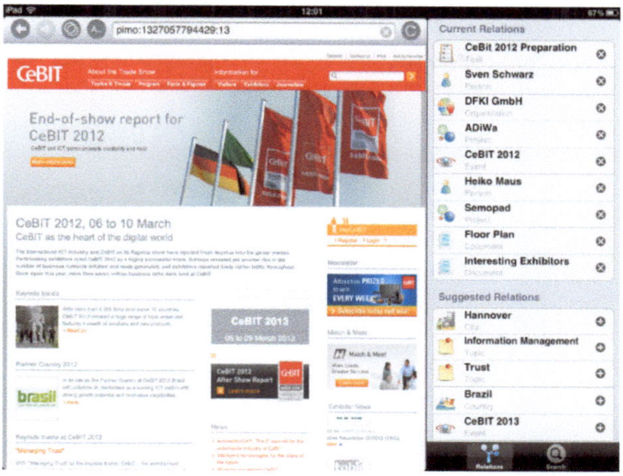

Fig. 6 Semopad: A web page with annotated and suggested things from the PIMO on the iPad

Mobile access to the Personal Knowledge Space: In the Semopad project[7] we bring the power of the Semantic Desktop paradigm to mobile devices. As Apple's iPad is a favorite and often used tablet device, we realized an iOS App for the iPad 2 (see Fig. 6). The software is still being developed and is not yet available via the App Store. Although accesses the PIMO and provides an interface similar to the sidebars we implemented so far such as for Mozilla Firefox. Users can browse the internet and get to see existing and suggested annotations next to the web page at hand.

The related PIMO concepts can be viewed and browsed. For searching things in the PIMO, Semopad realized a faceted search (see Fig. 7). That way, the Semopad App can be used in the mobile setting to quickly annotated found web pages, as well as, to efficiently look up information in the PIMO. As the user does not want to enter much text, the project focuses on use cases with minimal required interaction, particularly minimal taps on the (virtual) keyboard. Therefore, the PIMO provides an interactive search for things, requiring only the beginning or part of the concept's name and showing the options while the user is typing. This feature is, of course, also available in the desktop GUIs, but in the App it is crucial.

---

[7] https://www2.dfki.de/intranet/research/projects/Project_674

The App can also be run when being offline and uses a sophisticated caching mechanism which allows high performance in slow network scenarios, e.g., when having to use a bad or slow UMTS connection: The App always shows the cached information first while, in the background, the PIMO is queried for recent changes and respective updates are requested. As soon as an updated version of a displayed thing is available, the display is corrected accordingly. So, when a Semopad user is inspecting a thing and a remote colleague (e.g., in the company building) is adding a new interesting resource to the thing, the Semopad user sees the changes right away, respectively, as soon as he is online again (after lacking internet connection).

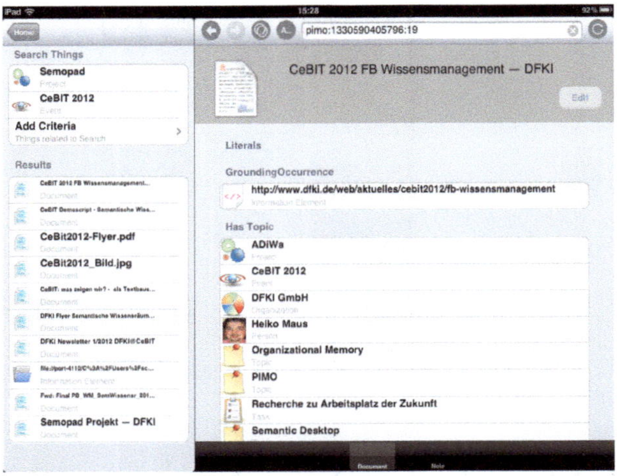

Fig. 7 Faceted search in Semopad and viewing a PIMO concept

4 Conclusions

The paper presented a new ecosystem of applications and plug-ins for the Semantic Desktop approach. The goal is to provide knowledge worker with one vocabulary – the PIMO – across application borders. Furthermore, starting from previous research and systems, the individual PIMOs are now embedded in a GIMO that will give the opportunity to follow a bottom-up approach for knowledge management in groups. We presented several new prototypes and referenced other systems belonging to the ecosystem and also gave outlook to further research that will be addressed in future work. We expect more insights into the question on how the ubiquitous availability will influence the PIMO build up, its usage, and the user benefit in knowledge work as well as the influence to cooperative construction of the GIMO. This insight will be especially complemented by the experience from our spin-off gnowsis.com with their product Refinder which follows a similar approach with individual PIMOs and sharing to friends.

Acknowledgments. Parts of this work were supported by the projects ADiWa (funded by German Federal Ministry of Education and Research - 01IA08006) and Semopad (funded by Stiftung Rheinland-Pfalz für Innovation - 961-386261/1001).

References

1. Sauermann, L., Bernardi, A., Dengel, A.: Overview and Outlook on the Semantic Desktop. In: Proc. of the First Semantic Desktop Workshop at the ISWC Conference 2005 (2005)
2. Dengel, A.R.: Knowledge Technologies for the Social Semantic Desktop. In: Zhang, Z., Siekmann, J.H. (eds.) KSEM 2007. LNCS (LNAI), vol. 4798, pp. 2–9. Springer, Heidelberg (2007)
3. Sauermann, L., van Elst, L., Dengel, A.: PIMO - A Framework for Representing Personal Information Models. In: Proceedings of I-SEMANTICS Conference, Graz, September 5-7. J.UCS, pp. 270–277. Know-Center, Austria (2007)
4. Dengel, A.: Six Thousand Words about Multi-Perspective Personal Document Management. In: Proceedings EDM, IEEE Int'l Workshop on the Electronic Document Management in an Enterprise Computing Environment, Hong Kong (2006)
5. Grimnes, G.A., Sauermann, L., Bernardi, A.: The Personal Knowledge Workbench of the NEPOMUK Semantic Desktop. In: Aroyo, L., Traverso, P., Ciravegna, F., Cimiano, P., Heath, T., Hyvönen, E., Mizoguchi, R., Oren, E., Sabou, M., Simperl, E. (eds.) ESWC 2009. LNCS, vol. 5554, pp. 836–840. Springer, Heidelberg (2009)
6. Sauermann, L., Heim, D.: Evaluating long-term use of the Gnowsis Semantic Desktop for PIM. In: Sheth, A.P., Staab, S., Dean, M., Paolucci, M., Maynard, D., Finin, T., Thirunarayan, K. (eds.) ISWC 2008. LNCS, vol. 5318, pp. 467–482. Springer, Heidelberg (2008)
7. Schumacher, K., Sintek, M., Sauermann, L.: Combining Fact and Document Retrieval with Spreading Activation for Semantic Desktop Search. In: Bechhofer, S., Hauswirth, M., Hoffmann, J., Koubarakis, M. (eds.) ESWC 2008. LNCS, vol. 5021, pp. 569–583. Springer, Heidelberg (2008)
8. Grimnes, G.A., Adrian, B., Schwarz, S., Maus, H., Schumacher, K., Sauermann, L.: Semantic Desktop for the End-User. i-com, vol. 8, pp. 25–32. Oldenbourg Verlag (2009)
9. Maus, H., Schwarz, S., Haas, J., Dengel, A.: CONTASK: Context-Sensitive Task Assistance in the Semantic Desktop. In: Filipe, J., Cordeiro, J. (eds.) ICEIS 2010. LNBIP, vol. 73, pp. 177–192. Springer, Heidelberg (2011)
10. Klinkigt, M., Kise, K., Maus, H., Dengel, A.: Semantic Retrieval of Images by Learning from Wikipedia. In: König, A., Dengel, A., Hinkelmann, K., Kise, K., Howlett, R.J., Jain, L.C. (eds.) KES 2011, Part IV. LNCS, vol. 6884, pp. 212–221. Springer, Heidelberg (2011)
11. Rostanin, O., al Agroudy, P.: SWord: Semantic Annotations Revisited. In: König, A., Dengel, A., Hinkelmann, K., Kise, K., Howlett, R.J., Jain, L.C. (eds.) KES 2011, Part II. LNCS, vol. 6882, pp. 410–419. Springer, Heidelberg (2011)
12. Schwarz, S., Marmann, F., Maus, H.: Extracting Personal Concepts from Users' Emails to Initialize Their Personal Information Models. In: König, A., Dengel, A., Hinkelmann, K., Kise, K., Howlett, R.J., Jain, L.C. (eds.) KES 2011, Part II. LNCS, vol. 6882, pp. 430–439. Springer, Heidelberg (2011)

13. Aschoff, F.-R., Schmalhofer, F., van Elst, L.: Knowledge Mediation: A Procedure for the Cooperative Construction of Domain Ontologies. In: Motta, E., Shadbolt, N.R., Stutt, A., Gibbins, N. (eds.) EKAW 2004. LNCS (LNAI), vol. 3257, pp. 506–508. Springer, Heidelberg (2004)
14. Rostanin, O., Maus, H., Suzuki, T., Maeda, K.: Using lightweight knowledge modeling to improve proactive information delivery. In: Proc. of the 2nd Int'l Conference on Agents and Artificial Intelligence (ICAART 2010), January 22-24, pp. 611–614. INSTICC, Valencia (2010)
15. Schwarz, S.: Context-Awareness and Context-Sensitive Interfaces for Knowledge Work Support. PhD-Thesis. University of Kaiserslautern, Verlag Dr. Hut (2010)
16. Sauermann, L.: The Gnowsis Semantic Desktop approach to Personal Information Management. PhD-Thesis. University of Kaiserslautern. Verlag dissertation.de (2009)
17. Sauermann, L., Dengel, A., van Elst, L., Lauer, A., Maus, H., Schwarz, S.: Personalization in the EPOS project. In: Proceedings of the Semantic Web Personalization Workshop at the ESWC 2006 Conference, Budva, pp. 42–52 (2006)
18. Kubo, S., Tsuji, H., Maus, H., Dengel, A.: Visualizing Personal Trend on Gnowsis Semantic Desktop. In: IEEE International Conference on Systems, Man, and Cybernetics (SMC 2008), pp. 2765–2771. IEEE (2008)

Simulation-Based Knowledge Management in Airport Operations

Saeid Nahavandi, Doug Creighton, Michael Johnstone,
Vu Thanh Le, and James Zhang

Centre for Intelligent Systems Research, Deakin University, Geelong, Australia
{Saeid.Nahavandi,Douglas.Creighton,Michael.Johnstone,
Vu.Le,James.Zhang}@Deakin.edu.au

Abstract. Capturing and retaining knowledge in any organization is a major challenge. This talk describes how these challenges have been addressed through simulation and modeling techniques for complex engineered systems. A series of case studies that focus on airport processes are used to demonstrate the concepts. Furthermore, the additional benefits that a simulation model can bring, through online control and decision-making support, are discussed.

Keywords: Simulation Modeling, Knowledge Management.

1 Introduction

An organization's knowledge is its intellectual capital, including both the tangible and intangible assets [1]. In the information economy, the proper management of knowledge will give a significant competitive edge to an organization [2, 3]. Knowledge management can be broken into several phases: acquisition (access and codification), generation and sharing. These phases interconnect to constitute a never-ending learning cycle for an organization.

To achieve the goals of good knowledge management, many knowledge methods have been proposed and implemented, such as mentoring, training and development, knowledge project, knowledge repository, communities of practice, intermediary role, story-telling, collaboration, social network analysis, scenarios, knowledge mapping and experiments [4]. The explosive developments of the Internet, ubiquitous computing and cloud technology in the last decades have ushered in many novel tools for knowledge management, such as discussion forums, weblogs and wikis [5] and Visual Wiki [6].

These novel methods are very helpful in capturing the implicit and tacit knowledge and text-based search can be used to extract knowledge from the data [7]. However, in terms of knowledge visualization and knowledge generation, computer simulation enjoys distinct advantages that have not been realized by many organizations. Recently, researchers have begun to specifically consider the connection between simulation and knowledge management [8, 9]. A simulation model can be viewed as a virtual process parallel to the real world and as such it encodes a great deal of knowledge about the real world. Furthermore, in simulation, the model building process drives knowledge acquisition; its final result typically accords understanding to the human agents in an intuitive way.

Simulation has several paradigms: agent-based simulation (ABS), system dynamic (SD) simulation and discrete event simulation (DES) [10-12]. In this paper we will focus on discrete event simulation, the most widely used technique of the three. A discrete event simulation can not only provide accurate prediction of the system behavior (hard system thinking), it can also be used to facilitate problem understanding and management learning (soft system thinking) [12, 13]. More specifically, a computer model provides a concrete platform on which people can base their discussion rather than relying on the manipulation of abstract concepts in mind, thus facilitating the generation of new knowledge, both tangible and tacit. As there is no generally agreed definition of knowledge, to further explore the interaction between simulation and knowledge, a specific perspective called computational information processing system [14, 15], is adopted here. This perspective is used in our previous study [8] to interpret the simulation efforts for a warehouse simulation model.

According to this perspective, knowledge is essential in interpreting or lifting raw data to the level of information that is semantically meaningful to a reasoning agent, either a computer system or a human agent (user). Combined with knowledge about the world entities, a running simulation model provides understanding to a human agent. By changing the scenario and utilizing case based reasoning (CBR) the human agent can make projections and obtain deep insight into the system in question. In the process the frame of reference changes from a computer program to a human agent. This transition makes sense as a decision support system needs to assist human experts in decision making rather than replacing them. It is possible to search simulation cases with similar parameters, characteristics and applied methods based on techniques of case-based reasoning (CBR), namely, indexing and retrieval of similar cases. When the simulation cases contain quantitative attributes that are hard to index, or inexact qualitative attributes, the fuzzy-set-based approach is very promising on this regard. Other artificial intelligence methods such as genetic algorithm have also been successfully applied to case-based reasoning, where the genetic search technique assigns relative importance of feature weights for case indexing and retrieving.

Distinct from other tools of knowledge management, a simulation model can also be used in the real time decision-making when coupled with the business control systems. In the online mode the communication between a business control system and the real system can be forwarded to the simulation model, and the simulation model then calculate the required Performance Indicators (PIs), which will serve as online feedback to the control system so it can adjust its decision making accordingly. This real time closed-loop decision- making system further extends the value of the simulation model and partly offsets the cost of simulation model development.

2 Simulation Model Development

Simulation model development is a process that involves producing computer models to approximate the behavior of real processes. Although specification and requirements alter amongst problems, the standard methodology applied in simulation model development process remains unchanged between projects. In some situations an established model might not follow any standard structure, procedure, specification, rules and regulation. However, automatism without thorough planning could lead to a number of issues. As requirements change through multi-project phases, making modifications to design and implementation is a challenging task. Under these conditions, having a formal structure to follow in the model building stage is an essential step. This ensures that transparent irrelevant factors are excluded from the process, which minimizes the overall development time and prevent accumulate errors.

The prominent phase in the modeling process involves developing an understanding of a system. This involves the development of design specification documentation that formulates the problem, defines constraints and describes the expected outcomes. This material is constructed and provided as a guideline to facilitate the model building process. Some benefits of the design specification include:

- Minimize development time and prevent anomalous operational behavior;
- Provide a set of clear objectives to ensure all requirements are met;
- Provide a clear set of important performance factors during data analysis for knowledge generation and sharing;
- Provide appropriate constraints to ensure accurate model behavior that resembles the real process;
- Provide appropriate model detail that accurately represents the system at macro and micro levels and communicates effectively to different audiences; and
- Enhance the quality of a project.

The model development phase also involves acquisition and interpretation of knowledge. This task usually can be accomplished through extracting and processing the information that is currently available and accessible to the project. This is an obligatory process, in order to determine what data is missing that required acquisitions in order to reformulate the problem to meets requirements and deadlines. The formal steps in the simulation model development process are described in the next section.

3 Steps in Simulation Study

The appropriate method to approach a simulation study has been well formulated in literature [16-18], however the focus is commonly applied to the act of modeling. Here we will revisit the common approach to simulation and apply a focus to the ability of the simulation study methodology to provide benefits not usually associated with such a study. Primarily, these benefits relate to knowledge, both the generation of new knowledge, and the collation, unifying and sharing existing knowledge.

The standard steps in a simulation have been defined previously [19], and are redescribed in the following sections.

3.1 Problem Formulation

This is where the exact problem that the simulation study should address is defined. The problem may not necessarily relate to a production issue, it may be defined as a lack of understanding of a given system, or the need to capture knowledge for a particular process.

3.2 Setting of Objectives and Overall Project Plan

The objectives typically indicate what questions the simulation model should be capable of answering. The project planning relates to typical project management items such as resources, timelines and costs. In terms of knowledge management, this step provides a summary of what knowledge the study will generate, that is, the output from the model will generate new knowledge by providing results to the questions put to the model.

3.3 Model Boundary Condition and Assumptions

Assumption needs to be made to ensure that the developed models operate within a predefined set of limitations. This is a mandatory process to prevent the model from operating beyond and outside its initial design capability, which would cause insensitive information being generated.

3.4 Model Conceptualization

This step, best guided by experienced simulation engineers, relates to abstracting the real world system into a conceptual model, which simplifies the system down to an appropriate level, strike a balance between ease of modeling and fidelity to the real world. This can be a difficult process and the knowledge generated here tends to deal with the current system operation and complexity.

3.5 Data Collection and Correlation Analysis

Data collection is driven by the model, and the data required as input to the model. This will usually become an iterative time consuming process, as the model complexity or requirements change, so too will the data requirements. Depending on existing data collection systems, this process may exceed the actual model build, but proves to be an indispensable aspect of the simulation study. Analysis of the data collected provides several knowledge gains. Firstly, it can verify the operation of the system, either backing or refuting previous descriptions of the system. Secondly, it can reveal trends that were previously unknown. Both points provide a deeper understanding of the system and may impact the model conceptualization by changing the understanding of the real world system, or reveal new problems that the model should address.

3.6 Model Translation

Model translation is the act of creating the software-based representation of the system using a modeling package. This step again requires experienced simulation engineers in order to proceed quickly and accurately. It is our belief that with the ease of modern simulation packages, models should now be built in 3D. This provides several benefits regarding knowledge management. A 3D representation of the system is easier to recognize and become familiar with during training. The 3D model can serve as a tool when discussing the system, and when the model is run, problems such as bottlenecks or inadequate buffer space are quickly identifiable.

In terms of collating knowledge of a particular system, this step tends not to contribute to the knowledge pool. If the aim of the study was to collect and unify knowledge, this step may be skipped entirely, however, the power of the model would never be realized. The model is able to generate new data by providing results to particular scenarios, and therefore can be used extensively once created. Not only can the model produce results to address the questions being asked of the system, it can be used for training, scenario planning and many other items.

3.7 Verification and Validation

To verify the model is to ensure that the model runs as intended without errors while validation ensures that the model is an accurate representation of the real system. Both steps tend to be an iterative process, requiring model and data analysis and add to the knowledge about the system.

3.8 Model Output Analysis

This is where new knowledge of the system can be generated. By analysing output data, and applying different input configurations and operating conditions it is possible to accurately determine how a system will behave. The measurable metrics used to evaluate system performance varies between systems under. Some of the most common performance metrics used in simulation output analysis for knowledge generation include:

- Determine transient and steady-state condition
- Determine system warm-up and cool-down period
- Determine system recovery time
- Throughput, travel time, inventory level, queue length, resource and space utilization
- Input output correlation
- Number of replications
- Duration of the model simulation time
- The effect caused by variability
- Describe significance of results using statistical analysis procedures

3.9. Documentation and Reporting

Documentation of the mode, including progression, input data, operating rules, output data and analysis provide a centralized source of knowledge about the current system. This documentation can be used to standardize user's knowledge of the system and to train new users.

Through each step in the simulation study, knowledge about the existing system is either gathered, corrected or created. The act of simulation pools this knowledge into a single repository that can be used to unify an organization's knowledge of a system, and provides training material for new employees as they enter the organization.

4 Simulation Case Study

In this section two case studies on airport operations are investigated. The first case study involves the knowledge generation in Baggage Handling System (BHS) and the second system is the Airport Security Checkpoint (ASC) operation.

4.1 Case Study 1: Airport Baggage Handling Systems

The baggage handling system (BHS) in airports plays an important role, ensuring bags are secure and delivered on time. It is the key component within major airports to ensure smooth transition of luggage and ensure a safe flying experience by preventing dangerous material from entering the plane. The performance of the baggage handling system is crucial to airport operation.

A BHS is generally complex and comprises of many thousands of resources. For a full description of the various components within a BHS we refer to our previous work [20-23]. In this study a small scale BHS model, illustrated in Fig. 1 is examined. This system has four input delivery conveyors, one transfer input conveyor, five level one screening machines and three level three screening machines, two automatic tag reader (ATR) machines, three make up output loops and one manual encoding station. The system performance is evaluated through investigating the key measurable metrics. This was done with the aid of an output analysis tool for knowledge generation and sharing. Input analysis results from captured data for security screening and other time taking operation is summarized in Table 1.

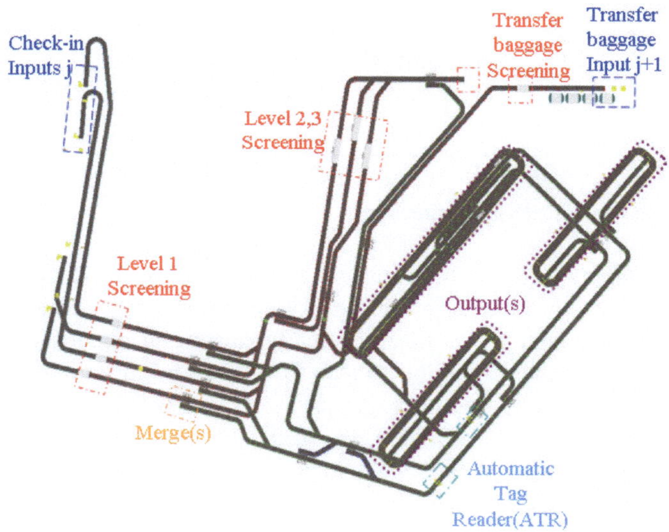

Fig. 1 Case study 1, a small scale Baggage Handling System simulation model

Table 1 System resource specification

Resource Type	Pass Rate (%)	Processing Time Distribution
Level 1 Screening	70	Constant (5)
Level 2 Screening	80	Constant (10)
Level 3 Screening	85	Normal (15, 3)
Level 4 Screening	90	Constant (20)
Level 5 Screening	90	Constant (6)
Automatic Tag Reader	Origin = 95, Transfer = 80, Group = 95	NA
Manual Encoding Station	100	Constant(12)

The output analysis tool used for knowledge generation and sharing through display charts has the heterarchical structure demonstrated in Fig. 2. The structure represents the predefined analysis information used for the graphical display of results. Each chart description for different analysis types is defined inside the chart definition collection. This predefined information is sent to the search engine according to the analysis type before display on a chart.

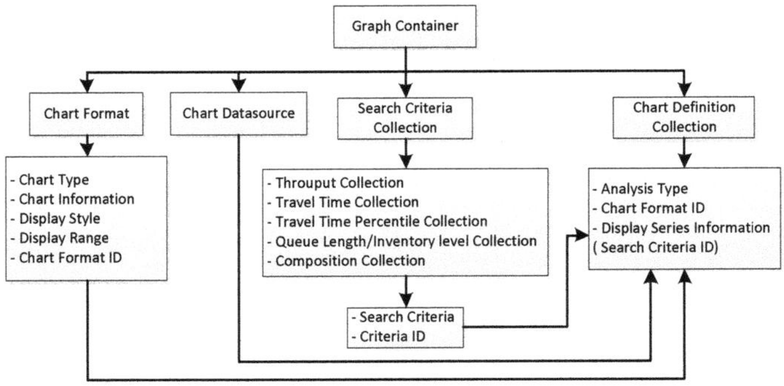

Fig. 2 Standard analysis charting structure

Typical outputs from the heterarchical structure in Fig. 2 are given in Fig. 3a and Fig. 3b. Fig. 3a shows the travel time of all systems bags through the system for different mean inter-arrival times. It clearly indicates that at least 85 percent of bags travel through the system in less than 17 minutes for mean inter-arrival times ranging from 2 to 10 seconds. Fig. 3b shows the travel time with respect to the system congestion level. This figure was generated by combining baggage travel time and in-system congestion level, which was analyzed using a one-minute bin interval on nine different loading inter-arrival time scenarios. The figure clearly indicates the increase in congestion level, as the loading rate intensifies toward the mean inter-arrival time of two seconds. Furthermore, this system also shows signs of a developing bottleneck when the loading rate inter-arrival time is below four seconds, due to data outliers that visible in Fig. 3b.

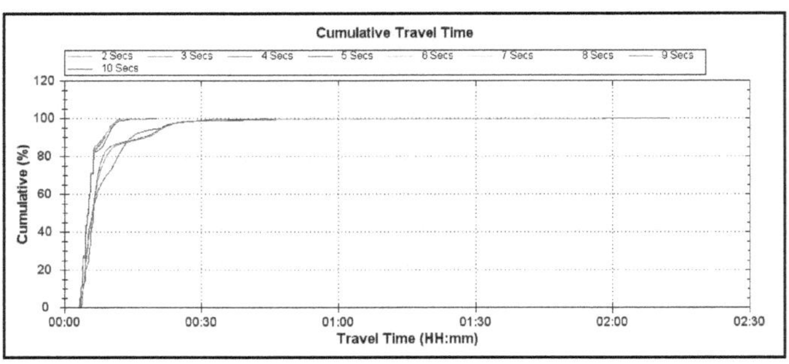

Fig. 3 a) Baggage cumulative travel time

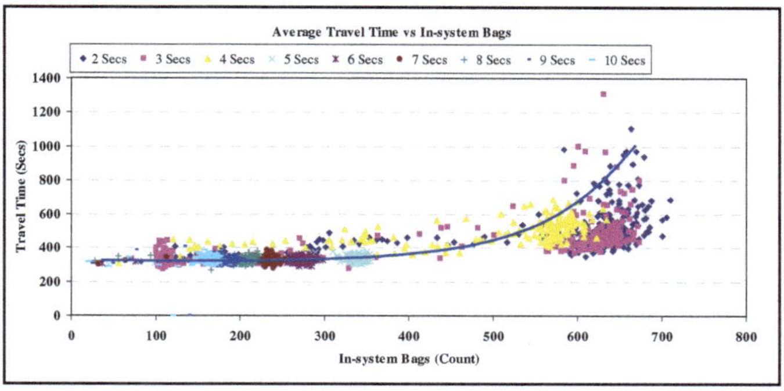

Fig. 3 b) Travel time on different system congestion level

4.2 *Case Study 2: Airport Security Checkpoints*

Airport security checkpoints act as a threat pre-emptive mechanism to enforce a safe traveling experience. It aims at identifying individuals that pose a threat to passenger and equipment. The process is controlled by security officers, who are responsible for screening all bags and scanning all passengers that pass through the checkpoint.

The security checkpoint comprises of many elements and sub-processes. In its primary operation, a passenger firstly divest their belonging, goes through the Walk-Through Metal Detector (WTMD), arriving at the composure area to collect their screened bags and may be selectively chosen to go through the Explosive Trace Detection (ETD) process before exiting the checkpoint. The checkpoint operation under investigation for case study two is illustrated in Fig. 4. Input analysis of captured data for selected sub-process pass rate and processing time information is summarized in Table 2.

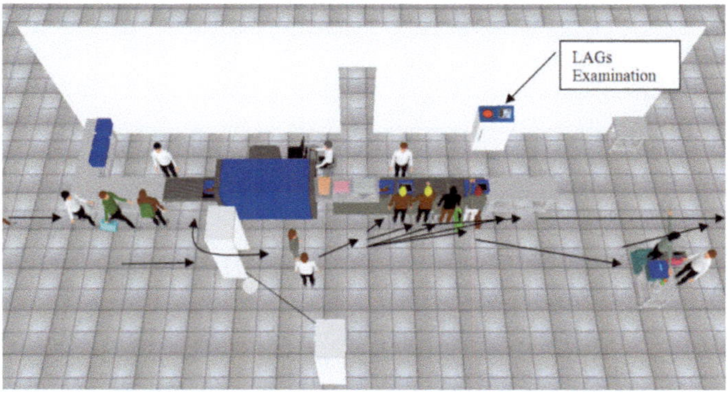

Fig. 4 Case study 2, a small scale Airport Security Checkpoint simulation model

This particular study involves approximately 600 different scenarios with different operating conditions and input variables. The magnitude of the problem makes it impractical to perform manual output result analysis. Therefore, a refinement to the existing analysis tool interface is reshaped, which enabled batch scenario processing and auto reporting feature. This is necessary to enhance and allow robust knowledge generation and sharing process. The interface of the auto-analyzer used for auto report generation is illustrated on Fig. 5.

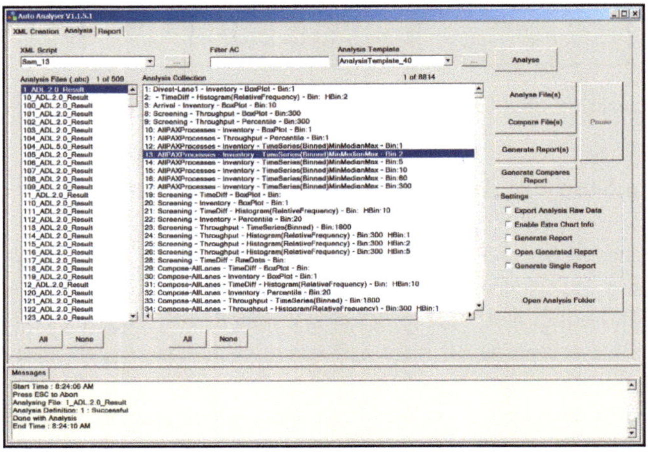

Fig. 5 Auto-analyzer with batch mode scenario analysis and auto report generation enabled

Table 2 Security checkpoint sub-resource specification

Resource Operation	Pass Rate (%)	Processing Time Distribution
X-ray Level 1	80	2.+WEIBULL(2.64, 0.81,1)
X-ray Level 2	90	2.+WEIBULL(2.64, 0.81,1)
X-ray Level 3	95	2.+WEIBULL(2.64, 0.81,1)
X-ray Level 4	100	2.+WEIBULL(2.64, 0.81,1)
ETD scanning	100	9+164*BETA(1.89, 6.14,1)
WTMD Level 1	81.70	0
WTMD Level 2	81.41	0
WTMD Level 3	92.44	0
Wand	92.44	20.4*(1./UNIFORM(0,1,1)-1.)^(-1./1.4)
Frisk	99.99	20.4*(1./UNIFORM(0,1,1)-1.)^(-1./1.4)

Using the updated analysis tool, which inherited a similar structure to Fig. 2, a sample output generated from the study is demonstrated in Fig. 6. Fig. 6 combines the throughput analysis of 49 different scenarios and displays them on a single chart. The figure shows the mean hourly throughput rate on varying number of liquid inspection machines and operators with respected to different liquid alarm rates. It is evident that the throughput decreases as the alarm rate increases. When there are two operators controlling two or more liquid inspection machines, the throughput does not alter greatly between alarm rates of 5 to 30%. A second set of results used in knowledge sharing is shown in Fig. 7, which was generated using the charting structure described in Fig. 2 and the auto-analyzer presented in Fig 5. The figure clearly illustrates system hourly throughput operating range under different liquid inspection operating policies.

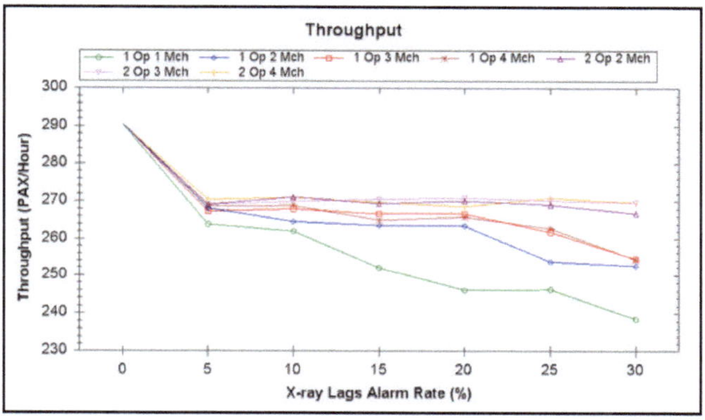

Fig. 6 Passenger throughput level on increasing in liquids inspection alarm rate and variability in number operators

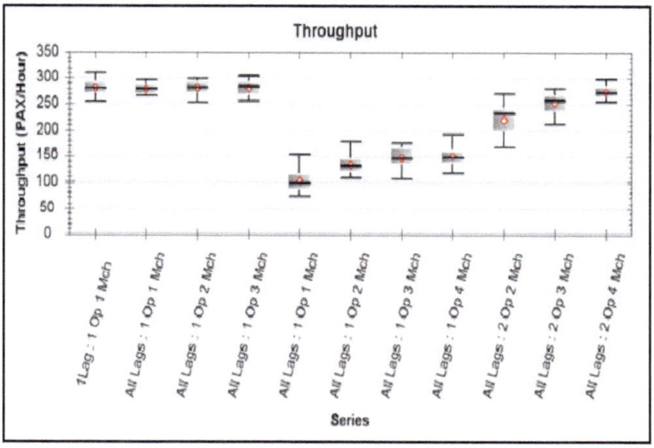

Fig. 7 Passenger throughput level on different liquid inspection scenario

5 Conclusion

In this work we demonstrated the utility of simulation modeling methodology in capturing existing knowledge and generating new knowledge. By collecting, analyzing and verifying data through the modeling process, knowledge pools are created and verified. These knowledge pools are able to unify perceptions about the system subjected to the study, and provide a detailed training induction for new employees. Knowledge can be created once the simulation model has been built. By applying different input configurations and operating rules, the systems' expected behavior can be determined. From this the model can be used for additional purposes such as training, process optimization and control system testing.

Two case studies were presented that highlight the knowledge that was obtained through data collection and subsequent analysis of that data, while the study of each systems' behavior led to an understanding of those systems and we were thereby able to determine what behavior would be expected for different conditions.

Acknowledgments. This research was supported by the Australian Research Council and the Centre for Intelligent System Research (CISR) at Deakin University.

References

1. Rus, I., Lindvall, M.: Guest editors' introduction: Knowledge management in software engineering. IEEE Software, 26–38 (2002)
2. Davenport, T.H., Prusak, L.: Working knowledge: How organizations manage what they know. Harvard Business Press (2000)
3. Lee, H., Choi, B.: Knowledge management enablers, processes, and organizational performance: An integrative view and empirical examination. Journal of Management Information Systems 20, 179–228 (2003)
4. Fahey, L., Srivastava, R., Sharon, J.S., Smith, D.E.: Linking e-business and operating processes: The role of knowledge management. IBM Systems Journal 40 (2001)
5. Wagner, C., Bolloju, N.: Supporting knowledge management in organizations with conversational technologies: Discussion forums, weblogs, and wikis. Journal of Database Management 16 (2005)
6. Hirsch, C., Hosking, J., Grundy, J., Chaffe, T., MacDonald, D., Halytskyy, Y.: The Visual Wiki: A new metaphor for knowledge access and management. In: 42nd Hawaii International Conference on in System Sciences, HICSS 2009, pp. 1–10 (2009)
7. Twietmeyer, G.A.G., Lyth, D.M., Mallak, L.A., Aller, B.M.: Evaluating a New Knowledge Management Tool. Engineering Management Journal 20, 10 (2008)
8. Zhang, J., Creighton, D., Nahavandi, S.: Toward a synergy between simulation and knowledge management for business intelligence. Cybernetics and Systems 39, 770–786 (2008)
9. Diaz, R., Bailey, M.: Building knowledge to improve enterprise performance from inventory simulation models. International Journal of Production Economics 134, 108–113 (2011)

10. Davidsson, P.: Multi Agent Based Simulation: Beyond Social Simulation. In: Moss, S., Davidsson, P. (eds.) MABS 2000. LNCS (LNAI), vol. 1979, pp. 97–107. Springer, Heidelberg (2001)
11. Venkateswaran, J., Son, Y.J.: Hybrid system dynamic—discrete event simulation-based architecture for hierarchical production planning. International Journal of Production Research 43, 4397–4429 (2005)
12. Robinson, S.: Discrete-event simulation: from the pioneers to the present, what next? Journal of the Operational Research Society 56, 619–629 (2005)
13. Robinson, S.: Soft with a hard centre: Discrete-event simulation in facilitation. The Journal of the Operational Research Society 52, 905–915 (2001)
14. Agnar Aamodta, M.N.: Different roles and mutual dependencies of data, information, and knowledge- an AI perspective on their integration. Data and Knowledge Engineering 16, 191–222 (1995)
15. Althoff, K.D., Weber, R.O.: Knowledge management in case-based reasoning. Knowledge Engineering Review 20, 305–310 (2005)
16. Banks, J.: Handbook of Simulation
17. Law, A.M.: Simulation modeling and analysis, 4th edn. McGraw-Hill, Boston (2006)
18. Chung, C.A.: Simulation modeling handbook: a practical approach (2003/August 11, 2008)
19. Banks, J.: Introduction to Simulation. In: 1999 Winter Simulation Conference Proceedings, pp. 7–13 (1999)
20. Johnstone, M., Creighton, D., Nahavandi, S.: Status-based Routing in Baggage Handling Systems: Searching verses Learning. IEEE Transactions on Systems, Man, and Cybernetics, Part C: Applications and Reviews 40, 189–200 (2010)
21. Le, V., Creighton, D., Nahavandi, S.: Simulation-based input loading condition optimisation of airport baggage handling systems. In: 10th IEEE International Conference on Intelligent Transportation Systems, Seattle, WA, USA (2007)
22. Khosravi, A., Nahavandi, S., Creighton, D.: Estimating performance indexes of a baggage handling system using metamodels. In: Proceedings of the 2009 IEEE International Conference on Industrial Technology, pp. 1–6 (2009)
23. Khosravi, A., Nahavandi, S., Creighton, D.: Interpreting and Modeling Baggage Handling System as a System of Systems. In: Proceedings of the 2009 IEEE International Conference on Industrial Technology, pp. 1–6 (2009)

Incremental and Interaction-Based Knowledge Acquisition for Medical Images in THESEUS

Daniel Sonntag

German Research Center for AI (DFKI),
Stuhlsatzenhausweg 3, 66123 Saarbruecken, Germany
sonntag@dfki.de

Abstract. Today, the major challenge in medical imaging is the so called knowledge acquisition bottleneck. We cannot acquire the necessary medical image knowledge that ought to be used in the software application easily as it is hidden in the heads of medical experts. In this article, we provide an example of how an incremental knowledge acquisition process for radiology images can be implemented to solve this problem. Thereby, we integrated Semantic Web technologies with a variety of automatic and manual annotation tools for radiology images. We developed the prototypes in the context of a large scale German research program for a new Internet infrastructure based on semantic technologies - THESEUS. According to the complex medical finding processes in the MEDICO use case, the different annotation tools should be used for very specific purposes. After four years of prototyping automatic and manual annotation tools (2009-2012), we developed a divide-and-conquer strategy for future knowledge acquisition processes. This divide-and-conquer strategy turns out to be very effective in the radiology domain, but produces many infrastructure requirements. It also relies on high-end intelligent user interfaces such as the Radspeech dialogue system which are not available in today's clinical environments.

1 Introduction

A prior usability analysis to identify the requirements for industrial applications, where image semantics play a role, is very useful. In many circumstances, different requirements have to be met during knowledge acquisition, refinement, and retrieval. In addition, work from the area of the Semantic Web should be integrated in such a way that the process of using image semantics, and relying on it, does not produce too much knowledge engineering overhead. Unfortunately, in many industrial domains such as medical radiology, a vast amount of images is produced and manual annotations are not feasible. In addition, these medical image annotations must be refined and augmented during a complex medical workflow.

Our clinical partner, the University Hospital Erlangen in Germany, has a total of about 50 TB of medical images. They are currently doing about 150,000 medical examinations producing 13 TB of data per year. Many 2D and 3D image series in radiology, and individual images in particular, require specific semantic annotations of the image contents which cannot be automatically provided (figure 1). These annotations are extremely helpful and increase the quality of patient treatment processes; in addition to satisfying the trend to store and organize all patient data, including health records, laboratory reports, and medical images in digital libraries, effective retrieval of images builds on the semantic annotation of image contents. In the medical domain, the proper selection of specific image contents can improve the treatment process to a large degree since the doctor can consult similar cases and other doctors' treatment plans. This case-based reasoning is very effective in the medical domain. At the same time it is crucial that clinicians have access to a coherent view of image data within their particular diagnosis or treatment context. Semantic annotations should provide the necessary image (region) information.

In order to address these issues, namely the knowledge acquisition bottleneck and different user interface requirements at different medical work we designed and implemented an incremental knowledge acquisition process for radiology images against the THESEUS project's background (section 2). This process relies on an integrated ontology-based approach of structured knowledge for medical images (section 3) and takes the special requirements of the radiology department into account. Based on these requirements, automatic and manual annotation frameworks can be constructed (sections 4 and 5) and combined, thereby implementing an incremental knowledge acquisition process (section 6). Section 7 provides a conclusion.

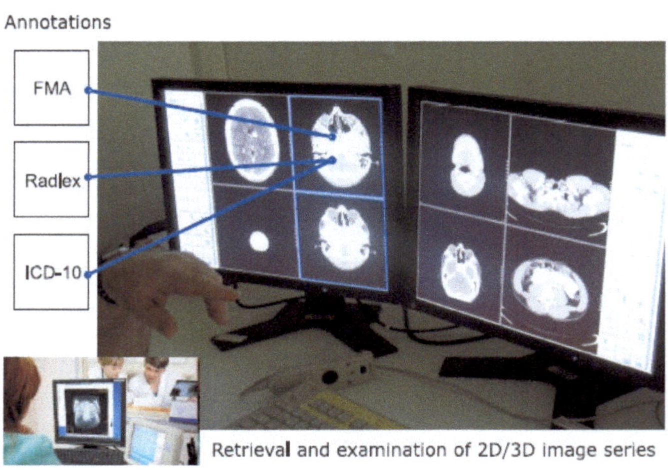

Fig. 1 Image series and semantic annotation requirements

2 THESEUS Background

Partners from the academic sphere and the business community are working together within the framework of THESEUS (currently Germany's largest IT research program with a total budget of more than 200 million euros) to meet the challenge of creating an Internet of services based on data semantics. Just as the legendary figure of Theseus in Greek mythology succeeded in escaping from the Minotaurs labyrinth, the research program of the same name has developed ways to navigate through the increasing quantities of data found on the Internet.[1] The technologies being developed within the THESEUS program are preparing the way for a future Internet of Services. This will make it possible for services that are now available on the Web only separately, such as online shopping, flight bookings and research support, to be combined and linked with one another. Several application scenarios show how the technologies can be used for innovative tools, services, and business models in particular application domains such as healthcare.

MEDICO[2] addresses the need for advanced semantic technologies in medical image and patient data search. The objective is to enable a seamless integration of medical images and different user applications by providing direct access to image semantics. Semantic image retrieval should provide the basis for the help in clinical decision support and computer aided diagnosis. During the course of lymphoma diagnosis and continual treatment, image data is produced several times using different image modalities. After semantic annotation, the images need to be integrated with medical (textual) data repositories and ontologies.

RadSpeech[3] aims to build the next generation of intelligent, scalable, and user-friendly semantic search interfaces for the medical imaging domain, also based on semantic technologies. Ontology-based knowledge representation is used not only for the image contents, but also for the complex natural language understanding and dialogue management process. RadSpeech shows a speech- based annotation system for radiology images and focuses on a new and effective way to annotate medical image regions with a specific medical, structured, diagnosis while using speech and pointing gestures on the go.

Our investigations throughout the MEDICO and RadSpeech research projects which, as described, focus on semantic medical image search and user interaction [16] respectively, have shown us that several types of structured knowledge are relevant for the annotation of the images. In addition, several types of knowledge acquisition processes are needed. The combination of those individual processes towards an incremental knowledge acquisition process for radiology images in THESEUS is the main focus of this article.

---

[1] Under the THESEUS umbrella, more than 60 research partners from academia and the business world have come together to develop new technologies and applications. Their goal is to facilitate access to information, combine data to form new kinds of knowledge and lay the groundwork for new services on the Internet.

[2] http://theseus-programm.de/en/920.php

[3] http://www.dfki.de/RadSpeech/

3 Structured Knowledge for Medical Images

Structured knowledge in radiology has a multitude of different aspects, which can be divided into different representational ontologies in RDFS and OWL. The annotations for medical images are based on the assumption that those elements at higher levels are more stable, shared among more people, and thus change less often than those at lower levels. For example, the *Upper Ontology* describes very general concepts like *time, space, organization, person,* and *event,* which are the same across all domains. *The Information Element Ontology* represents the information elements of the incremental knowledge acquisition process (figure 2). For the *Medical Ontologies,* a separation into mid- and low-level ontologies is not so clear since they usually cover a broad spectrum of concepts ranging from very abstract ones like "heart" (which are not very likely to change) to macromolecules (which are updated and added frequently). However, the medical ontologies are

- The Foundational Model of Anatomy (FMA) ontology [7] for anatomical annotations;
- The International Classification of Diseases (ICD-10)[4] for disease annotations; and
- Radlex to express visual features of the visual manifestation of a particular anatomical entity or disease [6].

On the images, any combination of anatomical, disease, and visual annotations is allowed and multiple annotations of the same image region are possible. As a result, all messages transferred between internal and external components which deal with image contents are then based on RDF data structures which are modeled in the respective ontology instances (also cf. [1, 4, 13]). This is only possible when all the annotation ontologies are available in the respective format. Especially for the most critical disease part, the ICD-10 was not available in OWL, although the biomedical ontology community has focused on establishing interoperability and data integration. Several country- and language-specific adaptations of ICD-10 exist which share the general structure of the WHO version but differ in certain details. We presented an approach for modeling the hierarchy of the ICD-10 using OWL so that we can easily use it in our ontology framework and have enough expressivity to convey special data relations (figure 2(1)). For example, specialties such as "Exclusion" statements, which make statements about the disjointness of certain ICD-10 categories, are modeled in a formal way. The important thing is that we crawled the necessary data from the language-specific ICD-10 web pages, and this procedure can be transferred to other image semantic domains, as far as the necessary image terminology is available online. Noy and Rubin have also presented an approach for translating the Foundational Model of Anatomy ontology (FMA) to OWL [10]. (From their approach we adopted the idea to split the generated ontology into an OWL-DL and an OWL-Full component.) The resulting integrated data model is a prerequisite to get accurate annotations on decision-relevant image contents in medical imaging.

[4] `http://www.who.int/classifications/apps/icd/icd10online`

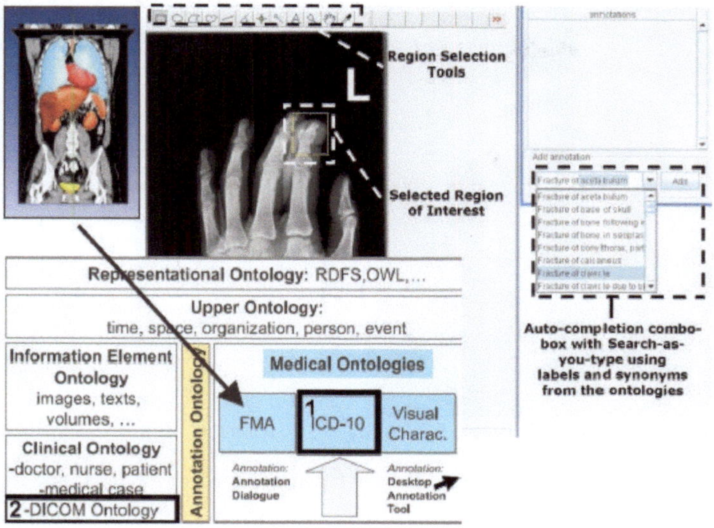

Fig. 2 Knowledge structure and automatic manual annotation

4 Automatic Annotation

Automatic annotation of medical images has three basic components. First, you can extract knowledge from metadata that is produced during the image generation process. Second, you can use image recognition software to detect anatomical concepts and landmarks. Third, you can try to reason about the plausibility of special configurations being detected while using ontological background knowledge.

DICOM Standard. The Digital Imaging and Communications in Medicine (DICOM) Standard (http://medical.nema.org/) ensures the interoperability of information on medical images. Manufacturers of imaging equipment and imaging information systems and manufacturers of peripheral equipment (e.g., computer monitors and image archives) conform to this standard. The Siemens computer tomography (CT) and magnetic resonance imaging (MRI), which we used to produce our image material, use this standard to encode a multitude of image metadata about the image generation process. Figure 3 shows the subset of these data we extract from the image headers in order to create ontology instances automatically (cf. the ontology model in figure 2(2)). As can be seen, we can extract about fifty image features in the context of the image study, the patient, and low level image characteristics. These metadata provide the necessary information to create the patient image instances to be augmented by the image content semantics of specific image regions.

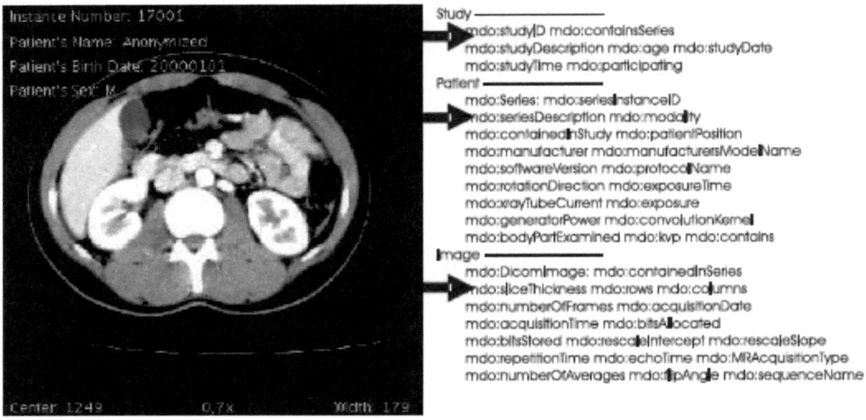

Fig. 3 DICOM data that can be extracted from the image header

Image Recognition. The CT and MRI systems produce detailed pictures of organs, soft tissues, bone, and virtually all other internal body structures. Today, organs of the chest and abdomenlincluding the heart, liver, biliary tract, kidneys, spleen, bowel, pancreaslcan be detected with great accuracy. But the automatic detection of image semantic of, e.g., malicious tissue in the context of cancer, is extremely difficult. Although we use state-of-the-art organ and land-mark detection software [12] with a special focus on organs, landmarks (also cf. top left of figure 2), and lymph node segmentation [2], many further reasoning and manual annotation steps are necessary.

The ontology knowledge structures become effective when axiomatic relations apart from subsumptions can be exploited. Spatial relations are a promising area of research in this automatic reasoning area since they complement well-known linguistic phenomena being put into the ontology context (e.g., Wordnet relations) and at the same time allow both the automatic modeling of special configurations and the human judgment/evaluation for plausibility. One idea we evaluated is the incorporation of a spatio-anatomical ontology for automatic plausibility checks of the found configuration of automatically detected organs [8]. We first learned a model of plausible organ constellations inductively from an annotated corpus of 3D volume data sets. The model, an ontology-based canonical representation of the spatial relationships of organs in the human body, can be used to check the results of a state-of-the-art medical object recognition system for 3D CT volume data sets for spatial plausibility.

The interesting thing is that, on a dataset of 1118 instances, the model produces only 76 false positives and 213 false negatives. This means that while precision is relatively high, the recall is moderate with 65.5%. As a result, a lot of further manual control is needed to find the erroneous automatic recognition results. This is one of the reasons why manual annotations are needed not only for the disease, but also for the anatomical level on medical images in radiology.

5 Manual Annotation

Manual annotation means that the radiologist must use a special human-computer interaction system to perform the required image annotations. This process reveals many usability issues. We will first describe what the desktop workstation and the special multi-touch installation in combination with a dialogue system looks like, before we discuss the usability issues in the context of the combined incremental process.

Desktop Workstation. For the manual semantic annotation on a regular desktop workstation we developed a new medical semantic annotation and retrieval tool RadSem [9]. It consists of a component that implements a method to annotate images and upload/maintain a remote RDF repository of the images and image semantics. In order to ease the task of finding appropriate annotations, we use *auto-completing* combo-boxes.

A screenshot of parts of the annotation tool is depicted in figure 2 (right) which shows a simple orthopedic example. The broken bone of the index finger can easily be annotated while using the auto-completion combo-boxes with a search-as-you-type functionality. The resulting annotation is accurate but very time-consuming.

Radiology Dialogue System. It is crucial that clinicians have access to a coherent view of image data within their particular diagnosis or treatment context (we experimented with a large touchscreen installation). These data include previous (rudimentary) annotations. A semantic dialogue shell should be used to ask questions about the image annotations and refine them while engaging the clinician in a natural speech dialogue at the same time. In the construction of a dialogue system for radiologists, we learned some lessons which we used as guidelines in the development of *semantic* dialogue systems [11, 14]; over the last years, we have adhered strictly to the developed rule "No presentation without representation." All the items presented on the touchscreen are basically surface representations of more complex ontological entities according to the described knowledge structure. This knowledge structure (section 3) allows a specific user to ask questions about the displayed image content and other region-based image elements. The domain-specific dialogue application for the radiology department (also cf. [15]), which uses a touchscreen (figure 4, upper right) to display the medical image windows, is able to process the following dialogue:

- **U:** "Show me the CTs, last examination, patient XY."
- **S:** Shows corresponding patient CT studies as DICOM picture series and MR videos.
- **U:** "This lymph node here (+ pointing gesture) is enlarged; so add the annotation: *lymphoblastic.*"
- **S:** Shows new annotation on the image and confirms database update.

The dialogue-based annotation can be done at a rate of approximately 6 annotations per minute (including the visual feedback phase) whereas the desktop-based annotation comes to a rate of approximately 3 annotations per minute. Most importantly, the prototype dialogue system delivers new semantic annotations instantly which are unavailable in the current clinical finding process so that the (senior) radiologist can directly detect errors visually.

6 Incremental Knowledge Acquisition Process

The incremental knowledge acquisition process (figure 4) relies on the structured ontological knowledge as introduced in the first section. Based on this prerequisite, we have been trying to formulate the process of automatic and manual image annotation. Hereby, two factors play a major role: the quality of automatic annotations and the usability of different intelligent user interfaces to control, correct, and add annotations. For us, usability means that people can use an Artificial Intelligence (AI) prototype easily and efficiently to accomplish their tasks. Prototypes that are usable enable clinicians to concentrate on their tasks rather than paying attention to the tools they use to perform their tasks. The prevalent interaction design issue that follows this definition is that the intelligent interfaces are

- efficient to use;
- quick to recover from errors; and
- visually pleasing.

To achieve all three of these a careful selection of involved components for manual annotation is vital. This can be substantiated by the current developments in clinical practice where *structured reporting* should be introduced. This means that the radiologists fill in special standardized forms. Radiologists feel restricted by these standardized forms and fear a decrease in focus and eye dwell time on the images [3, 17]. As a result, the acceptance for structured reporting is still low among radiologists while referring physicians and hospital administrative staff are generally supportive of structured standardized reporting since it can be used more easily for further processing. As a matter of course, the image semantics with RDF are a further step in this direction. These issues are explained in the context of industrial usability and our basic process steps for industrial dissemination.

6.1 Binocular View and Industrial Usability

As [5] point out, many research prototypes that use technically advanced but unimportant or unrealistic functionality for the specific domain or personal activities do not provide the AI support that users would appreciate most. This can, e.g., make a complex speech dialogue system languish as an infertile research prototype on demonstration computers which cannot be used in the context of industrial prototypes or real-world industrial dissemination. Accordingly, the binocular view of intelligent interfaces for industrial dissemination should study not only the suitability of a single algorithm and a component performance for a given user task, but also the industry user's interaction requirement in which the interaction will be used. In our specific radiology case, the feature that only a senior radiologist is responsible for the treatment plan, implicates that his or her interaction with the annotation system must be designed to be very effective. Although it is widely reductive to put it this way, a senior radiologist has three main goals: (1) access the images and image (region) annotations (a summary can also be synthesized), (2) complete them, and (3) refine existing annotations. These tasks can best be fulfilled while using a multimodal dialogue system. In contrast, less demanding manual annotation tasks, such as the correction of organ detection algorithms of image region selection can be done by, e.g., a first-year resident with the help of our desktop-based annotation tool. This tool can also easily be installed on virtually every computer in a hospital, whereas a speech dialogue system requires a specific hardware infrastructure.

6.2 Process Steps

The incremental knowledge acquisition process (figure 4) has four steps. First, the automatic metadata are extracted from the DICOM images and instantiated according to the structured/structural knowledge model. After that, a direct access to the RDF statements is possible while using, e.g., the query language SPARQL.

Second, the automatic image recognition software runs over the images to produce anatomical annotations according to the structural knowledge model. According to the spatio-anatomical ontology, automatic spatial plausibility checks can be executed. Hereby, the spatial reasoning process runs completely automatically and only the outlier configurations are presented to the medical experts.

Third, the experts can then use the manual annotation tool to correct or extend these configurations. At this stage, a very comprehensive set of image semantics, namely the study, patient, and low-level image feature information in combination with the automatically detected anatomical concepts and manual annotations with the desktop tool are available. These image model instances are not accurate enough for a proper diagnosis which results in a treatment plan, but accurate enough to be used in a semantic search and annotation system, the dialogue shell, which the senior radiologist can use.

Fourth, only when the images are retrieved and considered for a medical treatment plan, can accurate disease annotations be added by the senior radiologist while using the dialogue system which displays the image and patient data on a large touchscreen. It is even possible to search for similar disease annotations in other patients' contexts for a comparable study. Currently, we are trying to extend the high-level process of patient findings and image annotations to a mobile scenario, where we can use a special pen to recognize annotations on normal paper and/or used the iPad as a mobile dialogue system and touchscreen device for the senior radiologist (also cf. the project Radspeech, http://www.dfki.de/RadSpeech/).

Our hope is that the resulting process successfully supports the complex healthcare process in which radiology images are used. The development of automatic processing applications is as essential as the design and implementation of intelligent user interfaces for specific purposes. In our view, only this combination will produce successful decision support systems for industrial dissemination.

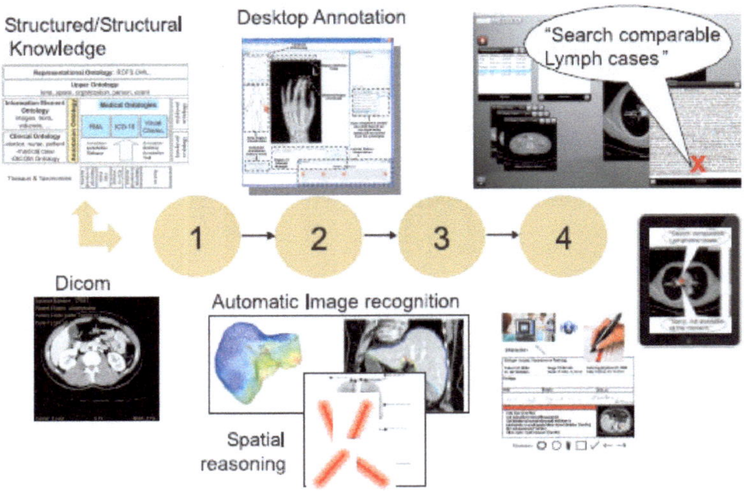

Fig. 4 Incremental Knowledge Acquisition Process

7 Conclusion

In discussions with radiologists we found out that three typical clinical scenarios are of interest for further analysis of clinical knowledge requirements and (incremental) knowledge acquisition: (1) the clinical reporting process; (2) the patient follow-up treatment (i.e., monitoring the patient's health condition and the development of the disease); and (3) the clinical disease staging and patient management. In this paper we have explained a process that takes structured medical knowledge as input and provides an incremental process for the patient follow-up and clinical disease staging process by addressing the bottleneck to annotate appropriate image semantics. The process can be applied to new patients and image data, but also to image time series on a given patient in order to monitor a patient over time, e.g., how a cancer evolves over time under medication/radiation which produces new annotations about changing characteristics.

In addition, we provided automatic and manual annotation scenarios and a MEDICO server architecture with several HCIs/dialogue systems to meet the requirements of a distributed software infrastructure and/or usability issues. These issues have been explained in the context of our basic architecture approach for industrial dissemination. An incremental knowledge acquisition process for radiology images seems to be adequate. But we produced many infrastructure requirements and relied on high-end speech-based dialogue systems which are not available in the industrial sector today.

The question of how to integrate the acquired image knowledge with other types of data, such as patient data, is paramount. In a further step, individual textual findings should be organized according to a specific body region and the disease context both of which can be interlinked to several text passages. (Because diseases can touch diverse regions, this organization only helps to visualize the data, but should not preclude linking various lesions.) Currently, we are evaluating the proper usage of information extraction technology for this purpose. The main problem is that the text processing tools cannot be easily adapted to the medical domain. Finally, educators may find our process can help trainees learn the important elements of reports and will encourage the proper use of radiology terms (structured reporting). We hope that structured reporting will also help to ease the task of text mining.

Acknowledgments. I would like to thank all colleagues at DFKI and all partners in the MEDICO and RadSpeech projects for their valable contributions to the iterative process described here. This research has been supported by the THESEUS Program funded by the German Federal Ministry of Economics and Technology (01MQ07016).

References

1. Fensel, D., Hendler, J.A., Lieberman, H., Wahlster, W. (eds.): Spinning the Semantic Web: Bringing the World Wide Web to Its Full Potential. MIT Press (2003)
2. Feulner, J., Zhou, S.K., Huber, M., Hornegger, J., Comaniciu, D., Cavallaro, A.: Lymph node detection in 3-d chest ct using a spatial prior probability. In: The Twenty-Third IEEE Conference on Computer Vision and Pattern Recognition, CVPR 2010, San Francisco, CA, USA, June 13-18, pp. 2926–2932 (2010)
3. Hall, F.M.: The radiology report of the future. Radiology 251(2), 313–316 (2009)
4. Hitzler, P., Krötzsch, M., Rudolph, S.: Foundations of Semantic Web Technologies. Chapman & Hall/CRC (August 2009)
5. Jameson, A.D., Spaulding, A., Yorke-Smith, N.: Introduction to the Special Issue on Usable AI. AI Magazine 3(4), 11–16 (2009)
6. Langlotz, C.P.: Radlex: A new method for indexing online educational materials. RadioGraphics 26, 1595–1597 (2006)
7. Mejino, J.L., Rubin, D.L., Brinkley, J.F.: FMA-RadLex: An application ontology of radiological anatomy derived from the foundational model of anatomy reference ontology. In: Proc. of AMIA Symposium, pp. 465–469 (2008)

8. Möller, M., Ernst, P., Sonntag, D., Dengel, A.: Automatic spatial plausibility checks for medical object recognition results using a spatio-anatomical ontology. In: Proc. of the International Conference on Knowledge Discovery and Information Retrieval (KDIR 2010), Valencia, Spain, October 25-28 (2010)
9. Möller, M., Regel, S., Sintek, M.: Radsem: Semantic annotation and retrieval for medical images. In: Proc. of the 6th Annual European Semantic Web Conference, ESWC 2009 (June 2009)
10. Noy, N.F., Rubin, D.L.: Translating the foundational model of anatomy into owl. Web Semant. 6, 133–136 (2008)
11. Oviatt, S.: Ten myths of multimodal interaction. Communications of the ACM 42(11), 74–81 (1999)
12. Seifert, S., Kelm, M., Moeller, M., Mukherjee, S., Cavallaro, A., Huber, M., Comaniciu, D.: Semantic annotation of medical images. In: Proceedings of SPIE Medical Imaging, San Diego, CA, USA (2010)
13. Sonntag, D.: Ontologies and Adaptivity in Dialogue for Question Answering. AKA and IOS Press, Heidelberg (2010)
14. Sonntag, D., Engel, R., Herzog, G., Pfalzgraf, A., Pfleger, N., Romanelli, M., Reithinger, N.: SmartWeb Handheld — Multimodal Interaction with Ontological Knowledge Bases and Semantic Web Services. In: Huang, T.S., Nijholt, A., Pantic, M., Pentland, A. (eds.) ICMI/IJCAI Workshops 2007. LNCS (LNAI), vol. 4451, pp. 272–295. Springer, Heidelberg (2007)
15. Sonntag, D., Möller, M.: Unifying semantic annotation and querying in biomedical image repositories. In: Proceedings of International Conference on Knowledge Management and Information Sharing, KMIS (2009)
16. Sonntag, D., Schulz, C., Reuschling, C., Galarraga, L.: Radspeech, a mobile dialogue system for radiologists. In: Proceedings of the International Conference on Intelligent User Interfaces, IUI (2012)
17. Weiss, D.L., Langlotz, C.: Structured reporting: Patient care enhancement or productivity nightmare? Radiology 249(3), 739–747 (2008)

Complex Decision Making to Support Urban Search and Rescue Operations

Lars Hildebrand[1] and Wolfgang Vautz[2]

[1] Dortmund University of Technology,
Computer Science Department, Chair 1, Otto-Hahn-Str. 16,
44221 Dortmund, Germany
[2] Leibniz-Institut für Analytische Wissenschaften - ISAS - e.V.,
Bunsen-Kirchhoff-Str.11, 44139 Dortmund, Germany
lars.hildebrand@tu-dortmund.de,
wolfgang.vautz@isas.de

Abstract. The project "Second Generation Locator for Urban Search and Rescue" or simply SGL, is an EC funded research project, which aims at the support of rescue people during the search of entrapped people or dead bodies in collapsed structures. One part of the project is the establishment of unattended, wireless monitoring devices, that are able to raise alarms in case of detected signs of life, danger, or death. This article gives a short overview of the systems structure, the concepts of sensor fusion, and the use of fuzzy logic as the central decision making mechanism.

Keywords: Decision making, sensor fusion, fuzzy logic, wireless sensors, urban search and rescue.

1 Introduction

The aim of the project "Second Generation Locator for Urban Search and Rescue" (SGL) is to advance the current state of the art for early location technology of entrapped people or dead bodies in collapsed structures. The approach is that of integrating multiple sensing elements into operational devices and improving communication management and data fusion techniques. The project considers the development of innovative portable devices and probes for continuously monitoring the conditions of voids and data measurements regarding vital medical parameters of the victims. This novel, integrated approach will be organized around multi-sensory localization systems and devices (Second Generation Locator) supported by a command and control framework. This framework will integrate location and monitoring methods with USaR (Urban Search and Rescue) logistics, energy management and other applications that will support managing USaR operations in the most reliable, efficient, safe and economic way. The SGL platform has a modular and scalable architecture, enabling evolutionary development. The remote early detection system consists of autonomous devices operating without the support of rescuers. Its mission is supporting safety of rescuer's teams by identifying potential risks on operation area and signs of life on those locations where they are placed. These devices have communication capabilities to send and receive information to/from Command and Control Center [1, 2].

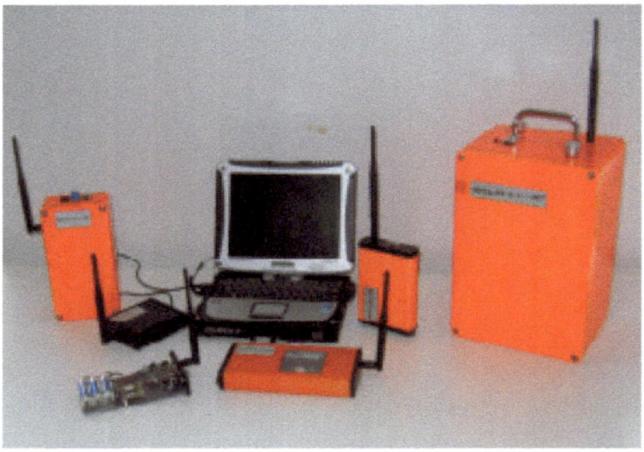

Fig. 1 REDs probes and controller, from left to right. Back: audio probe, gateway, REDs controller, accelerator, gas sensor system. Front: GPS node (open and closed).

2 Information Processing

Information processing is one of the basic tasks of information handling. The information processing should follow a general scheme, no matter which device is used for the information processing. The information processing is structured into the stages and is based on the well-established pipes & filters architecture that was already used for complex decision-making processes in the field of mobile robotics [3]. An overview of the pipes & filters architecture can be found in figure 2. The system consists of a set of sensors, processing hardware, and radio modems. Sensors can be as simple as temperature sensors or CO-sensors, but also as complex as ion-mobility spectrometers (IMS) [4]. The number is not limited. Simple sensors are more related to attended devices, due to the reduced computational power and the limited energy capacity of these devices. A combination of simple and complex sensors can be used on unattended devices. The sensor input is followed by data pre-processing stages. After data pre-processing is finished, the signals have to be characterized. Characterization involves continues monitoring as well as assessing the signal and assignment of critical values. If critical values are found, the matching stage has to identify the signals and send them to an aggregation stage. This stage collects information from all sensors and all devices to give accurate information of the measured scenario.

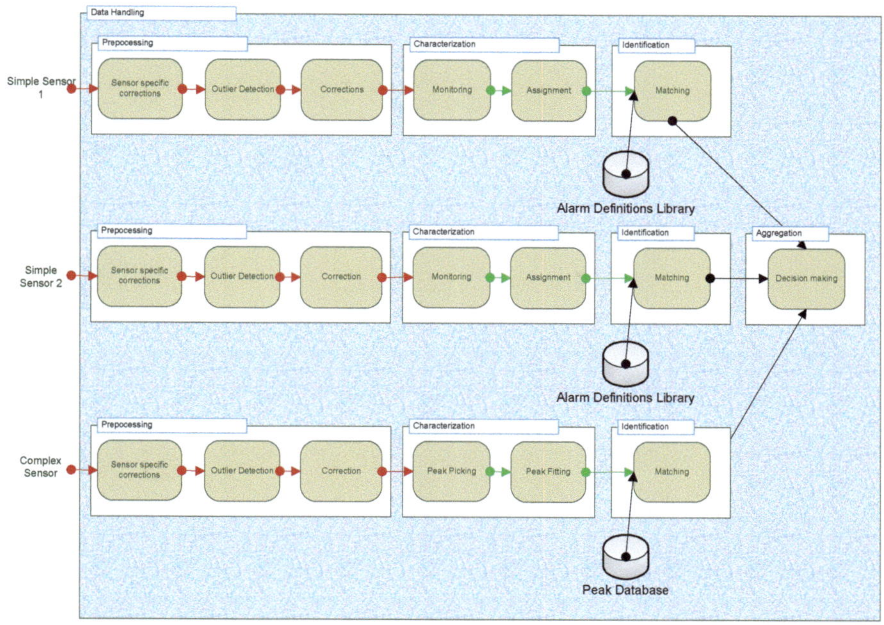

Fig. 2 Overview of the data handling system

Not all sensors need this complex information path. For very simple sensors the path can be reduced to e. g. outlier detection and assignment. If single sensors are used, the aggregation stage is not needed, if single devices are used, the aggregation stage has to be put into that device. If multiple devices are used, the aggregation stage is situated in the central controlling system or the Command and Control center (for the needed communication) an RF-based systems have to be used.

3 The Trapped Human Experiment

In September 2010, the Trapped Human Experiment took place in Loughborough, UK [5-7]. During ten experiments, each single one six hours long, exhaled air was analyzed using ion mobility spectrometers (IMS), CO and CO2 gas sensors, as well as O2, liquid petroleum (LP), and NH3 gas sensors. The whole experiment setup was designed to simulate a collapsed building. During one of the experiments the sample point, from which the exhaled air was sampled, was changed to get an impression of how the concentration of gases changes, if the depth of the entrapped human changes. The next figure shows the combination of all signals. The different experiments can be clearly distinguished based on the O2 and CO2 signal. The O2 signal drops to an amount of 17 – 19 %, if the void is occupied. The CO2 signal rises to about 2 % in this cases, it drops to 0 % if the void is empty.

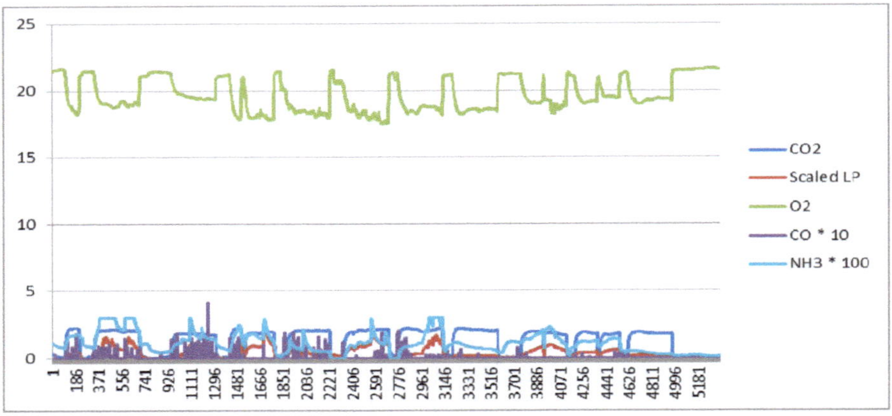

Fig. 3 Overview of the concentrations for CO, CO2, LP, NH3, and O2

All devices are equipped with CO and CO2 gas sensors. Both signals can be used to detect entrapped humans and presence of fire or dangerous environments. The experiments have shown that humans can be detected by analyzing the CO2 concentration. The CO concentration was nearly beyond detection limit and was amplified for graphing in the diagrams.

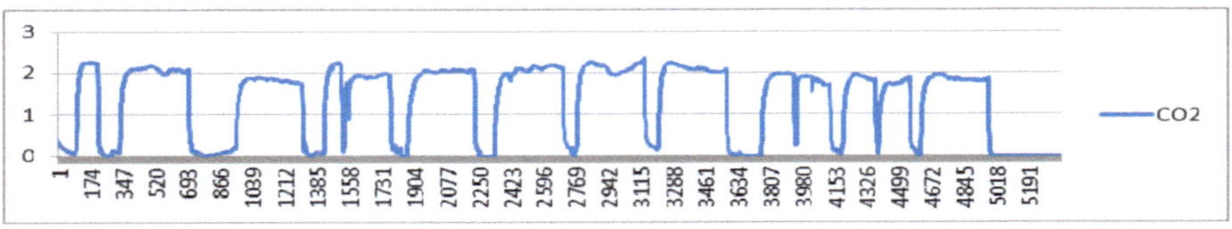

Fig. 4 Concentration of CO2 during the experiments

Typical concentrations of CO2 vary in the range of 1.8 – 2.2 %, depending on the person inside the vault. The O2 sensor shows a significant drop of the O2 concentration. Again the drop is not constant over all experiments, it varies between 17.5 % and 19 % instead.

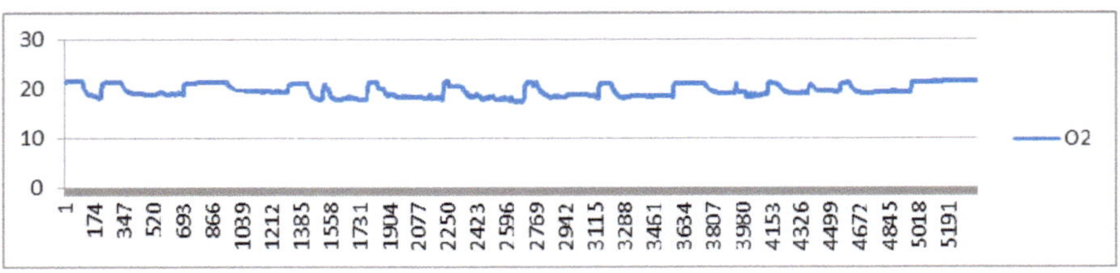

Fig. 5 Concentration of O2 during the experiments

The concentrations of NH3 and liquid petroleum (as an indicator for explosive gases) was also measured during the experiments. This data is not discussed in this work, but can be found in [7].

4 Decision Making

Decision making is a central point for automated, unsupervised devices like the here discussed probes and controllers. These devices are left unattended at places in the search area, where they monitor chemical and physical sensor, to detect signatures that correspond to special situation. The type of situation has to be detected and alarms have to be raised and send to the Command and Control Center. In the scenario of the SGL project, two main approaches for decision making have to be distinguished, decision making based on implicit knowledge and decision making based on explicit knowledge. Implicit knowledge results from learning methods that are able to make decisions, but are not able to explain, why they advised exactly that solution. Typical examples for the implicit decision making are algorithms like, artificial neural networks, self-organizing maps, networks, based on the adaptive resonance theory, and decision making based on principal component analysis. Knowledge is always stored in an implicit form, e. g. as a set of hundreds of weights that are used by an artificial neural network [8].

Explicit decision making systems are based on learning algorithms which are able to explain the decision taken. Knowledge is stored as a set of intervals, rules for decision trees, or in form of linguistic terms, if fuzzy logic is used. Typical learning algorithms for explicit decision making systems are decision trees, interval based logic, and expert systems, based on fuzzy logic. Especially for the SGL for USaR project, the chance to get reasons for an advised decision is very important. The knowledge used for the decision making will be revised often, e. g. new type of sensors or new signature for signs of life are available. From this point of view, a decision making system, based on explicit knowledge is necessary [9].

4.1 Fuzzy Logic to Cope with Sensor Information

From the study of the Loughborough data, it can be seen that critical values for the concentration of measured compounds cannot be defined in terms of fixed intervals. See CO2 as an example: Typical concentrations of CO2 vary in the range of 1.8 – 2.2 %, depending on the person inside the vault. Due to the fact, that not all humans of the world can be measured against their CO2 production, the setting of the interval limits to 1,8% and 2.2% would be wrong. These numbers are based on ten humans only. Fuzzy logic can be used in exactly these cases, because fuzzy logic is able to interpolate varying criteria. Fuzzy logic is also capable of handling imprecise or incomplete information. The practical use of fuzzy logic can be found in the non-linear control of technical systems, as well as in the rule based decision making. If fuzzy logic is used for obtaining numerical values, such in control applications, the Takagi-Sugeno approach is usually a suitable model. In the scope of the SGL project we propose the use of the Mamdani approach, due to its linguistic expression of premises and conclusions.

For the corresponding fuzzy rules computation, a standard max-min composition algorithm can be used.

The use of sensor fusion is a major principle of the whole SGL way to process information. The sensor fusion can be directly mapped to fuzzy logic rule based systems. Each sensor is used as a linguistic variable, with its own unique identifier. An arbitrary number of inputs can be used, no limits exist. Depending on the source of the input, two stages of decision making can be stated:

1. decision making on a single device

If all inputs are directly connected to the device, decision making can take place on the device. This way of decision making is typical for simple devices

2. decision making on an integrating device

If inputs from different devices has to be used, an integrating device, like controllers or Command and Control Centers has to be used. These device are connected with the input sending devices by RF-communications and are able to receive all necessary inputs.

4.2 Decision Making in Urban Search and Rescue Operations

The processing pipeline is defined as the processing pipeline for sensor fusion on different devices. The fuzzy rule based systems use rules to perform the decision making. This step is depicted as aggregation/decision making in the above diagram. The aggregation is calculated using fuzzy rules with more than one input source in the premise. In contrast to the decision making on simple devices, the aggregation uses distributed sensor input. The decision making is based on the evaluation of all fuzzy rules. An example for a fuzzy logic rule base, which can run on a controller device, is the following:

```
IF      probe1. input.CO2             is      high
AND     probe2.input.LP               is      high
THEN    output.danger_of_explosion    is      high

IF      probe1.input.CO2              is      low
AND     probe2.input.LP               is      low
THEN    output.danger_of_explosion    is      low
```

4.3 Application of Methods

The data gained during the Trapped-Human-Experiment (THE) are suitable to show how a fuzzy rule based system can be used to detect whether a human is inside a void or not. After thorough analysis of the recorded data by the University of Loughborough the fuzzy rule based system can be extended for field application. It can be derived, that an O2-signal in the range of 17% - 19% can be used as an indicator for an entrapped human. An O2-signal in the range of 20% - 21.5% is an indicator of an empty void. In case of an entrapped human the O2-signal drops quit fast (about ten minutes from 21.5% to 19.0%). The CO2-signal raises even faster (about five minutes from base level to 2%). The O2-level depends on the human being entrapped, so a fixed interval with hard limits is not suitable to model this. Instead fuzzy sets are used to indicate an uncritical and a critical level. The same happens to the CO2-signal; again two trapezoidal fuzzy sets can be used to distinguish between critical and uncritical levels.. The rule base connects both inputs to give an alarm if both signals are in a critical range. The rules are as follows:

Rule 1

IF	REDS.THE Experiment.CO2-signal	is	uncritical
AND	REDS.THE Experiment.O2-signal	is	uncritical
THEN	REDS.THE Experiment.Alarm	is	no

Rule 2

IF	REDS.THE Experiment.CO2-signal	is	critical
AND	REDS.THE Experiment.O2-signal	is	uncritical
THEN	REDS.THE Experiment.Alarm	is	no

Rule 3

IF	REDS.THE Experiment.CO2-signal	is	uncritical
AND	REDS.THE Experiment.O2-signal	is	critical
THEN	REDS.THE Experiment.Alarm	is	no

Rule 4

IF	REDS.THE Experiment.CO2-signal	is	critical
AND	REDS.THE Experiment.O2-signal	is	critical
THEN	REDS.THE Experiment.Alarm	is	yes

Evaluating the THE-data according to the fuzzy rule base, results in the following sets of generated alarms as depicted in the figure 6.

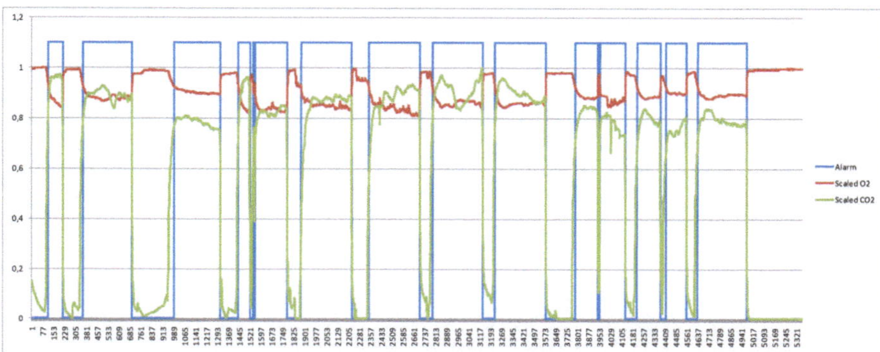

Fig. 6 Alarms generated by the rule based system

The next figure shows an excerpt of the data. The excerpt covers data from the experiments number 2, 3, and 4. It can be clearly seen, that the alarm is raised very early and canceled accurate to the end of the experiments.

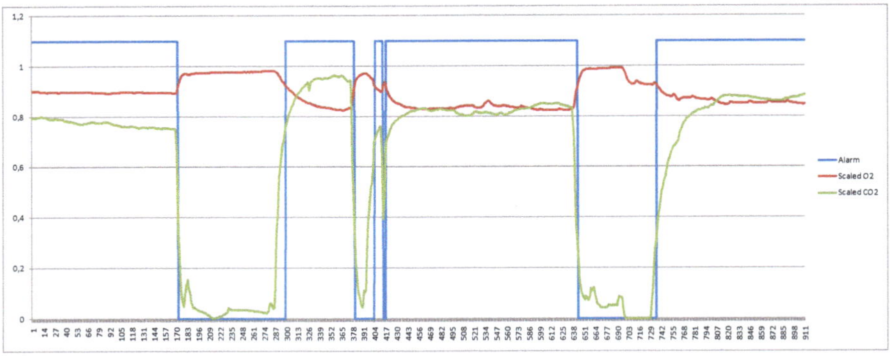

Fig. 7 Alarms generated by the rule based system during experiments 2, 3, and 4

5 Summary

The use of fuzzy logic based decision making is well suited for the detection of signs of life, danger, and death. The used rules are easy to understand and easy to change. The use of different fuzzy variables allows a modeling of the sensor fusion, typical for the demands of the search and rescue operations. The needed fuzzy rule sets are small and can be assigned to specialized parts of the decision making process. The use of fuzzy rules allows an explanation of the found decisions. In combination with the linguistic oriented way of defining rules, this allows an easy and fast adaption of the decision making process with respect to changing demands. The evaluation of fuzzy rules is fast in execution, even on simple hardware, which is typical for embedded sensor systems.

References

1. Statheropoulos, M.: Sensor network for search and rescue operations in collapsed buildings. ERCIM News (81), 49–50 (2010)
2. `http://www.sgl-eu.org/`
3. Hildebrand, L., Michalski, C., Wickrath, M., Valentin, H.: Strategy implementation for mobile robots using the pipes & filters architecture. In: Proceedings of the FIRA Robot World Congress 2003, Wien (2003)
4. Eiceman, G.A., Karpas, Z.: Ion Mobility Spectroscopy, 2nd edn. CRC Press (2005)
5. Vautz, W., Baumbach, J.I., Westhoff, M., Zóchner, K., Carstens, E.T.H., Perl, T.: Breath sampling control for medical application. International Journal for Ion Mobility Spectrometry 13(1), 41–46 (2010)

6. Bödeker, B., Baumbach, J.I.: Analytical description of IMS-signals. Int. J. Ion Mobil. Spec. 12, 103–108 (2009)
7. Huo, R., Agapiou, A., Bocos-Bintintan, V., Brown, L., Burns, C., Creaser, C.S., Davenport, N., Gao-Lau, B., Guallar-Hoyas, C., Hildebrand, L., Malkar, A., Martin, H., Moll, V.H., Patel, P., Ratiu, A., Reynolds, J.C., Sielmann, S., Slodzynski, R., Statheropoulos, M., Turner, M., Vautz, W., Wright, V., Thomas, P.: The Trapped Human Experiment. Journal of Breath Research 5 (2011)
8. Winter, J., Hildebrand, L.: Interactive Decision Trees and Artificial Neural Networks. In: Proceedings of the International Conference on Knowledge Management for Composite Materials, Germany (2007)
9. Hildebrand, L., Fathi, M.: Linguistic color processing for human-like vision systems. In: Proceedings of Electronic Imaging 2000, IS&T/SPIE 12th International Symposium, San Jose, USA (2000)

Integrated Modeling of Technical and Business Aspects in Service Networks

Frank Schulz[1], Simon Caton[2], Wibke Michalk[2], Christian Haas[2], Christof Momm[1], Markus Hedwig[3], Marcus McCallister[4], and Daniel Rolli[4]

[1] SAP Research, Karlsruhe, Germany
[2] Karlsruhe Service Research Institute (KSRI), Karlsruhe Institute of Technology, Germany
[3] Information Systems Research, University of Freiburg, Germany
[4] Conemis AG, Karlsruhe, Germany
frank.schulz@sap.com,
{simon.caton,wibke.michalk,ch.haas}@kit.edu,
christof.momm@sap.com, markus.hedwig@is.uni-freiburg.de,
{marcus.mccallister,daniel.rolli}@conemis.com

Abstract. The current trend towards a global services economy provides significant opportunities and challenges. For establishing complex services and delivering competitive advantages, several service providers have to work together. This collaboration creates a service network as an organizational form to be managed by a so-called service integrator. Within a service network, multiple dependencies between the resulting service and the contributions of the various service providers exist, on both technical and business aspects. In addition to the functional aspects, the non-functional service properties and respective service levels are of great importance. Successful joint management of the technical and business dependencies is a key prerequisite for the successful management of service networks.

This paper contributes an approach to the integrated modeling of dependencies in service networks. The relations between services and the relations between service levels are made explicitly and form the basis for effective decision support and management of service networks. The concept has been implemented and evaluated in various case studies and industrial settings.

Keywords: Service Value Network, Service Level Agreement, Cloud Computing.

1 Introduction

Services are becoming a major backbone and innovation driver of modern economy. IT services on the levels of computing resources (Infrastructure-as-a-Service), middleware (Platform-as-a-Service) and applications (Software-as-a-Service) are the main constituents of cloud computing. With respect to the production of complex services, service networks have emerged as a form of joint value creation within cloud service markets [2]. In a service network, several providers contribute their specific expertise to establish a complex service that would not have been possible before. The service network is coordinated by a service integrator who is responsible for selecting and combining component services and presenting one combined service to customers. This service integrator, also called service intermediary, is the only party facing the customers from contractual point of view. Hence, the integrator is responsible for creating, negotiating and delivering complex services. This includes the agreement on quality of service and price, and the management of associated risks.

Quality of service (QoS) is expressed with the help of metrics or key performance indicators (KPIs) like availability, response time, throughput, or support levels. Thresholds and tolerable ranges of these metrics are defined in service level objectives (SLOs). They are collected in a service level agreement (SLA) that is a contract between service provider and service consumer and defines the obligations of both parties. For the case of failing to keep the obligations, usually some compensation is defined. If the service provider misses a service level objective, typically a monetary penalty or a certain time of service usage without charging is agreed. This compensation is only due if the service consumer adheres to his obligations, for example not exceeding a certain rate of service requests. The identification of fulfilled or violated service level objectives requires a detailed monitoring and precise definitions of the underlying metrics and their interpretation.

As the service intermediary is the only customer facing party, and acting as a service provider towards the customer, the intermediary has to coordinate the underlying service network and to manage the service quality and associated risks. In particular he has to be aware of the dependencies between services and between service level agreements, in order to know the constraints during negotiation and the potential options during service delivery. Only the full understanding and visibility of dependencies of all service instances will allow an effective risk-aware management and optimization of service networks.

Related questions have been investigated before in the context of web service composition [3][11][16]. The combination of web services is a special case of the more general service networks considered here, which also cover infrastructure services and application services. We emphasize the explicit modeling of dependencies between all involved artifacts. This enables an evaluation and ranking of the dependencies as decision support for service intermediaries.

The key contribution of this paper is a concept and an implementation for modeling technical and business dependencies between services and between their quality characteristics. This enables effectively supporting the service intermediary in managing service networks and related service level agreements.

In section 2, the relevant background on service level agreements and service value networks is introduced. The main results are presented in section 3, and section 4 reports on the evaluation of the proposed concepts based on an implementation and usage within an integrated case study. Section 5 discusses related work and section 6 concludes with a summary and an outlook.

2 Foundations

Service level agreements provide an approach to combine technical and business aspects. The following subsections introduce SLAs and service value networks respectively, and provide the basis for the integrated modeling of dependencies.

2.1 Service Level Agreements

There are several approaches for formalizing **service level agreements**; and the WS-Agreement specification [1] is among the most important ones due to its wide adoption in practice and its extensibility. WS-Agreement specifies the high level format of an SLA document and requires the inclusion of application-specific KPIs and SLOs. The structure of an SLA document according to WS-Agreement is shown in Fig. 1(a).

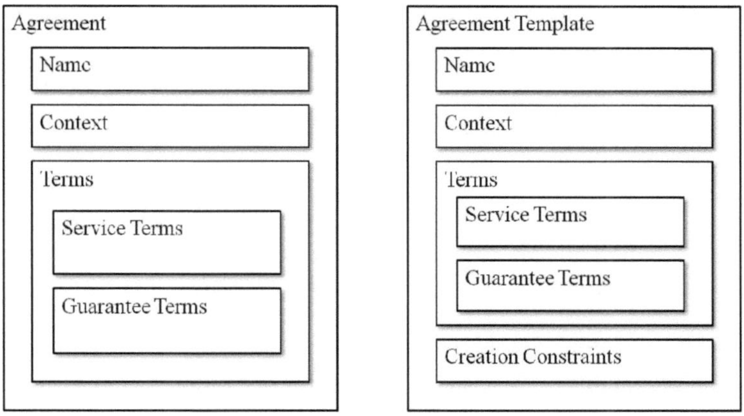

Fig. 1 (a) WS agreement document (b) WS agreement template

Following the document's name section, the context describes the participants, the expiration time and optionally the template from which the document was derived. The terms section contains the detailed information. Both the service terms and the guarantee terms are recursively defined with a term compositor. The service terms describe the service for example by linking to a WSDL document and by referring to a service endpoint. They also contain the specific metrics to be monitored. The guarantee terms express the service level objectives. An objective is a condition on one or more service metrics to be fulfilled by the obliged party. The guarantee terms also include business values like price and penalty in case of SLO violation.

Related to a service level agreement is a **service level agreement template** (SLAT), see Fig. 1(b). A template is used during service negotiation. It describes the offer of a service provider, while possibly leaving some parameters undefined.

During negotiation, a potential customer can choose these parameters, or select between several templates. SLA templates are also specified by WS agreement. Their structure is similar to SLA documents, and contains an additional section with creation constraints. These constraints restrict the potential values of parameters during negotiation.

When several SLA templates are offered for one service, they typically contain guarantee terms on the same metrics and differ only in the target values of the service level objectives. In such a case, the SLA (template) may be considered as a vector of SLOs. We introduce the following definitions for ranking SLAs and SLA templates.

Definition 1: SLO1 is called *better* than SLO2 if the guarantees are better from the consumers' point of view, for example higher availability or reduced response time for a service request. If two SLAs contain SLOs for the same metrics, SLA1 is called *better* than SLA2 if it is better for all SLOs. In other words, SLA2 is dominated by SLA1 in all vector components, and we write SLA1 > SLA2.

Example 1: Given the following three SLAs in informal notation:

SLA1: Availability > 95 % and Response Time < 2 sec
SLA2: Availability > 98 % and Response Time < 1 sec
SLA3: Availability > 99 % and Response Time < 1 sec.

Then SLA3 > SLA2 > SLA1. In business context, SLA1 may be labeled "bronze", SLA2 "silver" and SLA3 "gold". It is also possible that two SLAs cannot be compared according to vector dominance. For example, a service provider might offer a "high availability" SLA and a "fast response" SLA:

SLA1: Availability > 99 % and Response Time < 2 sec
SLA2: Availability > 95 % and Response Time < 1 sec.

In this case, neither SLA1 > SLA2 nor SLA2 > SLA1. The comparisons can be defined on service level agreements and on service level agreement templates in the same way.

2.2 Service Value Networks

Service value networks (SVNs), or service networks for short, are business networks that create value through the composition of services within an open service market [2][9]. A service network can be represented as a graph, as shown in Fig. 2. Each node refers to a service, and substitution groups classify equivalent services. Each path through the graph from the virtual source node *s* to the virtual target node *t* represents a valid service combination that provides the desired functionality of the complex service network.

Note that the graph does not necessarily represent a workflow with a defined sequence of services. While the order of services is important in some scenarios, only the set of services defined by a path is required for the following discussion. In Fig. 2, the providers of the services are omitted. They are not needed for the following discussion either, because the service level guarantees of each service instance are considered individually. For assessing joint risk distributions, this information would be useful though.

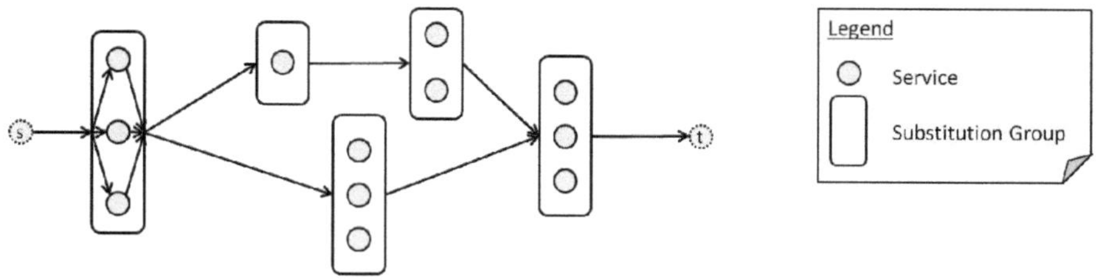

Fig. 2 Service Value Network

3 Integrated Modeling of Relations

This section presents the proposed model and several important use cases.

3.1 Modeling Dependencies

In this section, a model of the various relationships between services, SLA templates and SLAs in service networks is developed. A formal representation allows capturing the dependencies and enables decision support for the intermediary. Fig. 3 shows the involved entities. A *service* type describes the functional properties of a service. For one service, several *service level agreement templates* can exist and specify different potential quality levels. From one template, several *service level agreements* can be derived as the actual contracts between providers and consumers.

The model takes the perspective of a service intermediary. Hence it contains the *offered service* as the result of the service network, and the *required services* as the contributions by the participants in the service network. This functional relation is captured in a *service topology*. In other words, each path through a SVN graph like Fig. 2 corresponds to exactly one service topology instance.

The relation expressed in the service topology is continued to SLA templates and SLAs. For the offered and required services, similar topologies can be established for their SLA templates and SLAs. The *SLA template topology* describes the dependency between offered and required SLA templates, and the *SLA topology* represents the relation bet-ween active SLAs of provided and consumed services in a deployed service network.

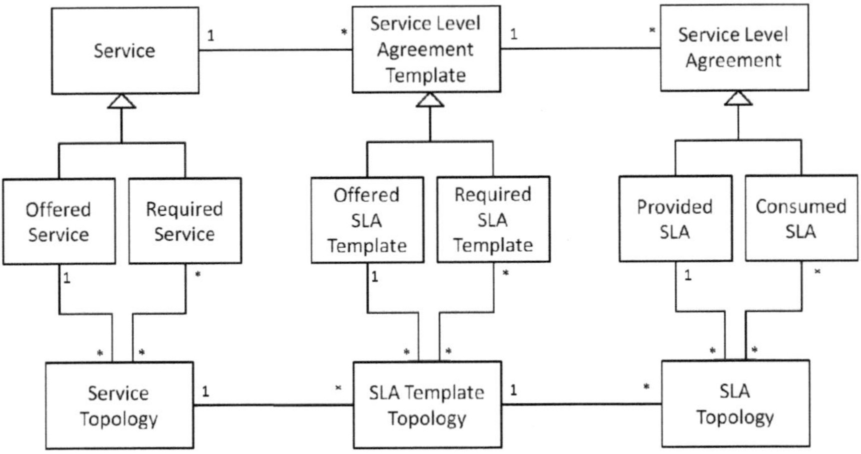

Fig. 3 Dependency Model

3.2 *Using the Model*

The proposed model provides the foundation for an effective service level management in service networks. Three main usages of the model are presented in the following sections.

3.2.1 Monitoring

A major application of the model is the immanent definition of a comprehensive structure for monitoring across a service value network. Even though many partners can collaborate and contribute to services in service value networks and those partners are usually independent, the consistent modeling delivers both the relationships between service providers as well as the detailed agreements between them as immediate input for the monitoring configuration. A monitoring solution that is prepared for interpretation of such models is automatically configured by this input.

The monitoring provides valuable feedback to both customers and service integrators. It assists in deriving useful proposals for future modifications and enhancements. Such modifications can mean switching supplying Service Providers or renegotiating service level agreements.

In the following it is explained how the monitoring of service metrics is realized on three different technical layers. On the infrastructure level, there are the infrastructure service resources like D-Grid and various cloud offerings such as Amazon EC2 and Amazon S3 that can be monitored. Each service is supervised by a dedicated agent that measures the actual state of the service, including availability and responsiveness. A full-fledged instance of a service value network model defines the monitoring down to the periodic time interval for the automatic measurement by the deployed agents. The collected data is sent to the monitoring server component and stored in the monitoring database. Web services are defined on a middle layer, for example intermediate services that typically provide building blocks of functionality but no complex process or application logic. Each monitored service is supervised individually, data is aggregated and the SLOs are calculated in the backend by using metrics for defined SLOs. The composite services of the upper layers represent full-fledged Software-as-a-Service applications. On this SaaS level, the services are also connected to the monitoring software. This builds the ground for reliable control over the entire service network.

On all layers, the monitoring system is responsible for collecting measurements and checking SLOs defined for the corresponding services such as the above mentioned availability, latency etc. Monitoring thereby lays the foundation for systematic service management across a whole service value network.

3.2.2 Planning of Services Capacities and Provisioning

During negotiation of the service offering with a potential customer, the service integrator has to ensure that it will be feasible to provision the requested service in the agreed quality. This task requires a suitable planning of service capacities. The service integrator has to consider regular service usage and irregular usage spikes, and to perform an appropriate sizing of the required component services. At this point, the topologies introduced in section 3.1 are used. For realizing a service delivery that meets the agreed service level objectives and optimizes the business revenue, the service integrator compares the potential options which are given by corresponding topologies. For one service to be offered, several service topologies can be considered, and for each of these service topology, several SLA template topologies can be compared. Given this candidate set of SLA template topologies, different strategies for selecting the most suitable ones are possible. One approach is to consider only the topologies that meet the customer's service level requirements, and then choosing the most profitable one among those. For efficient identification of the SLA template topologies fulfilling the requirements, the comparison scheme of definition 1 is used. Based on the partial order, only the SLA templates that are *better* than the customer requirements need to be taken into account, possibly including an appropriate safety margin. This means that other templates dominated according to SLO vector dominance can be discarded without affecting the result.

Another approach is to rank all available templates based on profitability with respect to the specific customer, and then to select the first template that meets the servive level requirements, again taking a safety margin into account.

The templates may contain a different set of metrics than requested by the customer. If the template contains a superset of the customer metrics, the additional metrics need not to be taken into account during vector comparison. If however the customer required additional metrics, only those templates that cover all the metrics are allowed in the candidate set. A contract with the customer will need additional negotiation, and potentially the extension of SLA templates by the missing metrics.

3.2.3 Risk Analysis

As discussed in section 3.2.2 above, an important task of service level management for service integrators is the best choice of service combinations and SLA combinations, i.e. the selection of the most suitable SLA template topology. As a more comprehensive alternative to the direct approach discussed above, the selection of the best template can be based on a risk measurement taking potential penalties in case of SLO violation into account.

For each SLA template topology and each combination of template topologies offered by the service intermediary, the associated risk is calculated, and the templates are ranked according to risk. This methodology described in [8] applies concepts from portfolio theory and asset management. The template selection based on economic risk is better suited for maximizing revenue, because all templates are considered and evaluated, instead of only those that strictly meet the customer requirements. Depending on the penalty in case of SLO violation, a template with small safety margin might be ranked better according to economic risk than a a template with larger safety margin in the service level metrics but with higher penalty. The methodology is explained and discussed in [7].

4 Evaluation

The proposed model of service networks and their SLAs has been implemented as the repository component within the ValueGrids project [12]. In this context, it provides the basis for all service level management features. In particular, the repository allows assigning and distributing collected data to the correct entities in order to assess past service behavior. The service repository provides SOAP-based web service interfaces for accessing and manipulating the content objects. The interfaces are used for data exchange with other components of the ValueGrids framework, in particular the monitoring component, the service planning component and the risk analysis component. The loose coupling based on open web standards allows an easy extension of the framework.

The implemented framework realizes a tool chain for addressing the key tasks of service level management in service networks. It has been applied to a case study that contains a three-level service hierarchy (infrastructure services, web services, application services). On each level, certain services are required, and candidate service offerings and with different service level characteristics are available. Hence the service level management tasks occur at each level within the scenario.

The repository represents the central data exchange component within the ValueGrids tool chain. The proposed model of dependency modeling provides the basis for and enables tasks like analysis, evaluation and planning of service delivery and service consumption.

5 Related Work

In the context of web service composition, several authors discuss the aggregation of service level agreements [3][11][15][16]. They focus on the identification and aggregation of web service QoS metrics, and do not investigate the service level management problem and the usage of SLA templates in this context.

The authors of [6] present a conceptual framework for service level management that addresses dependencies as well. However they do not model these dependencies explicitly or use them for further analysis and management.

The representation of a SLA as a vector of service metrics has been used in [5]. The authors define a scalar similarity function or scoring function for expressing the difference between two SLA vectors, but they do not use the concept of vector dominance for comparing SLAs or SLA templates.

In our approach, service dependencies are modeled on one level only. A multi-level hierarchical SLA aggregation has been proposed and developed in [14]. It depends on the application domain whether such a model with visibility of SLA contracts along a service chain is suitable. We believe that a restricted knowledge of SLAs of the direct business partner only is closer to current practice in many areas.

6 Conclusion and Outlook

This paper introduces a dependency model for service level agreements in a service network. The functional dependencies between services are extended to the non-functional service properties described in SLA templates and SLAs. The proposed dependency model delivers the technical foundation for monitoring, risk analysis and capacity planning. Hence it provides the basis for an effective service level management in service networks.

The work presented here will be continued in several directions. In addition to a finite list of SLA templates, the dependency model can be extended to support parameterized templates. In this case, the functions for translating parameters between offered and required SLA templates have to be added to the model. Future research will also analyse the application of dependency models to multi-tenancy settings, where services are shared between several consumers.

Acknowledgments. This work is supported by the German Federal Ministry of Education and Research under promotional reference 01IG09004 (ValueGrids).

References

1. Andrieux, A., Czajkowski, K., Dan, A., Keahey, K., Ludwig, H., Nakata, T., Pruyne, J., Rofrano, J., Tuecke, S., Xu, M.: Web Services Agreement Specification (WS-Agreement). Open Grid Forum (2007)

2. Blau, B., Krämer, J., Conte, T., van Dinther, C.: Service Value Networks. In: Proceedings of the 11th IEEE Conference on Commerce and Enterprise Computing, Vienna, Austria, pp. 194–201 (2009)
3. Cardellini, V., Casalicchio, E., Grassi, V., Lo Presti, F.: Efficient Provisioning of Service Level Agreements for Service Oriented Applications. In: 2nd International Workshop on Service Oriented Software Engineering (IW-SOSWE 2007), pp. 29–35 (2007)
4. Jaeger, M., Rojec-Goldmann, G., Mühl, G.: QoS aggregation for Web service composition using workflow patterns. In: Proceedings of the 8th Enterprise Distributed Object Computing Conference, pp. 149–159 (2004)
5. Laforenza, D., Nardini, F.M., Silvestri, M.: Collaborative Ranking of Grid-enabled Workflow Service Providers. In: Proceedings of the 17th International Symposium on High Performance Distributed Computing (HPDC), pp. 227–228 (2008)
6. Ludwig, A., Franczyk, B.: COSMA – An Approach for Managing SLAs in Composite Services. In: Bouguettaya, A., Krueger, I., Margaria, T. (eds.) ICSOC 2008. LNCS, vol. 5364, pp. 626–632. Springer, Heidelberg (2008)
7. Michalk, W., Blau, B.: Risk in Agreement Networks. Information Systems and E-Business Management (IseBM) 9(2), 247–266 (2011)
8. Michalk, W., Caton, S.: Service Level Management in Dynamic Value Networks. In: INFORMATIK 2010: Service Science - Neue Perspektiven fuer die Informatik. Band 1, pp. 126–131 (2010)
9. Mohammed, A.B., Altmann, J., Hwang, J.: Cloud Computing Value Chains: Understanding Businesses and Value Creation in the Cloud. In: Economic Models and Algorithms for Distributed Systems, pp. 187–208 (2009)
10. Momm, C., Schulz, F.: Towards a Service Level Management Framework for Service Value Networks. In: INFORMATIK 2010: Service Science - Neue Perspektiven Fuer die Informatik. Band 1, pp. 521–526 (2010)
11. Papazoglou, M.P., van den Heuvel, W.J.: Web Services Management: A Survey. IEEE Internet Computing 9(6), 58–64 (2005)
12. Schulz, F., Michalk, W., Hedwig, M., McCallister, M., Momm, C., Caton, S., Haas, C., Rolli, D., Tavas, M.: Service Level Management for Service Value Networks. In: 5th International Workshop on Service Science and Systems (SSS 2012) at the IEEE Conference on Computer Software and Applications, COMPSAC 2012 (2012)
13. Schulz, F.: Decision support for business-related design of service level agreements. In: 2nd IEEE International Conference on Software Engineering and Service Science (ICSESS), pp. 35–38 (2011)
14. Ul Haq, I., Schikuta, E.: Aggregation Patterns of Service Level Agreements. In: Proceedings of the 8th International Conference on Frontiers of Information Technology, FIT 2010 (2010)
15. Unger, T., Leymann, F., Mauchart, S., Scheibler, T.: Aggregation of service level agreements in the context of business processes. In: Proceedings of the 12th Enterprise Distributed Object Computing Conference, pp. 43–52 (2008)
16. Zeng, L., Benatallah, B., Ngu, A.H.H., Dumas, M., Kalagnanam, J., Chang, H.: QoS-Aware Middleware for Web Services Composition. IEEE Transactions on Software Engineering 30(5), 311–327 (2004)

TCP Traffic Classification Using Relaxed Constraints Support Vector Machines

Mostafa Sabzekar, Mohammad Hossein Yaghmaee Moghaddam, and Mahmoud Naghibzadeh

Department of Computer Engineering,
Ferdowsi University of Mashhad, Iran
sabzekar@wali.um.ac.ir,
yaghmaee@ieee.org,
naghibzadeh@um.ac.ir

Abstract. The traffic classification problem is critical for management, security monitoring, and traffic engineering in computer networks. It has recently taken into consideration by both network operators and researchers. It allows network operators to predict future traffics and detect anomalous behavior and also allows researchers to create traffic models. In this paper, we use a new architecture of support vector machines, namely relaxed constraints support vector machines (RSVMs), to present a traffic classifier that can achieve a high accuracy without any source or destination address or port information. We just use packet length to predict the application class for each flow. RSVM is an efficient and noise-aware implementation of support vector machines that assigns an importance degree to each training sample in such a manner that noisy samples and outliers are given a less degree of importance. Experimental results with UNIBS and AUCKLAND, two sets of traffic traces coming from different topological points in the Internet, show that the proposed classifier is more reliable and has better accuracy.

Keywords: Traffic classification, Support vector machines, Relaxed constraints.

1 Introduction

Accurate identification of network applications is a crucial part of network operation and management. Network operators need to know the application class of different flows over their networks to react quickly in support of their various business goals. Furthermore, traffic classification has been helpful in network management activities such as Quality of Service (QoS), security monitoring, among others. The reasons for more interests in this field can be summarized as follows:

- The ability of assigning traffic flows to relevant classes of service is very important for Internet Service Providers (ISP) because some governments are clarifying ISP obligations with respect to "lawful interception" of IP data traffic [1].
- Traffic classification is the heart of any intrusion detection system. Detection of the application behind a given IP flow is the core part of many anti-virus and anti-worm applications.
- From the view point of QoS, accurate traffic classification is very helpful in identifying the application utilizing network resources, and facilitate the instrumentation of QoS for different applications [2,3].
- It is very necessary for advanced network management and traffic engineering.

In general, there are two approaches for traffic classification of flows; traditional methods vs. traffic statistical properties based ones [4]. Traditional methods are based on the inspection of a packet's TCP or UDP port numbers (*port based classification*) or the reconstruction of protocol signatures in its payload (*payload based classification*). Traditionally, the port numbers were used widely to network traffic classification. But, nowadays it is shown that this method is not appropriate because some applications may not use their port numbers which registered with IANA [5] and this led to inaccuracy of its classification results [6,7]. Payload based classification method is used in many researches (such as [8,9]) to solve some of the drawbacks of port based classification methods. Although payload based inspection avoids reliance on fixed port numbers and can produce accurate results, its high resource requirements and limitations with encrypted traffic make its use unfeasible in current high-speed networks. So, it imposes significant complexity and processing load on the traffic identification device. So, the traditional techniques are suffered by their dependence on the inferred semantics of the information in packet content (payload and port number). To solve the problems with traditional techniques, new traffic classification methods use statistical properties of flows (e.g. the distribution of flow duration, flow idle time, packet inter-arrival time, packet lengths and so on). They assume that these characteristics are unique for an application and can be used to determine the class of different applications.

1.1 Related Works

Several recent papers have positively reported on the feasibility of machine learning (ML) techniques to traffic classification problem [10-12]. These approaches use the statistical characteristics of the flows to distinguish different source applications from each others. Nguyen et al. [4] survey and compare the complete literature in the field of ML-based traffic classification. Support vector machine (SVM) [13] is one of the most promising techniques in the field of machine learning. Among existing classification methods, SVMs provide several advantages such as adequate generalization to new objects, absence of local minima, and representation that depends on only a few parameters.

SVMs have been used in some traffic classification researches and promising results have been reported. The limitations of other ML algorithms (such as Bayesian neural networks, hierarchical clustering, C4.5, etc.) for traffic classification that motivate us to use SVM are:

- These methods have very complexity and therefore are not suitable for real-time purpose.
- They may trap into local optimum.
- Accuracy is highly dependent on samples' prior probabilities. The training and testing samples may be biased towards a certain class of traffic. For example, the WWW traffic constitutes the large majority of the sample in [14].

There are some efforts to use SVM for classification of flows in computer networks. Li et al. [15] trained an SVM classifier to distinguish seven classes of applications from each other. Also, they proposed an automatic algorithm for feature selection and reduction from nineteen features. The reported accuracy of their method is interesting. The authors in [16] developed SVM-based classifiers that can be very effective at discriminating traffic generated by different applications, even with reduced training set sizes. They use just packet size of flows as features of training and testing data for SVM classification. In another work [2], the authors suggested an SVM-based method that classifies the Internet traffic into broad application categories according to the network flow parameters obtained from the packet headers. An optimized feature set is obtained via multiple classifier selection methods. In [17], SVM is used for classification of flows based on the notion of *protocol fingerprints* that provide a statistical behavioral description of the corresponding protocol. They take into account the size and the direction of the packets that compose a flow and also inter-arrival times as features of training and testing samples. Also, there are many researches that use SVM for traffic prediction. For example in [18] SVM was used to predict short-term traffic flow. Also, rough set was combined with SVM and the proposed method was compared to back propagation neural networks and single SVM.

Recently we proposed an efficient structure for SVM, namely relaxed constraints support vector machines (RSVM) [18] that assigns an importance degree to each training sample in such a manner that noisy samples and outliers are given a less degree of importance. In this paper it is used for traffic classification problem and experiments with real data sets proved our claim.

The remainder of this paper is organized as follows. A brief review of the architectures of SVM and RSVM is described in Section 2. The proposed method is explained in Section 3. Experimental results using real-world data sets are given in Section 4. Finally, Section 5 summarizes this paper.

2 Support Vector Machines

2.1 SVM Structure

Support vector machines as originally introduced by Vapnik within the area of statistical learning theory and structural risk minimization have proven to work successfully on many applications of nonlinear classification and function estimation. The problems are formulated as convex optimization problems, usually quadratic programs (QP), for which the dual problem is solved. Within the models and the formulation one makes use of the kernel trick which is based on the Mercer theorem. With this strategy, input points are easily mapped into a high-dimensional feature space. Then, SVM finds a separating hyperplane that maximizes the margin between two classes in this space.

Suppose that we have a training sample set $\{(x_1, y_1),(x_2, y_2),\ldots,(x_n, y_n)\}$ and each sample belongs to either of two classes with given label $y_i \in \{-1,1\}$ for $i = 1,\ldots,n$. When the training samples are linearly separable, the SVM separates the two classes with maximum margin between them without any misclassification error. The optimal separating hyperplane (OSH) can be achieved by solving the following QP problem:

$$\begin{aligned} &\text{Minimize } Q(w,b,\zeta)=\frac{1}{2}\|w\|^2+C\sum_{i=1}^{n}\zeta_i \\ &\text{subject to } y_i(w^T x_i+b)\geq 1-\zeta_i \\ &\qquad\qquad \zeta_i\geq 0, i=1,\ldots,n \end{aligned} \tag{1}$$

where w is a weight vector of hyperplane and b is the bias term. The parameter C is a regularization parameter that makes a balance between maximization of the margin and misclassification error. In many practical situations, a separating hyperplane does not exist. To allow for possibilities of violating, slack variables $\zeta_i \geq 0$ are introduced. In order to classify nonlinearly, a solution is to map the input space into a higher dimension feature space and searching the OSH in this feature space. Therefore, the mapping function (x) is introduced. To solve the QP problem, one needs to compute the scalar products of the form $(x_i).(x_j)$. It is therefore convenient to introduce the kernel function $K(x_i, x_j)= (x_i).(x_j)$. By using the Lagrange multiplier method and kernel trick, the QP problem for finding the SVM is defined as:

$$\begin{aligned} &\text{Minimize } Q(\alpha)=\frac{1}{2}\sum_{i=1}^{n}\alpha_i\alpha_j y_i y_j K(x_i, x_j)-\sum_{i=1}^{n}\alpha \\ &\text{subject to } \sum_{i=1}^{n}\alpha_i y_i=0, \quad 0\leq\alpha_i\leq C \end{aligned} \tag{2}$$

where $\alpha = (\alpha_1,\dots,\alpha_n)$ is the vector of non-negative Lagrange multipliers and solution of the QP problem (2). The point x_i with $\alpha_i \geq 0$ is called support vector (SV). The decision function is

$$D(x) = sign(f(x)) = sign(\sum \alpha_i y_i K(x_i, x) + b) \tag{3}$$

2.2 RSVM Structure

TABLE I. Taking another point of view, there are some problems with the standard SVM. Since the classifier obtained by SVM depends on only a small part of the samples (support vectors), it is very easy for it to become sensitive to noises or outliers in the training set. Another problem is that the contributions of all training samples to training the classifier are identical. Relaxed constraints SVMs (RSVM) instead, considers the fuzzy membership values in the constraints of the SVM formulation. As we discussed in [18], the RSVM is an efficient extension of the SVM algorithm that deals with these problems. It considers an importance degree for each training sample. It has been shown that it is robust against noisy data and outliers in data sets. The constraints of RSVM have more relaxation and flexibility because of their fuzzy inequalities. So, the problem of SVM (1) is reformulated as:

$$\begin{aligned} &\text{Minimize } Q(w,b,\zeta) = \frac{1}{2}\|w\|^2 + C\sum_{i=1}^{n}\zeta_i \\ &\text{subject to } y_i(w^T x_i + b) \underset{\sim}{>} 1 - \zeta_i \\ &\qquad\qquad \zeta_i \geq 0, \quad i = 1,\dots,n \end{aligned} \tag{4}$$

TABLE II. Considering a linear membership functions for fuzzy greater than or equal inequality, the RSVM formulation is as follows:

$$\begin{aligned} &\text{Minimize } = \frac{1}{2}\|w\|^2 + C\sum_{i=1}^{n}\zeta_i \\ &\text{subject to } y_i(w^T x_i + b) \geq 1 - \zeta_i - d_i(1-\alpha) \\ &\qquad\qquad \zeta_i \geq 0, \quad i = 1,2,\dots,n \end{aligned} \tag{5}$$

TABLE III. Here, d_i is the importance degree of sample x_i. A greater value for parameter d_i of the sample value x_i means violation from the constraints for this sample is higher and the effect of this sample on training of the classifier is lower. In other words, x_i is regarded as a noisy data. If the same d_i is assigned to all constraints, the system can equally tolerate crossing over any sample. The parameter α is the level at which the membership degree of the fuzzy inequality of constraints is cut. A larger value for α means our certainty in the whole set of data is higher and vice versa. Note that, if we have high certainty in the training samples, we should not permit constraint violations. With these two parameters, we can handle the tolerance and uncertainty for classification in a data set (for more details see [18]).

2.3 One-Class RSVM

RSVM can be used to solve one-class classification problems. In this kind of classification, one class of data is assumed as the target class and the rest of data as outliers. The trained classifier separates the target class from the others. This method tries to construct a boundary around the target data by enclosing the target data within a minimum hypersphere. The hypersphere is specified by its center *a*, and its radius *R*. The data description is achieved by minimizing the error function:

$$\begin{aligned} &\text{Minimize } = R^2 + C\sum_{i=1}^{n} \zeta_i \\ &\text{subject to } \|x_i - \alpha\|^2 \underset{\sim}{<} R^2 + \zeta_i \\ &\zeta_i \geq 0, \quad i = 1, \dots, n \end{aligned} \tag{6}$$

Such as the RSVM structure, the fuzzy inequality constraints in (6) will be transformed to a crisp one with defining a linear membership function. As described in [18], one-class RSVM has better results than one-class SVM (SVDD).

2.4 Multi-class RSVM

TABLE IV. The basic SVM is designed to separate only two classes from each other. However, in many real applications such as traffic classification, a method to deal with several classes is required. A solution is to decompose a multi-class problem into several two-class classification problems. We can use RSVM for multi-class classification problems. To do this, modifications should be applied to one-against-all, pairwise, and DAG SVM classifiers. In the one-against-all RSVM, we train *m* RSVM, where *m* is the number of classes. RSVM_i separates class *i* from the remaining classes. A testing sample x_t is assigned to the class with the maximum decision function value. In the pairwise RSVM and DAG RSVM, $m(m-1)/2$ RSVMs are trained. RSVM_ij is the optimal separating hyperplane (OSH) between class *i* and class *j*. In the pairwise RSVM, a testing sample x_t is assigned to class with maximum decision function. The DAG RSVM uses a decision tree in the testing stage (see more details in [18]).

3 The Proposed Method

3.1 Flow Representation

In this paper we used just size of the packets that compose a flow. Also, we consider only TCP flows. A flow is composed by packets as they are seen by the network device that collects them. Each flow is a bi-directional ordered sequence of packets that are exchanged between a pair of connected endpoints, each one identified by an IP address and a TCP port. Since the packets inside a flow that have no any payload do not introduce any additional information for the classification, we ignored them. These packets have the size equal to zero. So, each flow is mapped to an ordered sequence of feature values based on each packet's length. In order to distinguish the direction of the packets, we add to each packet length a constant value $\varepsilon=1000$, and change the sign of the obtained number when the packet is traveling from the server to the client:

$$\text{If packet}_i \text{ sent by client: } \delta_i = \text{size}(\text{packet}_i) + \varepsilon,$$
$$\text{If packet}_i \text{ sent by client: } \delta_i = -\text{size}(\text{packet}_i) - \varepsilon.$$

So, each flow is represented by a vector $f = (\delta_1 , \delta_2,..., \delta_n)$ and it is ready for training a classifier. By this relations the feature values for each flow lying in the intervals (1040,2500] and [-2500,1040). Note that the minimum size for IP and TCP headers is 40 bytes.

3.2 RSVM Classification

After preparing the training set, a traffic classification scheme is proposed. Figure 1 shows the structure of the proposed method for traffic classification.

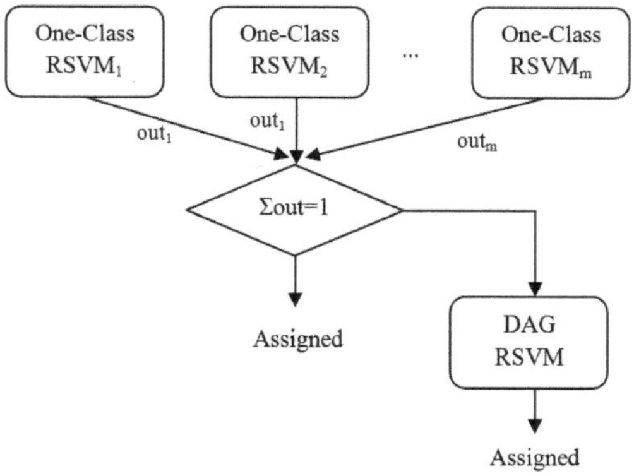

Fig. 1 General overview of the proposed method

As shown in the figure, m one-class RSVMs are trained to describe each application layer class. Also, $m(m-1)/2$ RSVMs are trained between each pair of classes for RSVM multi-class classification. In testing phase, when a test flow is applied to the proposed architecture, each one-class RSVM produce an output. If the test flow is inside the surface of the i-th class, the output of one-class RSVM_i (out_i) will be equal to one and otherwise it will be zero. If only one of the outputs of one-class RSVM is equal to one, the test flow is assigned to corresponding class. For example, if only out_i is equal to one and the others are zero, the test flow is assigned to i-th class. If more than one one-class RSVMs produce output equal to one, we will use multi-class DAG RSVM to determine the class of the test flow.

In the next section, we evaluate the proposed method using two traffic traces collected from two different network positions.

4 Experimental Results

In this section, we evaluate our proposed method using two real data sets and compare it with similar work that is presented in [3]. The authors in [3] used SVM for the classification of flows.

4.1 Data Sets

In this paper we use two traffic traces collected from two different network positions. The packet traces of UNIBS data set were collected at the border router of Brescia University. The second (AUCKLAND data set) is collected from the border router of the University of Auckland. Table 1 and Table 2 reports the protocol classes for each data set and the percentage of flows that it has generated.

Table 1 Composition of the traffic gathered at UNIBS and AUCKLAND border routers

UNIBS		AUCKLAND
Protocol	Flows (%)	Protocol
http	79.8	
ftp	0.1	http
smtp	4.9	https
pop3	1.7	smtp
bitTorrent	0.94	pop3
msn	0.51	ssh
edonkey	7.9	ftp
ssl	11	imap
other	0.1	

As shown in Table 1, we have nine classes of traffics in UNIBS data set. The "other" class contains the flows belong to *smb*, *gnutella*, *imap*, *aim*, *nntp* and *ssh* protocols. In AUCKLAND data set, we consider seven classes that the most of flows are assigned to them.

4.2 Results

In this subsection, we evaluate our proposed method and compare it with the similar work that is presented in [3]. The overall recognition rates of the proposed method are summarized in Table 2. In this experiment we used 70% of data points in each dataset for training SVM and RSVM classifiers and then in test phase, the remaining of data points is used for testing. Then, the recognition rate for each classifier in each datasets is calculated.

Table 2 Recognition rates of SVM vs. RSVM-based classifiers

UNIBS Recognition Rates		AUCKLAND Recognition Rates	
SVM	RSVM	SVM	RSVM
91.87%	**95.22%**	88.29%	**96.59%**

Table 3 Classification accuracy of SVM vs. RSVM-based classifiers

UNIBS					**AUCKLAND**			
	SVM		***RSVM***			***SVM***		***RSVM***
Protocol	***TP***	***TN***	***TP***	***TN***	***Protocol***	***TP***	***TN***	***TP***
http	100	98.12	100	99.37	http	95.00	100	97.50
ftp	80.00	100	100	100	https	100	100	100
smtp	90.00	97.5	90.00	99.5	ftp	100	95.12	100
pop3	80.55	99.42	88.89	100	Imap	61.75	98.52	100
bitTorrent	70.34	98.42	80.25	100	pop3	86.67	99.43	96.67
msn	99.46	88.12	99.46	96.00	Smtp	61.00	97.05	88.57
edonkey	92.18	99.47	85.34	98.94	Ssh	100	100	100
ssl	97.14	98.86	97.14	99.43				
other	94.44	98.96	100	98.96				

As shown in Table 2, the experimental result of the proposed method for AUCKLAND is very promising. A common way to characterize a classifier's accuracy is through metrics known as *False Positives (FP), False Negatives (FN), True Positives (TN) and True Negatives (TN). TP* is the percentage of correct detection of malicious behavior, whereas, *TN* is the percentage of correct detection of normal behavior. Similarly, *FP* is the percentage of incorrect classification of normal behavior (which is 100% -TP) and *FN* is the percentage of incorrect classification of malicious behavior (which is 100%-TN). Table 3 show the results of true positive (*TP*) and true negative (*TN*) metrics for our data sets. As shown in Table 3, for UNIBS data set, the proposed method has produced better results (except *edonkey* class). For AUCKLAND data set, the results of the proposed method are interesting and promising.

5 Conclusions

Accurate Internet traffic classification is a crucial part of network operation and management. It has recently taken into consideration by both network operators and researchers. In this paper, we used relaxed constraints support vector machines (RSVMs), an efficient and noise-aware structure of SVMs, to predict the application layer protocol for each TCP flow. The main advantage of this method is its robustness against noisy data and outliers. This cause more reliability for network operators to determine the class of each flow with more precision. Also, it allows the classifier to do correctly with as little training as a few hundred samples. Furthermore, we used only the size of packets of a flow as feature vectors. For evaluation, two sets of traffic traces are used. In almost all cases, the accuracy of the proposed classifier is very good.

References

1. Baker, F., Foster, B., Sharp, C.: Cisco architecture for lawful intercept in IP networks. Internet Engineering Task Force, RFC 3924 (2004)
2. Yuan, R., Li, Z., Guan, X., Xu, L.: An SVM based machine learning method for accurate internet traffic classification. Information Systems Frontiers 12(2), 149–156 (2010)
3. Este, A., Gringoli, F., Salgarelli, L.: Support Vector Machines for TCP traffic classification. Computer Networks 53, 2476–2490 (2009)
4. Nguyen, T., Grenville, A.: A Survey of Techniques for Internet Traffic Classification using Machine Learning. IEEE Communications Surveys & Tutorials 10(4), 56–76 (2008)
5. Internet Assigned Numbers Authority, IANA (2008), `http://www.iana.org/assignments/port-numbers`
6. Carela-Español, V., Barlet-Ros, P., Cabellos-Aparicio, A., Solé-Pareta, J.: Analysis of the impact of sampling on NetFlow traffic classification. Computer Networks 55, 1083–1099 (2011)
7. Karagiannis, T., Broido, A., Faloutsos, M.: Transport layer identification of P2P traffic. In: Proceedings of ACM SIGCOMM IMC (2004)
8. Sen, S., Spatscheck, O., Wang, D.: Accurate, scalable in network identification of P2P traffic using application signatures. In: WWW 2004, New York (2004)
9. Moore, A., Papagiannaki, K.: Toward the accurate identification of network applications. In: Proc. Passive and Active Measurement Workshop (2005)
10. Williams, N., Zander, S., Armitage, G.: A preliminary performance comparison of five machine learning algorithms for practical IP traffic flow classification. ACM SIGCOMM Comput. Commun. Rev. 36(5) (2006)

11. Erman, J., Mahanti, A., Arlitt, M., Cohen, I., Williamson, C.: Offline/realtime traffic classification using semi-supervised learning. Performance Evaluation 64, 9–12 (2007)
12. Auld, T., Moore, A., Gull, S.: Bayesian Neural Networks for Internet Traffic Classification. IEEE Transactions on Neural Networks 18(1), 223–239 (2007)
13. Vapnik, V.: Statistical Learning Theory. In: Adaptive and Learning Systems for Signal Processing. Communications, and Control. Wiley, New York (1998)
14. Moore, A., Zuev, D.: Internet traffic classification using Bayesian analysis techniques. Performance Evaluation Review 33, 50–60 (2005)
15. Li, Z., Yuan, R., Guan, X.: Accurate classification of the internet traffic based on the SVM method. In: International Conference on Communications, pp. 1373–1378 (2007)
16. Este, A., Gringoli, F., Salgarelli, L.: Support Vector Machines for TCP traffic classification. Computer Networks 53, 2476–2490 (2009)
17. Crotti, M., Dusi, M., Gringoli, F., Salgarelli, L.: Traffic classification through simple statistical fingerprinting. ACM SIGCOMM Computer Communication Review 37(1), 5–16 (2007)
18. Sabzekar, M., Sadoghi Yazdi, H., Naghibzadeh, M.: Relaxed constraints support vector machine. Expert Systems (2011), doi:10.1111/j.1468-0394.2011.00611.x
19. Liu, P., Chen, P., Jiang, Q., Li, N.: Short-term traffic flow prediction based on rough set and support vector machine. In: International Conference on Fuzzy Systems and Knowledge Discovery, pp. 1526–1530 (2011)

Chapter 2

Semantic Technologies for Industrial Management and Process Controlling

Next Generation Product Lifecycle Management (PLM)

Michael Abramovici and Youssef Aidi

Ruhr-University Bochum, Universitätsstr. 150 D44801 Bochum, Germany
{Michael.Abramovici,Youssef.Aidi}@itm.rub.de

Abstract. Product Lifecycle Management (PLM) is an established industrial approach constituting an enterprise integration platform for engineering processes, data, systems, and the involved actors throughout the entire lifecycle of a product. Based on comprehensive literature and own surveys, as well as on the experience gathered in current research projects, the paper in hand summarizes the main development trends in PLM. Due to the shift of most enterprises from selling physical products to offering sustainable Product Service Systems (PSS), and influenced by different economic and social drivers, the PLM approach will be extended to manage the growing complexity of processes and data. Furthermore, next generation PLM systems will lead back knowledge about the use of PSS to earlier development phases to make the development process more efficient and to continuously improve the performance of PSS offerings. Therefore, next generation PLM will consider new data models, PLM processes, and knowledge-based methods. Finally, new PLM systems will adopt a more flexible IT architecture. The main PLM development trends are illustrated by examples from different PLM research projects.

Keywords: Product Lifecycle Management, Product Service Systems, Data Management, Feedback Management.

1 Introduction

Due to rapid technological developments, products are becoming more complex and their development less and less manageable. The lifecycle of modern products is characterized by an interdisciplinary interaction of a large number of actors, concerning both providers and customers. These use a variety of domain specific IT tools such as CAD, CAE, CAM, etc., which create a highly heterogeneous data landscape. Over the last decade, Product Lifecycle Management (PLM) has become the central management approach in engineering, where it is used as a company-wide integration platform. PLM is an integrated approach including a consistent set of models, methods, and IT tools for managing product data, engineering processes, and tools throughout the product lifecycle [1]. The main components of the PLM approach are illustrated in Fig. 1.

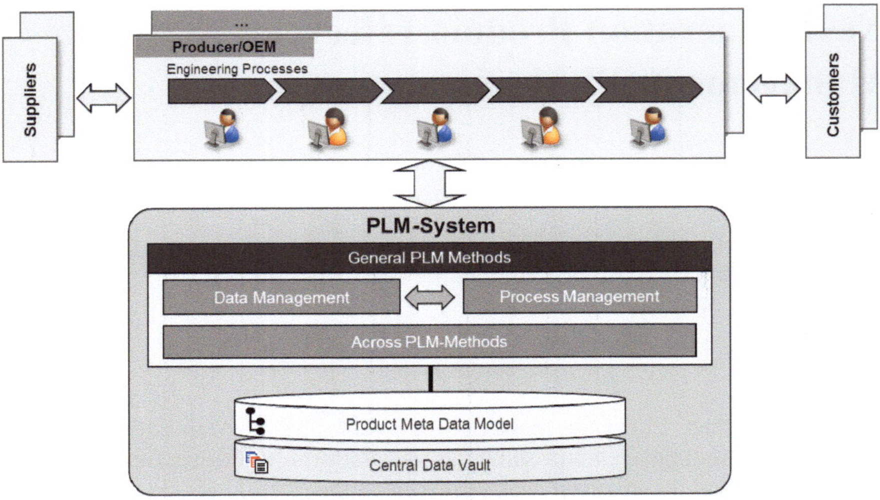

Fig. 1 The current Product Lifecycle Management approach

The current PLM Methods include data management, process management, as well as across PLM methods of the PLM approach. The PLM data management covers methods for analysis, organization, modeling and management of product data (e.g. meta data, product structures) as well as for document management (e.g. check-in / check-out procedures of documents). The PLM process management covers methods for analysis, modeling, simulation, controlling and documentation of PLM-specific administrative processes like release or change management. These two categories of methods are complemented by across PLM methods such as access management, engineering collaboration management or decision support methods.

2 Drivers for the Future PLM Developments

Due to emerging global markets, companies must develop suitable products, which meet the regional requirements of the different customers in each country. At the same time, the product compliance of the local regulations has to be guaranteed. In addition, rapid technological developments, unpredictable social and political changes and highly dynamic markets require a more flexible and adaptive engineering with regard to the development of individualized and sustainable products and of related services. Furthermore, companies have to deal with more unpredictable constraints like environmental disasters or resource scarcity (Fig. 2).

Today's industrial companies operate in a rapidly changing environment where a paradigm shift from product-centered business to customer value orientation is ongoing. The core of this shift is the offering of both products and the related

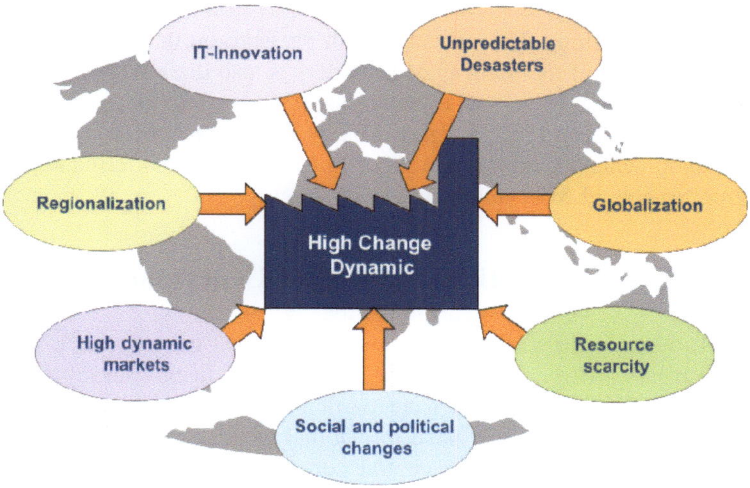

Fig. 2 Short and long time pressure on industrial companies

services. Those offerings, so called Product Service Systems (PSS), are defined as "integrated product and service offerings that delivers value in use" [2]. In this respect, it is the lasting satisfaction of the customer benefit that constitutes the central aspect of the PSS approach [3].

Generally, the PLM development is influenced by three groups of players: the PLM users, the PLM providers and the PLM researchers. Fig. 3 left shows the gap between those players, which was increasing during the last years, but is now converging. PLM providers are working together closely with users and research institutions for the development of new powerful solutions. At the same time, due to the high economic potential of PLM, to the stronger market pressure, the higher complexity of products and engineering processes, to new laws, guidelines

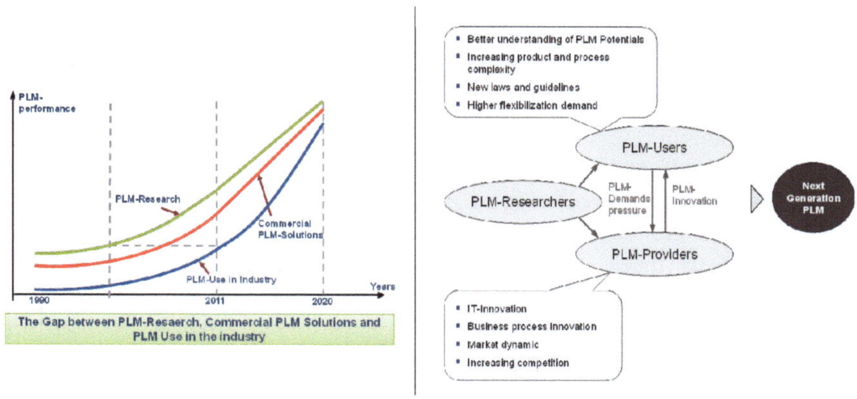

Fig. 3 The strong PLM development´s interaction between different PLM Players

and flexibility requirements the most industrial companies aim at improving existing PLM solutions. They create a high pressure on the PLM providers. The PLM development of the main PLM providers is also driven by emerging IT and business innovation (e.g. social networks, cloud computing) as well by an increasing competition (Fig. 3 right). Finally PLM researchers consider existing PLM developments but explore advanced PLM capabilities offering an orientation for future feasible PLM developments (Fig. 3 right).

3 Development Directions of the Next Generation PLM

In order to identify the main expectations for the next generation PLM a Delphi-study has been carried out by the ITM Bochum [4]. In this survey 40 industrial experts from different domains have participated and 20 use cases have been analyzed. Results of the study have shown that more than 71% of the experts are confident, that PLM will be the central company-wide integration platform for managing engineering data, processes, and IT tools in the next decade.

Fig. 4 shows, that according to these experts, next generation PLM has to cover not only planning and development processes, but the whole product lifecycle including the manufacturing, product use, as well as the product optimization and reconfiguration phases. In addition, next generation PLM solutions have to consider not only the management of a single lifecycle but of multiple lifecycles of products, wherein the used products can be remanufactured or reconfigured to meet sustainability goals. Current PLM solutions mainly consider the development of products [5] without taking into account service aspects. They have been developed to manage product classes or families of similar products. A single product instances and related services can change during the use phases [6]. This has to be considered by next generation PLM. Furthermore, current PLM systems focus only on formalized processes, like release and change management. The next generation PLM should consider new types of lifecycle processes like information feedback processes and decision support processes for gathering and leading back use information into earlier development phases. That way, knowledge about the own offerings and processes should be generated and used by developers and decision makers for improvement of products, services and their engineering processes. Most industrial companies collaborate within a large network of suppliers and service companies. As an enterprise integration platform, next generation PLM have to integrate all the network partners and the customers in order to enhance the offering of PSS, where providers are responsible for the use of products and for the execution of the required services.

In order to meet the above mentioned requirements and the experts' expectations following the dimensions featured in Fig. 4, an extension of the current PLM approach has been proposed and explored by ITM Bochum within several basic research projects such as (SFB/TR29:"Engineering of Product Service Systems", WirPro:"Leading back Product Use Knowledge into the Product Development"). The next sections of this paper describe in more detail the main expected PLM development trends concerning the next generation PLM data models, processes, methods and infrastructures.

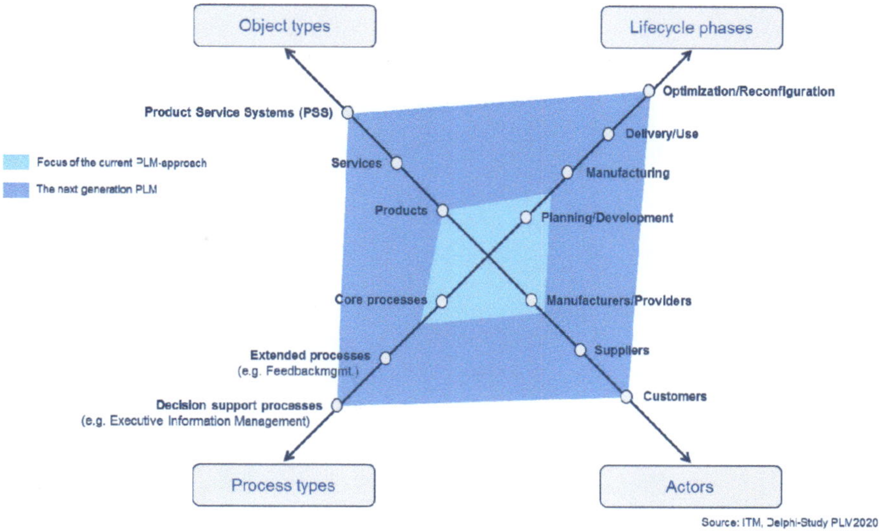

Fig. 4 The main development directions of PLM solutions [4]

4 Extension of the Covered Products and of PLM Data Models

Next generation PLM will manage highly heterogeneous products and services over the entire lifecycle. Thus, future PLM data models will be based on multidisciplinary, semantic, linked product and production models (Fig. 5). The Product engineering process includes the planning, development, and manufacturing phases. For those different phases, a lot of data and many models are employed which reflect all the relevant aspects of that phase. Fig. 5 shows the different digital partial models existing within product engineering, which will be integrated into a common PLM meta data model. In the planning phase, requirements and function models define the characteristics of the future product and constitute the framework for the subsequent engineering processes. In the development phase, both virtual and physical lifecycles of the products have to be integrated. Current PLM solutions focus mainly on the virtual lifecycle, especially classes of mono-disciplinary products (mechanical components) wherein the employed data models contain poor semantics (attributes only). Prototyping as well as testing models of a product represent the physical lifecycle within the development phase. The integration of the prototypes, testing results, and other simulation data with virtual product data opens new perspectives for a closed, efficient engineering. In the manufacturing phase, resources, equipment, and other manufacturing data will be managed and linked to the different product and manufacturing models.

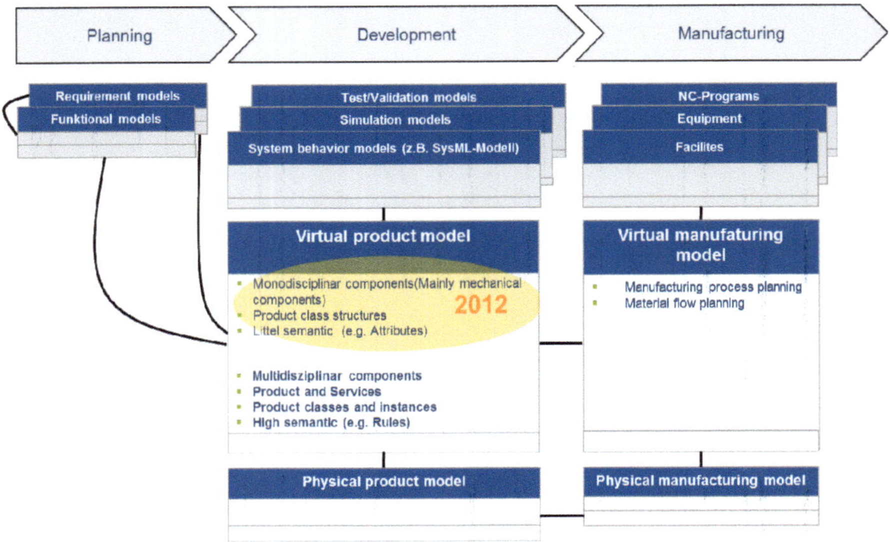

Fig. 5 The next generation PLM should integrate all partial models over the entire life cycle

5 Extension of PLM Processes and Methods

Current PLM solutions are focused only on the support of administrative information flow processes. Next generation PLM will improve the existing administrative processes but will also support further process type like planning, optimization, decision support and information feedback processes. Futre PLM solutions will also consider collaboration processes as well as material flow processes integrated with information flows (Fig. 6).

The main features of future PLM solutions covering these different process types are described in the following sub sections.

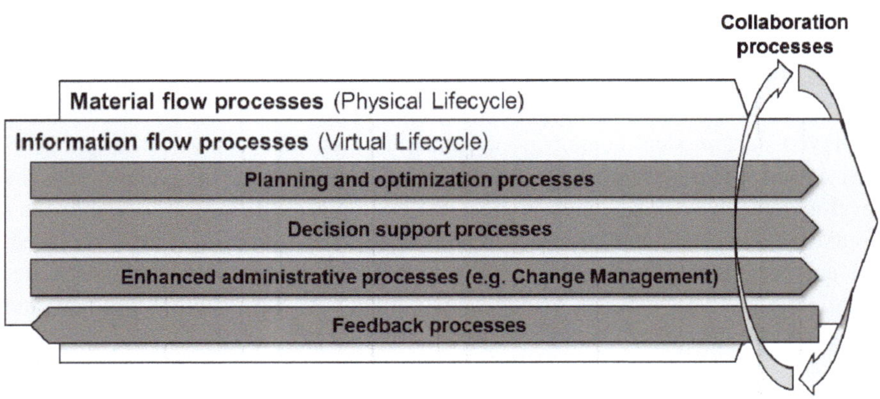

Fig. 6 The Extended PLM-Processes and methods within the next generation PLM

5.1 Enhanced PLM Administrative Processes

Competitive companies must be able to adapt their share of product and services to quickly respond to unforeseeable changes in the environment throughout the lifecycle caused by the different drivers listed in Fig. 2 [7]. Hence, the prompt reaction to these unpredictable changes along the overall lifecycle of products has a significant impact on the economic success of the companies and their network partners. This challenge can only be met by adaptive administrative processes, especially of Engineering Change Management (ECM) processes. Current PLM solutions employ existing change management methods which focus exclusively on the development and manufacturing phases and neglect the delivery, use, and reconfiguration phases. They cannot consider the complexity of PSS, which arises from the networking and mutual influence of products and services, as well as change dynamics during the delivery and use phases [3] [7]. In contrast to existing deterministic and fixed planed ECM processes, next generation ECM will support a real-time definition of executable ECM process-activities and their execution priorities depending on ECM contents, context, objectives, and the current conditions (i.e. adaptive process design and management). This allows a prompt configuration and immediate start up (e.g. continuous real-time plan-and-execute rather than static plan-and-execute) [8], taking into consideration the great uncertainties which arise in the development and use phases during the ECM process execution.

An example of such an adaptive ECM is a prototype developed by ITM, based on a new goal-oriented process management approach defined by Daimler AG and Whitestein Technologies [9]. The aim of this new goal-oriented management method is to replace processes that are planned in a fixed and sequential way and a priori, with dynamic and adaptive processes. When executed, the latter allows for near independent, real-time response in specified situations. Fig. 7 illustrates how these goals can be reached:

- First and foremost, the processes shall capture and characterize the defined business goal, independent of the solution. Goals can be split into further sub goals.
- Each goal is assigned to a generic implementation plan, which is merely made of independent tasks or activities without any predefined execution sequence or priorities.
- The specifications of tasks and activities and the order in which they are carried out are determined during the processes execution, in real time and depending on the main process issues and the current situation (rules) of the process.

Within the process, the tasks or activities are defined as intelligent agents. They represent the appropriate road to the (sub) goal, appropriately, independently, and subject to the rules [10]. Together with other decision support tools, they will also provide a process manager with recommendations for decision-making in view of the occurrence of further process steps.

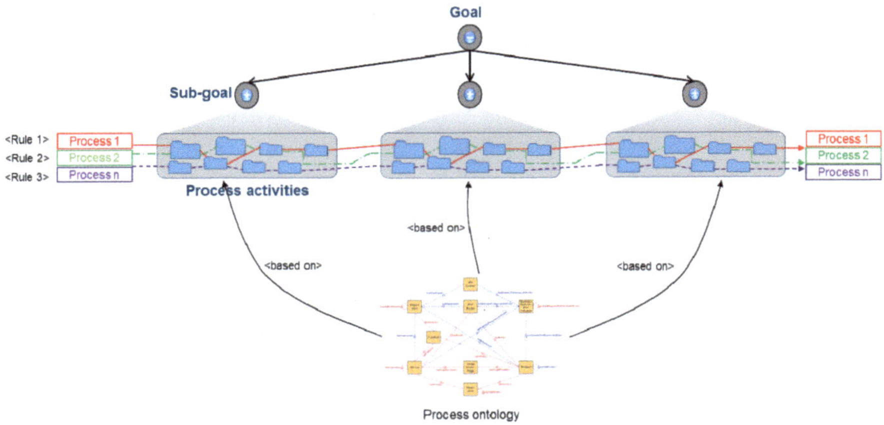

Fig. 7 Goal-oriented process modeling

Such a goal-oriented process modeling constitutes the basis for a new approach for an agile change management for PSS, which was developed and validated by ITM Bochum within the basic research project "SFB/TR29" [10]. Further research activities within the same project are conducted to use feedback information to develop an agile and knowledge based change management.

5.2 Enhanced Product and PSS Planning and Optimization Processes

The planning phase is considered the key phase of the product lifecycle that defines the framework for the next development activities as well as the main properties of the future product and the derived costs. In the permanently changing market situation, future products will have a shorter lifetime and should be adapted and improved continuously to the changing customer requirements. Hence, next generation PLM will more focus on the earlier planning phase. For instance, transparent cost management will be established, which allows an effective estimation of targets costs and efficient cost controlling. To reach this goal, all the financial information and data, as well as the different requirement functional and product structures throughout the entire lifecycle will be embedded in the PLM solution and linked to each other (Fig. 8) in order to enhance the transparency of the impact propagation of different changes in each structure.

Appropriate visualization tools will provide different developers and decision-makers a transparent overview of the arisen costs and cost drivers during the development and manufacturing phase and to give them an optimal estimation for the costs in later phases. Thus, planning and controlling activities will be carried out more efficiently. Furthermore, in global markets, many national and international norms and regulations must be considered. Compliance management will be a central component in next generation PLM with regard to giving developers and planners an integrated instrument to evaluate the compliance of products to the different legal issues at any given time. For instance, the PLM Provider PTC has developed the solution "Insight" which considers some of the aspects mentioned above.

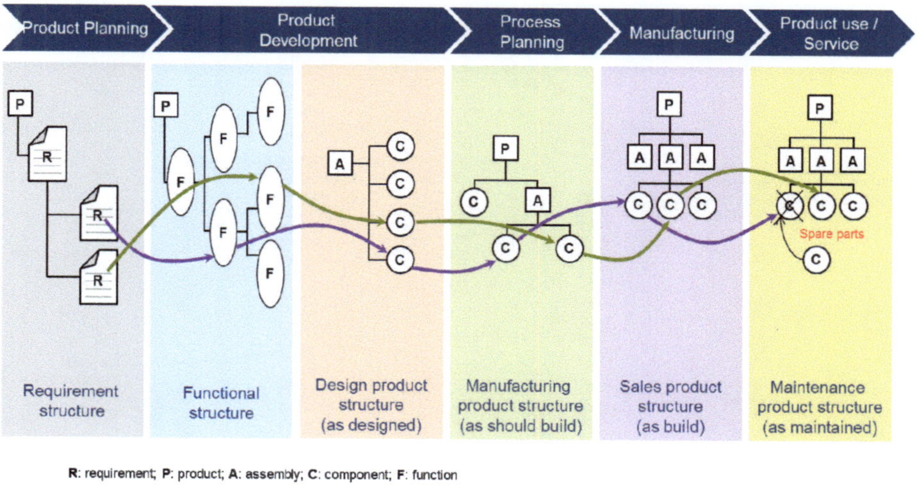

Fig. 8 Impact propagation through the linking of all lifecycle product structures within the integrated next generation PLM

5.3 PLM Embedded Information Feedback Processes

The majority of current industrial products are modular, mass-customizable systems, based on standard components. These mature standard components are subject to periodical design improvements (new releases or generations of a basic product/components type). This improved design of new product generations uses only isolated and unsystematic feedback from customers, retailers, or service partners, which mainly refer to warranty cases, complaints or product recalls. As product use information is not exploited systematically, new product generations are still suboptimal or over-engineered [11] [12] [13]. Product use feedback for the design of future product generations can be either subjective or objective. Some marketing-driven approaches (i.e. customer surveys, the Kano method, or Quality Function Deployment) facilitate a systematic acquisition and analysis of retrospective subjective customer data [14]. However, objective field data (i.e. sensor, operation or service data) is hardly available, collected or used [15]. Driven by progresses and a price digression of micro product embedded sensors, an increasing number of companies use product field data for condition monitoring solutions to facilitate preventive maintenance of critical parts [15]. Fig. 9 shows how such a feedback management can be embedded into a PLM solution.

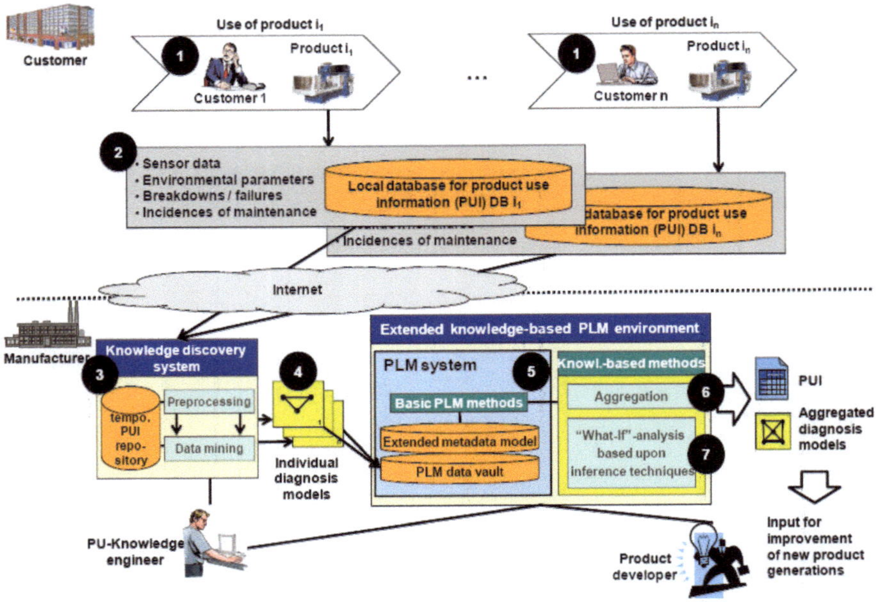

Fig. 9 PLM embedded Information Feedback Provision [13]

The shown concept above will be realized within the current research project "WirPro" by ITM-Bochum. A similar approach based on product tracking using smart embedded systems was followed within the project PROMISE, which is funded by the European Union.

5.4 PLM Support for Decision Processes

The increasing interdisciplinary of modern products and the introduction of product service offerings (PSS), as well as the related close interaction between providers, suppliers and customers, pose new challenges to managers and decision makers [16] In a single day, decision makers like managers are involved in a variety of tasks e.g. meetings, appointments, business negotiations, and report-reading [17]. Their experience cannot be fully used to manage such interdisciplinary and highly complex decision processes. First, as the most technical aspects of products and characteristics of engineering processes influence the economic end result. On the other hand, the required information within the decision processes is distributed on many systems and there is no transparency about its origin. Furthermore, current commercial systems like Business Intelligence solutions mostly focus on financial information and business operations, and thus cannot fully meet the requirements of decision makers in modern companies [18]. The employed applications (such as ERP, SCM, CRM,

PLM, etc.) at tactical, and strategic levels contain a lot of valuable data and information for decision making at each respective level. However, just a few provide special modules to meet the information demand of decision makers and managers. Those modules can also analyze the data stored in the single system. Next generation PLM as an enterprise integration platform will provide the opportunity of extracting and aggregating data from the different employed systems and to relate them with regard to providing a holistic and transparent support solution for decision processes along the entire lifecycle of products and PSS offering (Fig. 10).

To support decision processes, the following functions are required:

- **Monitoring** to offer integrated information on a dashboard for controlling projects, processes, products and available resources.
- **Analysis** by means of mathematical methods and based on existing internal and external data with regard to making trend analysis and forecasts which are used as important references for future decisions.
- **Reporting** to automatically provide decision makers with the right reports at the right time and according to their issues at the appropriate decision level and current situation. That way, unnecessary reports are avoided and the time for preparing the reports is greatly decreased.

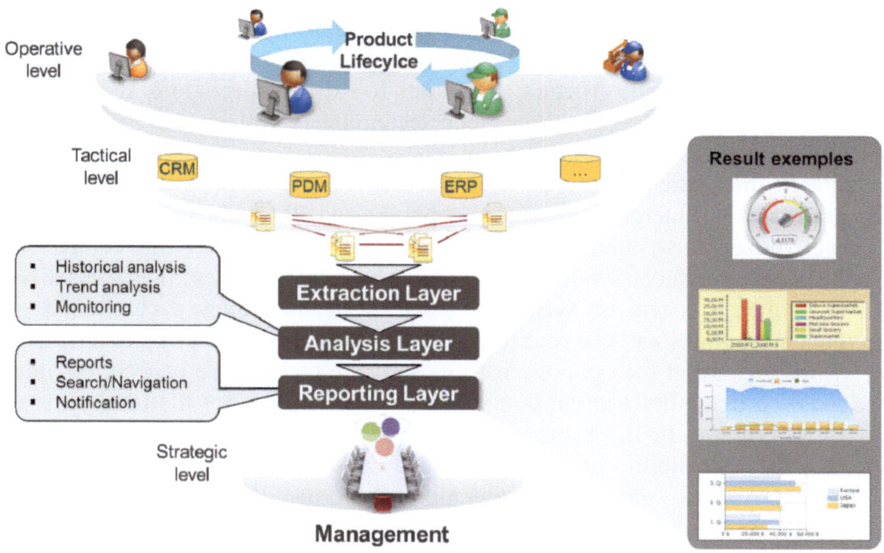

Fig. 10 Decision Support System within an enterprise PLM solution

5.5 *Enhanced Collaboration Processes*

Figure 4 shows the enlargement of the body of PLM users. In order to integrate all the actors within the lifecycle of products including customers, next generation PLM will support flexible collaboration. Engineering work is an ingenious and creative process which is made more difficult due to the increasing complexity of products. The different actors from the different disciplines have to cooperate in a closed way. Usually, engineers develop new ideas and concepts in think tanks which are supported by continuous knowledge sharing. However, the development process is more and more decentralized. Next generation PLM will provide company-wide knowledge bases, so that developers at different locations as well as development partners, customers, and suppliers can share their knowledge and experience and work closely to discuss and solve existing challenges. To support synchronous collaboration, many tools like 3D video conferences, application sharing, and meeting management will be provided within a PLM solution. For the support of asynchronous collaboration, project team task management, document, and knowledge management will be part of the next generation PLM.

6 The Next Generation PLM Infrastructure

Throughout the last decade, a considerable development of information technologies has taken place. Next generation PLM will use the newest information technologies. Knowledge management, flexible architecture and semantic data bases are the main streams in IT development which will have a big impact on the PLM infrastructure. According to expert expectations, the Service-Oriented Architecture SOA using cloud technologies will realize the expected enterprise integration platform.

51% of the participating experts in the mentioned Delphi-study state that an integrated ERP (Enterprise Resource Planning) and PLM implementation is the best enterprise system configuration for the IT integration of industrial companies. Sophisticated interfaces for data exchange for both systems as well as specialized modules in each system are available, so that they can complete each other. Such an implementation should be realized using the SOA architecture. More than 70% of the experts are of the opinion, that this software architecture concept provides the required flexibility and the best standardized interfaces for the integration of different systems. In the same direction, portal and web technologies should increase the accessibility of applications and data. All of the managed data and information within enterprise integration platforms should be assisted by knowledge base systems for a systematical generation and use of enterprise and product knowledge. Data Mining is one of many other technologies which open new opportunities to explore both structured and unstructured information. Finally, the emerging RFID technologies will facilitate the identification and tracking of products and product parts, in order to assist service activities or to enhance reconfiguration and disposal processes. They can also be employed for the identification of original parts with regard to product piracy.

7 Conclusion and Outlook

Product Lifecycle Management is the central approach for the integration of engineering data, IT tools, processes and actors involved in the entire lifecycle of a product. The vision presented in this paper is based on a Delphi-Study and several research projects and provides a valuable orientation for future PLM research and development activities. Driven by different economic and technological drivers, and according the existing political, market, and environmental constraints, next generation PLM have to deal with an increasing complexity of products and engineering processes. The paper in hand has presented the main extensions of the current PLM approach which will cover an enlarged product range and enhanced engineering processes. These extensions take into account the continuous development of information technologies which open new opportunities to make companies and their products more flexible and adaptive to the changing market conditions.

References

1. Abramovici, M., Schulte, S.: Study "Benefit of PLM – The Potential Benefits of Product Lifecycle Management in the Automotive Industry". ITM Ruhr-University Bochum, IBM BSC, Detroit (2007)
2. Datta, P.P., Roy, R.: Cost Modelling Technique for Availability Type Service Support Contracts: A Literatur Reivew and Empirical Study. In: Proceedings of the 1st CIRP IPS^2 Conference, Cranfield, UK, pp. 216–223 (2009)
3. Abramovici, M., Bellalouna, F., Goebel, J.C.: Towards adaptable industrial product service systems (IPS^2) with and adaptive change management. In: Proceedings of the 2nd CIRP IPS^2 Conference, Linköping, Sweden (2010)
4. Abramovici, M., Bellalouna, F., Neubach, M.: Delphi-Study PLM2020 – Experts expectation about the future development of the Product Lifecycle Management. Industrie Management (2010)
5. Eigner, M., Stelzer, R.: Product Lifecycle Management – Ein Leitfaden für Product Development und Lifecycle Management. Springer, Berlin (2009)
6. Främling, K., Chong, B.B., Brusey, J.: Globally unique product identifiers – requirements and solutions to product lifecycle management. In: Proceedings of the 12th IFAC symposium on Information Control Problems in Manufacturing (INCOM), Saint-Etienne, France, pp. 47–51 (2006)
7. Abramovici, M., Bellalouna, F., Goebel, J.C.: Adaptive change management for industrial product service systems (IPS^2). In: Proceedings of the TMCE 2010, Ancona, Italy (2010)
8. Kernland, M., Hoeffleur, O., Felber, M.: The Agility Challenge in Business Process Management. Product Data Journal 1, 40–42 (2008)
9. Burmeister, B., Arnold, M., Copaciu, F., Rimmassa, G.: BDI-Agents for Agile Goal-Oriented Business Processes. In: Proceedings of the 7th International Conference on Autonomous Agents and Multiagent Systems (AAMAS), Estorial, Portugal (2008)
10. Roa, A.S., Geogreff, M.P.: BDI Agents: From Theory to Practice. In: Proceedings of the 1st International Conference on Multi-Agent Systems, San Francisco, USA (1995)

11. Abramovici, M., Lindner, A., Krause, F.L.: Providing Product Use Knowledge for the Design of improved Product Generation. CIRP Annals - Manufacturing Technology 60(1), 211–214 (2011)
12. Abramovici, M., Lindner, A., Walde, F., Fathi, M., Dienst, S.: Decision Support for Improving the Design of Hydraulic Systems by Leading Feedback into Product Development. In: Proceedings of the 18th International Conference on Engineering Design (ICED), Copenhagen, Denmark, vol. 9, pp. 1–10 (2011)
13. Abramovici, M., Neubach, M., Fathi, M., Holland, A.: Knowledge-based Feedback of product Use Information into Product Development. In: Proceedings of the 17th International Conference on Engineering Design (ICED), Stanford, CA, USA, vol. 8, pp. 227–238 (2009)
14. Meier, H., Kortmann, D.: Leadership From technology to Use; Operation Fields and Solution Approaches for the Automation of Service Processes of Industrial Product-Service-Systems. In: Proceedings of the 14th CIRP Conference on Life Cycle Engineering
15. Uhlmann, E., Langmack, M., Geisert, C., Stelzer, C.: Herstellung von Rotationsbauteilen für die Mikroproduktion. In: Wt Werkstatttechnik, pp. 950–954. Springer-VDI-Verlag, Düsseldorf (2008) (online 11-12/2008)
16. Abramovici, M., Michele, J., Neubach, M.: Erweiterung des PLM-Ansatzes Für HybrideLeistungsbündel. In: ZWF, vol. 103, pp. 619–621. Carl Hanser Verlag (2008)
17. Wang, S.Z.: Research on Executive Information System (EIS) for Enterprises. Jianton University Xi´an, Doctoral Thesis (1999)
18. Leidner, D., Elam, J.: Executive Information Systems: Their Impact on Executive Decision Making. Journal of Management Information Systems 10, 139–155 (1994)

The Role of Semantic Technologies in Future PLM

Detlef Gerhard

Vienna University of Technology, Mechanical Engineering Informatics and Virtual Product Development Research Group, Getreidemarkt 9/307, Wien/Vienna, Austria
detlef.gerhard@tuwien.ac.at

Abstract. The paper describes research efforts to address three core problems of the complete process of product creation from an IT perspective and especially focusing on Product Lifecycle Management (PLM). Today's major challenges are the support of globally distributed and multi-disciplinary teams, complexity, and enterprise knowledge management. Essential building block for new approaches to cope with these challenges is the use of semantic technologies for the supportive IT environments. There are different aspects to be considered, integration of social software and community development in the context of open innovation and crowd sourcing as well as context driven enrichment of data and adaptive application integration. The aim of this contribution is to give a comprehensive overview of the requirements resulting from today's challenges and to introduce an approach to future IT working environment for engineers in the sense of organic computing (OC). Furthermore, the paper shows how Semantic Web technologies can contribute to more individual and flexible engineering application integration and knowledge management.

Keywords: PLM, Semantic Technologies, Organic Computing, Enterprise Mashups, Flexible IT-Architectures.

1 Introduction

PLM approaches and especially Product Data Management Systems (PDMS) are the major means for today's engineers to organize the exploding amount of product related information. Within the last 15 years of the development of this category of enterprise information systems they have become quite mature. Nevertheless, PLM in implemented practice - despite vendors' high gloss brochures - mainly encompasses the management of mechanical engineering design data including BOM management and exchange with Enterprise Resource Planning (ERP). PLM deployment in the very majority of cases ends with Start of Production (SOP) though a lot of companies gain good revenues from their service business. PLM supported multi-disciplinary project work as well as inter-enterprise project control within the different supply chain partners is truly an exception.

The strengths of today's PLM solutions are storing and managing metadata based information sources. Indexing and full text retrieval of textual information has also been implemented recently in most of the available systems. Nonetheless, weaknesses can be discovered in areas such as acquisition, discovery, aggregation, organization, correlation, analysis and interpretation of data and information. Currently PLM does not cover the whole available information source relevant for specific tasks and delivers too much or irrelevant results on queries. Results are not sufficiently in a form that users can navigate through and explore. PDMS and other enterprise business software systems have not delivered on its promise to fully integrate and intelligently control complex business and engineering processes while remaining flexible enough to adapt to changing business needs in co-operative product development tasks. Instead, most IT system environments are patchworks installed and interconnected by poorly documented interfaces and slovenly customized processes [1]. IT systems that were supposed to streamline and simplify business processes instead have brought a considerable level of complexity, containing tenth of databases and hundreds of separate software modules installed over years. Rather than agility, they have produced rigidity and unexpected barriers to flexibly adapt to organizational and process oriented changes.

2 PLM – Challenges

Today's engineers are to a great extend information workers. Engineers are heavily dependent on retrieving and using documents and existing models in order to fulfill various engineering design tasks. Exploring design concept alternatives during the early stage of the development is as important as learning from the original design process and understanding the rationale behind the decisions made (handling change requests) or searching for past designs when working on a similar product or problem (design reuse). Various studies show that engineers spend conservatively one third of their time retrieving and communicating information.

Applications in the context of PLM, Supply Chain Management (SCM) and Enterprise Resource Planning (ERP) have multiplied the number of database systems containing essential business information. Bridging these systems, e.g. via enterprise application integration or data warehousing, is complex, costly and of limited efficiency. But still, only 20 to 30% on relevant information for product development, production and other value adding core processes resides within those "big" enterprise IT systems. A huge amount of knowledge is generated in order to describe a product and corresponding processes. Internal information is captured in form of documents and models, not only 2D/3D CAD models and drawings, FEA and simulation files but also office documents for specifications, reports, spreadsheets, technical documentation, and informal communication documents like notes, memos, and e-mails. This user generated content reflects to a large extend valuable enterprise knowledge, but as it resides outside corporate databases, remains mostly unexploited. This unstructured information has to be described by metadata attributes in order to be retrievable within business systems like for instance PDMS which is in most cases the biggest barrier for making them available. Surveys in industry companies show that roughly 80% of a typical IT budget is spent for the management of 20% of the relevant data. Freely available web data is not available in enterprise applications. Information which is getting increasingly important and relevant for product creation processes are external resources, e.g. legal issues, regulations, patent information, international standard documents and certifications, supplier and product catalogs. With the emergence of the usage of Web 2.0 technologies like wikis or blogs - internet and intranet based - even more sources of unstructured information are building blocks of the enterprise knowledge.

Links between structured information fragments and unstructured contents of web pages are hardly available. Significantly more effort has to be spent to externalize internal knowledge (knowledge capturing), to create semantic enriched and context oriented data relationships leading to better information retrieval. One example characterizing this phenomenon is the aspect that engineering documents and models are different to other domains' information resources due to syntax variations and semantic complexities of their contents. Abbreviations, e.g. SLA (stereo lithography, service level agreement) reflect company or domain specific naming conventions. Acronyms and synonyms are widely used and depend on enterprise or international standards, e.g. "Steel" = "St37-2" = "S235JRG2" = "S235JR+AR" = "1.0038". There is also a wide range of domain specific peculiarities and the relationships among these, e.g.:

- customer requirements and specifications
- functions and performances
- structure design and materials
- manufacturing process selections

Current engineering practices are too weak in terms of reuse of previous knowledge. A tremendous amount of time is wasted reinventing what is already known in the company or is available in outside resources. Redundant effort per employee is increasing and causing enormous cost. One reason behind this is that engineers in general do not make sufficient efforts to find engineering content beyond doing mere keyword/metadata searches within the PLM environment. But this is also a matter of implemented procedures and regulations within the companies. Engineers too much stick to tools like spreadsheets or "private" desktop databases. Certainly this is a problem of acceptance and benefits. If one has developed a spreadsheet which contains data for individual demands and covers needed functions, the effort for data maintenance directly pays off for the individual or the team using the spreadsheet. Furthermore, developed functions directly correspond to the tasks and duties. Enterprise IT systems on the other hand are often intricately to use and literally seem absorb maintained data like a black hole without delivering a direct benefit to the user or only for other users down the road of the process chain. It is necessary to implement incentive system to motivate individuals to contribute to a co-operative knowledge by sharing information and eliminate islands of isolated information. Solving this issue is a matter of further developing enterprise IT systems in general to provide the demanded usability that common spreadsheet programs offer and that can be configured or customized to individual or role based needs.

Additionally to the inherent data oriented challenges, complexity is the main issue. Complexity is characterized by networked structures, nonlinear behavior and means multi-causation, multi-variability, multi-dimensionality, interdependence with the environment, and openness. In the area of product creation we are facing the three facets of complexity (in causal order):

- Product and system complexity
- Process and organization complexity
- IT landscape and tool complexity

Complexity of products is caused by multiple instances and variants of a base product to meet requirements with respect to customization demands and differentiation of the target markets. Furthermore, nearly all products consist not only of mechanical components including pneumatic and hydraulic parts but increasingly use electronics, automatic control parts, and contain firmware/software. Because there are many different domains of expertise involved in product development tasks, process complexity also increases through dissemination over locations (countries, cultures) and distribution within the supply chain (organizations). The biggest challenge here is not only the integration of different technical disciplines all of them using specialized IT tools but the diversity and dynamics of the relationships between project partners, manufacturers, vendors and suppliers. Whereas a decade ago manufacturing distribution was dominant distributed engineering and collaborative design require even more extensive use and support of advanced IT systems which again leads to more complex IT landscapes with docents of data formats and interfaces relying on heterogenic system platforms.

3 Related Work and Concepts of Semantic Technologies for PLM

To face the aforementioned challenges, semantic technologies a wider sense on different integration levels of industrial IT systems in the context of PLM have been analyzed, especially:

- Organic Computing Concepts
- Server Based Mashups
- Semantic Web Technologies

As complexity increases, the formal description and modeling of systems becomes more difficult hardly manageable within the given time constraints. The term "Organic Computing" (OC) [2] describes an architectural approach for biologically inspired IT systems with "organic" properties. OC is based on the insight that every technical system is surrounded by large collections of other more or less autonomous systems, which communicate freely and organize themselves in order to perform the actions and services that are required. Main characteristic property of an OC system is the capability of self-organization. Hence, an OC system is a technical system which adapts dynamically to the current conditions of its environment. Considerable work has been spent on OC within the informatics research community, e.g. a Priority Research Program of German Research Foundation [3] from 2006 to 2011. Most of the developed approaches refer to complex technical (mechatronic or so called cyberphysical) systems which are considered as autonomous communicating units. Within those systems it is impossible to predict a priori all eventualities of behavior and therefore the goal is to develop adaptive systems that can adapt flexibly to changes in self-organized manner.

Self-organization in the sense of IBMs autonomic computing concept [4] comprises a couple of "Self-X" characteristics

- Self-Awareness: Ability of a system to be aware of its state and its behaviors.
- Self-Configuring: Ability to configure and reconfigure itself under varying and unpredictable conditions.
- Self-Optimizing: Ability to detect suboptimal behaviors and optimize itself to improve its execution.
- Self-Healing: Ability to detect and recover from potential problems and continue to function smoothly.
- Self-Protecting: Ability to detect and protect its resources from both internal and external attack in order to maintain overall system security and integrity.

These Self-X capabilities can be summarized as means to achieve robustness of a system without intervention, be it manually or by a superior control instance. Flexibility and adaptivity means openness to function in a heterogeneous and dynamically changing world. Consequently, a flexible IT system must be built on standard protocols and interfaces in order to be portable across multiple hardware and software architectures. Self-organization is a concept that is known in a variety of fields. A self-organizing system must be endowed with certain basic capabilities as preconditions for self-organization. They must be able to communicate, to sense the environment and other agents, and to trigger a reconfiguration if changing conditions require it. Di Marzo et al. [5] define three types of systems with self-organization capabilities

- Physical systems, where a system changes into another state due to certain conditions or upon reaching some critical value.
- Living systems, whereby, e. g. an organ features special functionality that is way beyond the functionality provided by each cell it is made of.

- Social systems, whereby insects, communicate for example indirectly via their environment and are therefore capable of more sophisticated actions than any single insect.

A so called “Observer/Controller” architecture for self-organization is proposed by Müller-Schloer, Schmeck et al. [6]. The observer evaluates all data from the observed systems which characterize the current system state and combines all the relevant values for the controller. The controller evaluates the system behavior with respect to given or specified targets and influences the parameters of the controlled system. This so called directed self-organization needs to learn appropriate actions for identified system states in advance but also while the dynamic change of the system or environment in the course of its work.

The main challenges in terms of enterprise IT system architectures in the context of PLM are to cope with complexity and dynamic changes. The complexity of today’s business and engineering IT environments embedding PDMS in enterprise spanning configurations which are spread over different locations is not far from that of biological organisms. In fact, a PLM environment can be considered as socio-technical ecosystem. It is notoriously difficult to handle in software systems. Therefore, it is reasonable to explore OC approaches for future PLM development and architectures on middleware level [7] keeping in mind that PLMS are the major means for today’s engineers to organize the exploding amount of product related information created with the different CAD/CAE applications, manage collaboration, and control virtual product development processes. The fundamental insight here is that the complexity problem of products and processes cannot be solved through increasingly complex IT-Systems and connected development projects. An emphasis on simplicity, flexibility and efficiency is necessary.

Traditional communication and integration approaches of enterprise IT solutions follow a hierarchical or centralized structure with some master or backbone system and others depending thereon. This proposition is also subject to change when project partners change or mergers & acquisitions lead to structural changes. Integration and interfacing projects for PLMS or other business critical IT environments are far away from being able to follow the pace of these changes. They are ineffective in this context, since they fail to address several required features, e.g. flexibility for project based collaboration, heterogeneity in node capabilities, and management complexity of such systems. Approaches to build applications from service providing independent application modules - Service Oriented Architectures (SOA) - are not likely to do much better. The "Lego" idea behind this, the notion of reusable software modules, works on a small scale but as software grows more complex, reusability in terms of services on a variety of granularity levels and with different configurations mirroring individual processes of companies or departments becomes difficult if not impossible. A unit of software code is not similar to other software code in terms of scale or functionality, as Lego blocks are. Instead, software-code is widely various, semantic commonalities are rare and interfaces are heterogeneous.

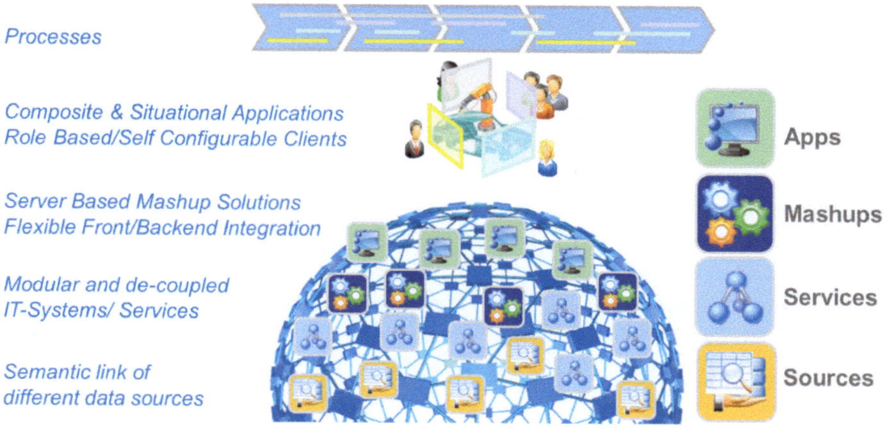

Fig. 1 Concept of future IT architecture for PLM

Figure 1 shows the concept of a future IT architecture for PLM. Self-organization in the sense of OC means that systems intelligent assistants rather than like rigidly programmed robots. It can be established e.g. through use of agent based software systems but also – on a higher level – involving users as a part of the system configuring and organizing portions of applications. The major goal of the OC approach in a wider sense is to master complexity and establish human machine interfaces oriented accordingly to user needs. The computerization of our environment opens a wide range of new applications in which the problem of controllability is the main issue. A user's claim is to be always in control, and control is based on predictability and transparency. Predictability means that the system behaves as expected. Predictability is closely related to confidence in a system environment. Transparency means that the user can always gain sufficient insight into the internals of the system, without being overwhelmed by details. A strong orientation of IT system environments such as PLM towards the human or user needs can be achieved by context awareness, i.e. the capability to anticipate to the extent possible system needs and behaviors and those of its context and the ability to manage itself proactively. Examples from Web 2.0 show how user behavior can be monitored and needs anticipated.

Flexibility can be achieved by making use of an infrastructure providing a technology for a flexible networking of different information fractals called Enterprise Mashups [8]. This is a strategic technology and will be the dominant model for the creation of composite enterprise applications according to various analysts. A mashup in the context of Web 2.0 is a hybrid WWW application that combines complementary elements from two or more sources to create one integrated experience. Content used in mashups is generally sourced from a third party via an API or from WWW feeds (e.g. RSS). Basically, the idea is to take multiple data sources or WWW services and compose them into something new or combined. In contrast to EAI (enterprise application integration) concepts, the unique point of this approach is that mashups allow business users and engineers to address their own information needs, to self-connect the data fragments in order to create information that answers their questions providing dynamic, user-specific views and customized filters. Mashups let users share their resulting services, making them a part of a services network in the sense of self-organization and self-optimization.

Advantages of enterprise mashup technology are:

- Mashups are user driven, users are able consume public (enterprise spanning or WWW based) and local services and contents on demand.
- Users are able join in data from outside the enterprise to include external data in their work whereas SOA efforts are largely inwardly focused.
- The granularity of services and can be right-sized by the consumer without having the IT department to guess or make time consuming analyses
- Composite and situational applications (role and/or project based) can be generated using configuration functions of the mashup server platform
- Interfaces/Adaptors to emerging data sources like Wikis, Blogs, and RSS are generally available.

Disadvantages of enterprise mashup technology:

- A consumer of a mashup service is not in control of the primary source of data. When sources are offline, a mashup is offline as well.
- Public accessible APIs for mashup services will limit the number of requests an application can make within certain period of time to avoid response problems.

The most important building block of a future IT working environment for engineers is based on semantic web technologies, also referred to as Web 3.0 technologies. Web 3.0 can be defined as a set of technologies that offer efficient new ways to help computers organize and draw conclusions from data whereas the term Web 2.0 is typically used to refer to a combination of

- improved communication between people via social-networking technologies
- improved communication between separate software applications (mashups) via open Web standards for describing and accessing data
- improved Web interfaces that mimic the real-time responsiveness of desktop applications within a browser window (Rich Internet Application – RIA technologies).

Web 2.0 technologies focus on social interaction and information acquisition. The intention of Web 3.0/Semantic Web is to improve the quality and transparency of data through provisioning of semantic relationships and ontologies. Web 3.0 - as seen by analysts like Gartner - is likely to decentralize enterprise wide information management. Self-optimization and context-awareness in the sense of OC can be accomplished by making use of ontologies within the PLMS data sources as well as improving usability on PLMS client side. Given a three tier architecture as implemented in most of today's PLMS, additional layers between data layer and application layer comprise the means for semantic enrichment of managed information, i.e.:

- consolidated and contextualized heterogeneous engineering documents and models,
- representation of knowledge in a more explicit, structured and navigable manner,
- user-centric computer-aided tools and methods for a shift from text based/metadata based towards visual information retrieval.

The aim is to achieve high quality information retrieval of structured and unstructured knowledge assets. Several Technologies and research domains (Information Retrieval, Language Engineering and Natural Language Processing (NLP), Text Mining) have led to a multitude of commercially or public available software systems for this purpose. They have to be assembled to a coherent system. Ontologies can be used as a sophisticated mechanism in order to structure an information repository mainly built from unstructured documents and to achieve better results in information retrieval systems.

One good example for heterogeneous and multiple data sources which cannot be integrated and maintained by one single authority is the use of material information in engineering design processes. The process of integrating data from multiple independently developed databases containing material information presents significant challenges to engineers that include resolving differences in metadata terms as well as data structures, formats, and metrics.

Y. Li [9] proposed a data warehouse approach aiming to address the issues of low precision and recall of retrieved data during materials selection processes. Data is extracted from databases, identified and data differences resolved. Adjusted and cleansed data is loaded into a centralized repository where end users can search data quickly and easily. This concept improves the recall and precision rates, and ensures the availability of deposited data. However, data update, cleansing, and maintenance is costly and this approach cannot easily adapt to changes in back-end database schemas or the addition of new databases.

A more flexible approach based on XML schemas is MatML (Materials Markup Language) which was especially developed to facilitate the exchange of materials information. MatML is the result of a NIST initiative to develop a data format specifically for the interchange of materials information [10]. It is the only materials schema that is proposed as a standard representing materials property data to resolve syntactic and structural heterogeneity. MatML provides a shared vocabulary and formalized constraints over data structures. Like any other XML schema it is simple, flexible, and human understandable, but provides little support for the semantic knowledge necessary to enable flexible dynamic mappings between vocabularies. Several commercial software applications, including Materiality, Granta MI, and ANSYS Workbench support import/export of material data as MatML files.

Ashino [11] developed the Material Ontology which comprises materials information, substance, property, environment, process, unit dimension, and physical constant. The vocabularies are from two sources: Matdata [www.matdata.net/index.jsp] and the Japanese National Institute for Materials Science's MatNavi [http://mits.nims. go.jp/db_top_eng.htm]. Zhang et al. [12] developed a Semantic Model for Materials scientific data (SMM) which includes knowledge such as materials classification, materials property, structure & composition, processing. The developed two types of ontologies are domain-specific and mapping ontologies. The domain-specific ontology encapsulates the high level structure of the materials science knowledge, whereas mapping ontologies define the mappings between terms in the SMM ontology and local database schemas. They also developed inference rules for defining new concepts such as "Corrosion Resistant Material". However, the authors do not discuss how they ensure SMM quality, particularly in terms of coherence.

Cheung et al. [13] have developed MatOnto as an extensible model for the exchange, re-use and integration of materials science data and experimentation. As such it enables representation of the relationships between a material structure, properties and processing steps involved in its composition and engineering. It also provides a common basis for the integration of materials data from heterogeneous, disparate databases. In contrast to XML schemas, ontologies enable semantic mapping between domain-specific knowledge structures.

RDF (Resource Description Framework) is a data model which was originally designed for providing metadata for Web resources. It provides for encoding structured information to a universal machine-readable format. With the URI (Uniform Resource Identifier) concept it is possible to denote any resource in a world-wide unambiguous way, i.e. any object possesses a clear identity within the context of a given application. RDF language features allow for modeling semantic aspects of a domain, hence, RDF can be seen as a lightweight ontology language though negative information cannot be specified as well as cardinality and disjunction. RDF is used to represent information and to exchange knowledge in the Web. OWL (Web Ontology Language) is used to publish and share sets of terms supporting advanced web search, software agents and knowledge management. RDF and OWL work together to form ontologies, OWL allows for declaring binary relationships between nodes, and enable the inferencing of new relationships between nodes via reasoning engines.

In the course of an ongoing sustainable design engineering project our aim is to include material information from different sources for Life Cycle Assessment (LCA) integrated into a PDMS [14] making it possible to allow for environmental reporting within an environment of different CAD solutions. Therefore, we defined for the use case of gear boxes an ontology containing the top level classes (concepts): *Life Cycle Assessment* (in order to assign life cycle stages to it), *general Information* (including all design parameters which cannot be assigned directly to a particular life cycle of a product), *Material, Manufacture, Distribution, Use*, *End of Life*. Relations were defined by using terminologies such as: *"consists of"*, *"is used for determination"*, *"determined by"*, *"has effect on calculation"*, *"has action"*, *"influences"* or *"effected by"*. For example, the class

Life Cycle Assessment consists of subclass *Material* which contains information about the life cycle stage *Material* (not the used material). *Weight* in combination with the *selected material* can be directly linked to environmental data containing environmental indicator values for materials. This will already allow for the evaluation of the life cycle materials, e.g. CO_2 equivalent from existing databases. The idea is now to use and extend the MatOnto ontology to link and derive LCA relevant material data on a semantic level. E.g. for LCA considerations all low alloyed steel products are considered equal in terms of CO_2 equivalent (See Figure 2). So the information needed is on a coarse granularity compared to the detailed material information provided by CAD systems.

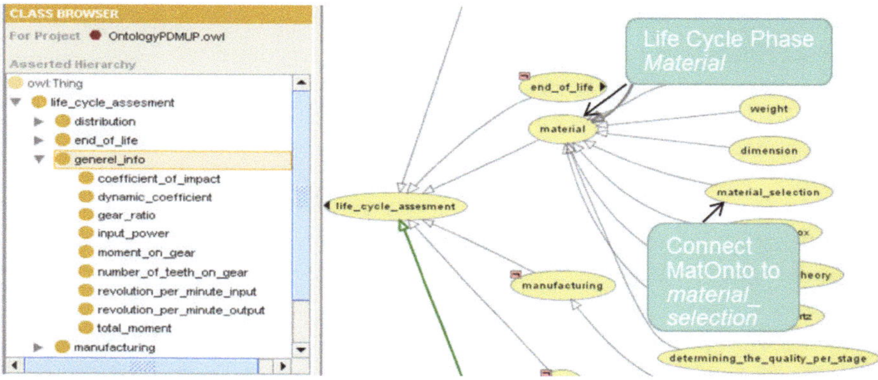

Fig. 2 Screenshot of part of the ontology (generated with Protégé OWLViz plug-in)

A huge amount of data is stored in (relational) databases. Building RDF triple stores out of them is impossible but there are bridges available that define a layer between RDF and the relational data. So relational database tables (eg, Oracle, OpenLink, …) can be mapped to RDF graphs and be used as well as triple stores. Advantages of semantic technologies and linked data bring new agility and expanded scope to enterprise applications, are:

- Enrichment of structured data with qualitative data from vast 'unstructured' sources like email, blogs, chat, and social Web pages
- Reconciliation of formats, structures and terminologies
- Identification of embedded meanings and relationships within and across resources
- Natural language processing to interpret imprecise requests and offer spelling corrections, close matches, and related content.
- Creation of innovative 'mashup' applications that seamlessly merge content and functionality from diverse sources such as databases, mapping services, and WWW resources.
- Dynamically Extraction of metadata to transform unstructured data into a fully classified resource and synthesize it with existing structured data

4 Conclusion and Outlook

Despite data explosion the individual knowledge decreases. Effects that can be observed are that the increasing dataset leads to reduced intake capacity of the human brain with consequences [15]:

- stronger information filters
- selective perception
- less-informed decisions
- only exciting information is noticed

Data integration is also a huge problem internal to companies. It is the highest cost factor in IT budget of large companies which operate a considerably large amount of databases. Traditional middleware improves and simplifies the integration process, but it misses the sharing of information based on the semantics of the data. Ontologies can rationalize disparate data sources into one body of information without disturbing existing applications, by:

- creating ontologies for data and content sources
- adding generic domain information

Sharing engineering information and providing access to relevant knowledge is a vital resource for enterprises. WWW and related technologies have had a tremendous momentum in the recent past. Web protocols, technologies, and middleware are well supported by various products evolving in the software industry and open source communities. Concentrating on the perspective of an engineer as a user of a sophisticated IT environment, PLM, Enterprise Content Management (ECM) and Personal Information Management (PIM) are converging or even merging. Semantic (Web) technologies offer new opportunities for enterprise IT aiming at establishing new approaches of handling information as resource and integrating different information sources. OC, Mashups, and Semantic technologies strive for a paradigm shift in looking at enterprise information systems: Hierarchical structures and integration approaches cannot be established in the same pace as changes of processes and organization occur. Instead networked relationships among different nodes of an IT environment and a demand (user) centric principle of looking at information resources are necessary.

References

1. Rettig, C.: The Trouble with Enterprise Software. MIT Sloan Management Review 49(1), 21–27 (2007)
2. Rochner, F., Müller-Schloer, C.: Emergence in Technical Systems. In: Special Issue on Organic Computing, vol. 47, pp. 188–200. Oldenbourg Verlag, Jahrgang (2005)

3. Priority Program 1183 Organic Computing (challenges for informatics regarding creation of technical systems from 2015) funded by the German Research Foundation (DFG) (2006-2011), `http://www.organic-computing.de/spp` (last viewed April 30, 2012)
4. Horn, P.: Autonomic Computing: IBM's perspective on the State of Information Technology. IBM Corp. (October 2001), `http://www.research.ibm.com/autonomic/` (last viewed April 30, 2012)
5. Di Marzo, G., Foukia, N., Hassas, S., Karageorgos, A., Mostéfaoui, S.K., Rana, O.F., Ulieru, M., Valckenaers, P., van Aart, C.: Self-organisation: Paradigms and Applications. In: Di Marzo Serugendo, G., Karageorgos, A., Rana, O.F., Zambonelli, F. (eds.) ESOA 2003. LNCS (LNAI), vol. 2977, pp. 1–19. Springer, Heidelberg (2004)
6. Richter, U., Mnif, M., Branke, J., Müller-Schloer, C., Schmeck, H.: Towards a generic observer/controller architecture for Organic Computing. In: Hochberger, C., Liskowsky, R. (eds.) INFORMATIK 2006 – Informatik für Menschen! GI-Edition – Lecture Notes in Informatics (LNI), vol. P-93, pp. 112–119. Köllen Verlag (2006)
7. Roth, M., Schmitt, J., Kiefhaber, R., Kluge, F., Ungerer, T.: Organic Computing Middleware for Ubiquitous Environments. In: Organic Computing—A Paradigm Shift for Complex Systems, pp. 339–351. Springer (2011)
8. Bitzer, S., Schumann, M.: Mashups: An Approach to Overcoming the Business/IT Gap in Service-Oriented Architectures. In: Nelson, M.L., Shaw, M.J., Strader, T.J. (eds.) AMCIS 2009, Part IV. LNBIP, vol. 36, pp. 284–295. Springer, Heidelberg (2009)
9. Li, Y.: Building The Data Warehouse for Materials Selection in Mechanical Design. Advanced Eng. Materials 6(1-2), 92–95 (2004)
10. Varde, A.S., Begley, E.F., Fahrenholz-Mann, S.: MatML: XML for Information Exchange with Materials Property Data. In: Proceedings of the 4th International Workshop on Data Mining Standards, Services and Platforms, Philadelphia, Pennsylvania, pp. 47–54 (2006)
11. Ashino, T., Fujita, M.: Definition of a Web Ontology for Design-Oriented Material Selection. Data Science Journal 5, 52–63 (2006)
12. Zhang, X., et al.: Material Scientific Data Integration for Semantic Grid. In: Proc. 3rd Int'l Conf. Semantics, Knowledge, and Grid, pp. 414–417. IEEE Press (2007)
13. Cheung, K., Drennan, J., Hunter, J.: Towards an Ontology for Data-driven Discovery of New Materials. In: AAAI Spring Symposium, Semantic Scientific Knowledge Integration (2008)
14. Ostad-Ahmad-Ghorabi, H., Rahmani, T., Gerhard, D.: Integrating LCA into PDM for Ecodesign. World Academy of Science, Engineering and Technology 7(81), 223–228 (2011)
15. Sternemann, K.-H.: Role based Clients in a Service Oriented Enterprise Architecture. In: Proceedings of the ProSTEP iViP Symposium, Berlin, Germany, April 9/10 (2008)

Use Case of Providing Decision Support for Product Developers in Product Improvement Processes

Michael Abramovici[1], Andreas Lindner[1], and Susanne Dienst[2]

[1] Ruhr-University Bochum, IT in Mechanical Engineering, Universitätsstr. 150, D-44801 Bochum, Germany
[2] University of Siegen, Institute of Knowledge Based Systems and Knowledge Management, Hölderlinstrasse 3, D-57072 Siegen, Germany
{michael.abramovici,andreas.lindner}@itm.ruhr-uni-bochum.de, dienst@informatik.uni-siegen.de

Abstract. Industrial goods like pumps, engines, and gears are subject of cyclical improvements. An important input for such product updates is information from the use phase, especially information about failures that occurred during the use of current products. In a current research project the authors have developed a concept and a prototype for a Feedback Assistant System. The first realized modules of this assistant provide filtered and condensed product use information to the product developer. The paper in hands presents an extended concept and use case to assist the product developer in choosing the most suitable alternative in product improvement processes. The decisions are based on hard facts using diverse criteria.

Keywords: Product Lifecycle Management (PLM), Decision Support, Product Improvement.

1 Introduction

During the product use phase of industrial goods, various, valuable information is generated. Industrial goods are produced in very large numbers and are subject to permanent improvement processes. For product improvement it is of special interest to identify critical components and their weaknesses. Hence, industrial products have embedded sensors that generate product use information (PUI). Additional data is only generated during unforeseen events (e.g. machine failures) or services events (e.g. regular inspections). Still, the data is not led back into product development and used for maintenance purposes only. By analyzing the PUI, the product developer can identify weaknesses and develop product alternatives.

Up to now, the leading of PUI into product development has not been addressed consistently. In fact, there are some approaches that address feedback. Nonetheless, these approaches address only specific feedback data sources (feedback from product-embedded information devices [1]) or specific feedback data receivers (e.g. feedback for quality improvement [2]). Therefore, the holistic conceptualization and development of a distributed Feedback Assistant System (FAS) for the acquisition, management, processing and the visualization of PUI is required.

In a first step the authors have developed and implemented FAS for the analysis and diagnosis of industrial goods and their failures. The two developed modules for analysis and diagnosis let the product developer identify weak spots. The concept is described in detail later on. In a second step, the concept is to be extended to support the product developer in improving current products. The concept and the use case for the implementation is the focus of the paper in hand.

2 Feedback Assistant System for Product Improvement

2.1 A Feedback Assistant System Solution

The authors have developed a concept for leading PUI into product development and implemented FAS for some of the above-mentioned tasks. The foundation of the concept is the PLM approach (see Fig. 1). Thus, basic PLM methods (e.g. user management) can be used further. In PLM, the PUI is usually not managed at all. The products themselves are managed in generation layers and not on a product instance level on which the PUI needs to be placed. Hence, the PLM product data model has been extended by a product instance layer that stores instance master data (e.g. serial numbers).

The PUI cannot be stored inside the PLM as this might lead to poor performance. Thus, the PUI is stored in a separate storage, a Data Warehouse (DWH). The DWH data model is based on the PLM data model [3]. The PUI stored in the DWH is acquired in the product use phase of industrial goods in the form of objective data. It is transferred from the heterogeneous customer databases (e.g. ERP, CRM) into a central database [4] [5]. Inside the DWH, the data is aggregated and pooled using Extract, Transformation and Loading (ETL) methods [6]. These are necessary steps to provide crucial information to the product developer later on [7]. Based on the central database, the following methods have been realized [6]:

- ***Analysis methods*** provide statistical data analysis methods, e.g. for managing key indicators, generating use profiles and their visualizations.
- ***Diagnostic methods*** provide knowledge-based methods (Bayesian Networks) to detect causes of failure. A Bayesian Network is a probabilistic graphical model that represents a set of variables and their probabilistic dependencies [8].

The product developer should be able to use these analysis and diagnostic methods without leaving the familiarity of his PLM environment as a graphical user interface has been implemented. Therefore, FAS has been coupled with a Product Lifecycle Management (PLM) system, which is the main data and process management system the product developer works with.

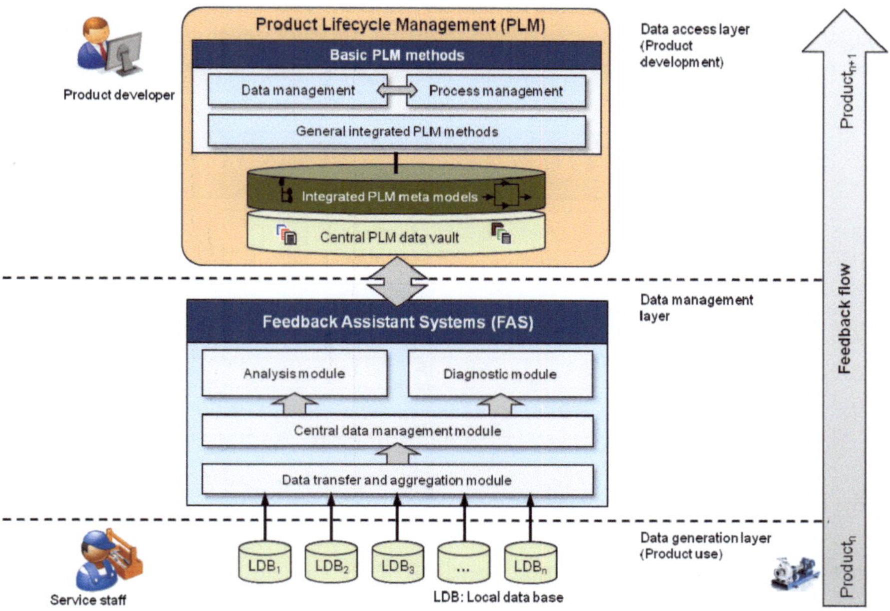

Fig. 1 Concept for the Feedback Assistance System

2.2 *Extension of the Existing Feedback Assistance System by Decision Support and Prediction Modules*

Current FAS support a product developer in analytical and diagnostic tasks such as identifying faulty components. The detection of faulty parts leads to the initiation of a product improvement process. However, the product developer himself is not supported in product improvement and has to identify several alternatives, define the criteria his decision is based upon, and finally choose an alternative that is mainly based on his personal experiences and expert knowledge. This is the second step of the research.

For this reason, the functions of the existing FAS must be extended by methods to support the product developer through multi-criteria decision-making during improvement processes. The product developer, though, still has to deal with the risk of making a wrong decision. In product improvement, there is more than one possible solution, and various criteria must be met in decision-making. To support these processes, a **Decision Support Module** as well as a **Prediction Module** will be developed. The developed use cases for these modules constitute the focus of this paper. These modules render decisions objective and comprehensible. Additional functions such as simulations, which are based on existing PUI, help to increase the decision quality. The motivation for this task is to enhance the quality of a developed product and to minimize incurring costs.

The decision support and prediction modules are integrated into the existing prototype within this project, which currently supports the product developer in finding weaknesses of existing products in use. The existing analytical and diagnostic methods are integrated into the usual product developer graphical user interface; the decision support and prediction methods are used as well (cf. Fig. 1).

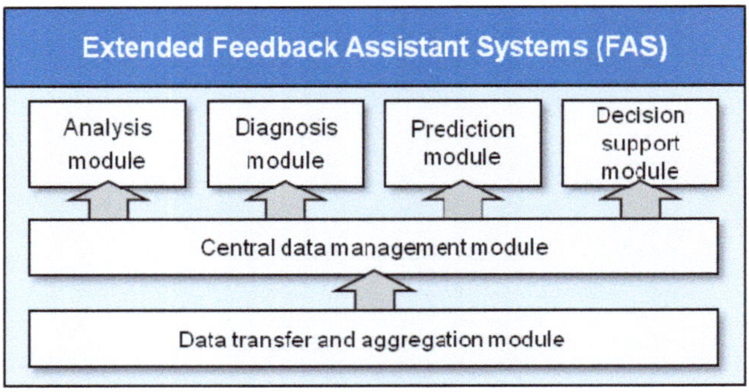

Fig. 2 Framework of the extended FAS

The designed framework thus consists of the already known modules. In addition, the modules for decision support and prediction will be implemented. These modules will be embedded in the FAS and connected to the already existing modules via high-performance interfaces (cf. Fig. 2).

2.3 *Considered Product Model*

Feasibility will be shown on the model of a sample industrial good. Industrial goods are usually built for long production times and, therefore, it is always a crucial task to improve the use phase [9]. In the first phase of the research, centrifugal pumps have been selected to serve as a model, as they fulfill all requirements for a sample product (See Figure 3) [6]. Centrifugal pumps will also serve as a model in the second phase, due to the fulfillment of the requirements and the data the authors have already gathered and analyzed [6]. The paper in hand presents the clutch of a centrifugal pump that is used to illustrate the use case of the Decision Support and Prediction module. The sample pump used here is equipped with sensors for collecting objective PUI [6].

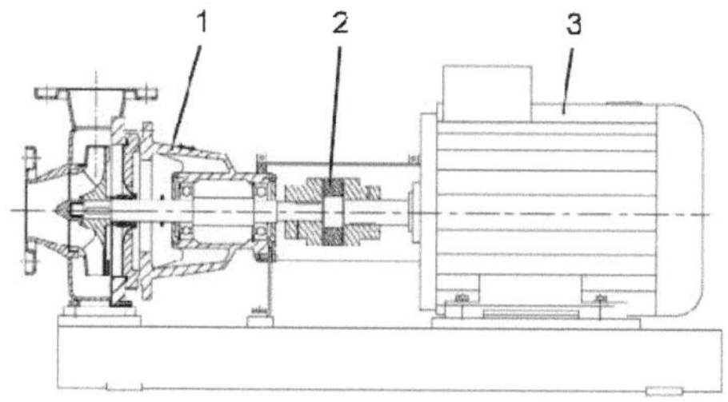

Fig. 3 Sample product: centrifugal pump

To illustrate the use case within this product improvement scenario , the clutch originally used in the sample product is a coupling disk. The advantages of this type of coupling are that it is easy to install and can balance torque oscillation and angular misalignments. Coupling disks are free of maintenance to a high degree. Only the connection component positioned between two metal parts and an elastomer need to be checked periodically and replaced if necessary. In the present improvement scenario, the coupling disk had to be replaced more often than expected, which led to high costs for spare parts and maintenance, so that an improvement process has been initiated.

3 Decision Support and Prediction Module

Decision-making is a crucial task in product improvement. This chapter introduces the process of solving problems in product development as a basis for decision-making, and presents the use case for the later implementation of the Decision Support and Prediction modules.

3.1 Problem-Solving in Product Improvement Processes

The main task of the decision support module is to support the product developer in choosing alternatives for improving existing products, and thus a decision support. A decision is the process of choosing one alternative. The following chapter describes the process of solving problems according to VDI guideline 2221 as a basis for decision making in product development and therefore the subsequently designed use case. The process of solving problems can be repeated several times while the steps taken remain equal.

The process is divided into six consecutive steps, but steps backwards are possible at any time throughout the improvement process (See Figure 4). The improvement process itself is a shortened product development process, as the early phases of the product development process can be omitted (cf. [10] [11]).

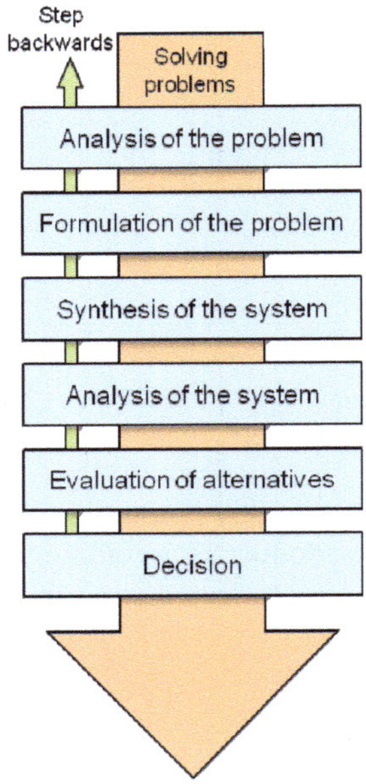

Fig. 4 Process of solving problems [10]

Analysis of the Problem: First, the problem has to be analyzed. The product developer requires information about the current state of the problem. Here, it is important to know which failure leads to the product improvement and if possible why the failure has occurred. In this case, the analysis and diagnosis modules of the FAS can be used.

Formulation of the Problem: Secondly, any information that is not relevant to find alternatives is dismissed. The problem can now be formulated in a short and precise way.

Synthesis of the System: After that, alternatives are generated. It is advisable to generate at least two alternatives to choose from, but not exceed the number of six alternatives. These alternatives must have different focuses so that the alternatives cannot replace each other. At this state, the alternative has the status of an idea. It is not a fully worked-out solution.

Analysis of the System: At this stage, quantifiable criteria for the evaluation are generated. There are different realization opportunities such as minimizing one factor while maximizing the opposite factor in the form of an explicit utility function or by decision rules [12] [13].

Evaluation of the Alternatives: relies on the criteria generated earlier. These are set together with their alternatives by an assessment method. Based on this, the alternatives are weighed so that the best solution can automatically be selected.

Decision: A decision is made by selecting the most suitable alternative. The product developer can check it by simulating the success. In the end the alternative is realized.

3.2 Use Case

The designed use case is the basis for the implementation of the later prototype and generated according to the requirements mentioned earlier. The use case is additionally founded on an intensive literature research and numerous interviews with product developers. Fig. 5 illustrates the use case, the arrows indicate the hierarchy of the sub-use case. The use case is embedded into the FAS. To execute the use case, two new modules, i.e. the decision support and the prediction modules need to be implemented.

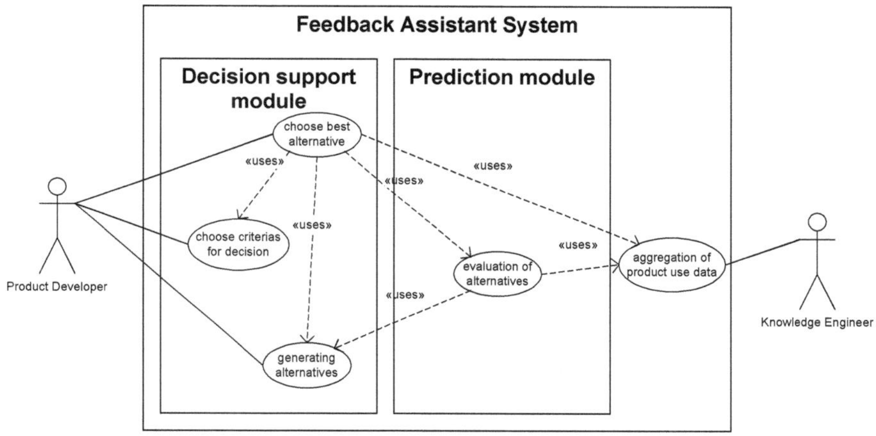

Fig. 5 Use case diagram for decision support of the product developer

The main use case is ‘choose best alternative’, which includes different sub-use cases that support the product developer in his decision-making. A part of this is to select the criteria for the decision and generation of alternatives. The ‘simulation of alternatives’ use case is part of the prediction module, but still supports decision-making. The simulation is based on aggregated product use information, which is part of the previously implemented FAS [6] [14]. A knowledge engineer supports the aggregation of the product use information.

3.2.1 Choose Criteria for Decision

Within the product lifecycle, there are always three main factors that have to be considered with every decision. They are: time, costs, and quality. These three factors compete with each other, which means that e.g. high quality leads to high costs and long production time. Cost reduction is one of the main goals in every step of the product lifecycle. The optimization of time and quality is not as easy, as it can be accomplished either by a reduction or a raise depending on the sub factors.

Taking a closer look at industrial goods, which is the designated group of products this use case has been developed for, the product use phase is the most important phase within the product lifecycle [9]. Because of the long lifetime, i.e. periods of more than 10 years, the costs of the use of a product exceed those of development and manufacturing by far. Hence, criteria that focus on the product use phase have been derived from the factors time, costs, and quality (cf. Table 1, below). They focus either on the effects an industrial good has on its environment (e.g. the owner), or on the effects the environment has on the product itself. Most importantly, maintenance has a huge effect on the actual use of the product. Most key performance indicators (e.g. availability, reliability) are based on factors influenced by the quantity and quality of maintenance events. Hence, the sub criteria of time include maintenance time and intervals, as well as standstills and lifetime. The sub criteria of quality include e.g. reliability, availability, the degree of standardization, accessibility, and manageability. The sub criteria of costs are divided into two groups: costs arising before product use (e.g. purchasing, installation, transportation costs), and costs incurring during the actual use (e.g. energy, staff, commodities, spare parts).

Table 1 Evaluation criteria for improvement alternatives

Time	Costs	Quality
• Short maintenance time	Costs before use:	• High reliability
• Long maintenance intervals	• Purchasing	• High standardization
• Reducing standstills	• Installation	• High availability
• Extending lifetime	• Introduction	• Easy verifiability
	• Transportation	• Simple installation
	Costs during use:	• Easy exchange of parts
	• Energy	• High accessibility
	• Staff	• Easy manageability
	• Commodities	• Strong connection component
	• Disposal of waste	• Eco-friendliness
	• Spare parts	
	• Variable staff	
	• Maintenance	
	• Cancellation	

The sub criteria presented in Table 1 are essential constraints that need to be taken into account in every product improvement process. They determine the most important boundaries every product developer must adhere to. Although more constraints can be found, the authors have reduced the amount as they aim at discussing only the most important criteria, as well as better handling and manageability for the product developer.

With reference to the clutch of the sample product, some criteria do not meet the actual case, e.g. 'energy costs' are not relevant, so that this particular criterion can be eliminated. The same applies to the strong 'connection component'. On the contrary, some criteria are more important than others. As high costs have incurred due to additional spare parts and maintenance, these aspects need to be taken into closer consideration.

3.2.2 Generating Alternatives

The generation of alternatives is the most important task during the improvement of existing products. The product developer is free to choose how grave the changes should be. Changes which address the early stages of product development (e.g. functional or active principle) might lead to a new layout of the product. In this particular case, subsequent simulations are no longer possible.

For this reason, the product developer is tied to boundaries when creating solution alternatives. These boundaries exist because the product he wants to optimize is an actual industrial good and currently in use. Therefore the changes cannot result in a completely new layout of the product; this would only be possible under new development. The main boundary is the available space. In addition, there are used materials, the question of statics, etc. This limits the product developer so that improvements can be fulfilled in the form of

- new dimensioning (e.g. thicker material)
- different material (e.g. harder material)
- different parameters (e.g. surface treatment)
- changed parts (e.g. to optimize the distribution of forces)
- new parts (e.g. to make assembly stiffer).

Additionally, all factors linked to installation and operations have to be taken into consideration when creating alternatives.

For the sample product a change of dimensions (larger connecting part) and materials (using thermoplastics instead of elastomer) have been taken into consideration. Here it is obvious that the solutions are within the existing boundaries, and thus a simulation of the changes is possible.

3.2.3 Evaluation of Alternatives

The evaluation of the above-developed alternatives depends on the degree of the carried-out changes. The degree of changes correlates with the phases of the product improvement process as changes in functional and active principles. Thus, the early phases have stronger effects on the product than changes in later phases and affect e.g. only the used material. The reason for this is that changes in the early phases of product improvement can hardly be determined in detail as there are too many unknown variables, which have to be taken into consideration. Building a computer model to simulate the characteristics of the different solutions would be very imprecise and lead to vague results.

The simulation of alternatives is based on detailed knowledge of the current product use phase. This information is available as objective data stored inside the prototype's data warehouse. Until now, that data has only been used for analytical and diagnostic purposes. Now, it will also be used to develop a loading case of the used product. This loading case describes the typical use of the product to be improved.

For the simulation of the different alternatives, different simulation methods can be used, e.g. FEM analysis to analyze the load on one or several components. At this point, however, it is more relevant to analyze the characteristics of the whole product. For that, as well as for diagnostic tasks, Bayesian networks can be used. These methods are limited to changes affecting the later phases of product improvement due to their needs of product use data.

Changes that cannot be simulated are evaluated by methods such as Failure Modes and Effect Analysis (FMEA). The evaluation is based on the product developer's expert knowledge and one expects him to be objective in his evaluation. In this framework, e.g. score evaluations can be used (cf. Table 2).

Table 2 Score evaluation of different clutch solutions (fictitious values)

		Original coupling disk (elastomer)	Thermoplas tic coupling disc	Full metal coupling	Hydro-dynamic coupling	Separating can coupling
Time	Short maintenance time	5	5	5	2	4
	Long maintenance intervals	4	4	5	4	3
	Reducing standstills	4	4	5	3	3
	Extending lifetime	4	4	5	3	4
Costs	Purchasing	5	5	4	1	2
	Installation	5	5	4	1	2
	Introduction	5	5	4	2	2
	Transportation	5	5	4	2	3
	Staff	5	5	5	3	3
	Commodities	5	5	5	3	4
	Disposal of waste	4	4	5	4	4
	Spare parts	5	5	4	2	2
	Variable staff	5	5	5	2	2
	Maintenance	5	5	4	2	2
	Cancellation	5	5	4	3	4
Quality	High reliability	1	3	4	3	3
	High standardization	4	4	4	2	2
	High availability	4	4	4	3	2
	Easy verifiability	3	3	3	3	3
	Simple installation	4	4	3	2	4
	Easy exchange of parts	4	4	3	2	3
	High accessibility	4	4	3	2	3
	Easy manageability	4	4	4	3	3
	Eco-friendliness	4	4	4	2	3
	Sum:	103	105	100	59	70

5: good; 1: bad

4 Conclusion and Outlook

The use case, as presented in the paper in hand, constitutes the foundation of an extension of the designed FAS. The decision support and prediction modules will be implemented according to the use case, which has been exemplified on the model of the clutch of a centrifugal pump. The decision support and prediction modules are a powerful tool to identify promising solutions for the product developer's individual problems, and present an objective and resilient benchmark. They are excellent add-ons to the FAS prototype, which is already a powerful tool to support the product developer in improving existing product generations. The analytical and diagnostic methods used to identify weaknesses and find first hints for solutions have proven feasible.

In the near future, researches will address the proper implementation of the presented prototype. Subsequent steps are the implementation of methods and algorithms for text mining, as well as methods for the visualization of the results and solutions found by the FAS within the product developers' context in an easy to handle and clearly arranged user interface. In this framework, special information and knowledge visualization methods will be developed, adapted, and integrated.

Acknowledgments. We express our sincere thanks to the German Research Foundation (DFG) for financing this research within the project 'Product Lifecycle Management Extension through Knowledge-Based Product Use Information Feedback into Product Development'.

References

1. Rostad, C., Myklebust, O., Moseng, B.: Closing the product lifecycle information loops. In: 18th International Conference on Production Research, Fisciamo, Italy (2005)
2. Edler, A.: Nutzung von Felddaten in der qualitätsgetriebenen Produktentwicklung und im Service. Verkehrs- und Maschinensysteme der Technischen Universität Berlin, Berlin (2001)
3. Dienst, S., Fathi, M., Abramovici, M., Lindner, A.: A Conceptual Data Management Model of a Feedback Assistance System to support Product Improvement. In: IEEE International Conference on Systems, Man and Cybernetics (IEEE SMC 2011), Anchorage, Alaska (2011)
4. Bauer, A., Günzel, H.: Data Warehouse Systeme - Architektur, Entwicklung, Anwendung. dpunkt Verlag, Heidelberg (2009)
5. Gluchowski, P., Gabriel, R., Pastwa, A.: Data Warehouse & Data Mining. W3L GmbH, Herdecke (2009)
6. Abramovici, M., Lindner, A., Walde, F., Fathi, M., Dienst, S.: Decision support for improving the design of hydraulic systems by leading feedback into product development. In: Proceedings of the 18th International Conference on Engineering Design (ICED), Copenhagen (2011)

7. Meier, H.: Ganzheitliches, regelbasiertes Verfügbarkeitsmanagement von Produktionssystemen (VeraPro). Apprimus Verlag, Aachen (2009) ISBN: 978-3-940565-99-0
8. Abramovici, M., Neubach, M., Fathi, M., Holland, A.: Competing Fusion for Bayesian Applications. In: Proceedings of the 12th International Conference on Information Processing and Management of Uncertainty in Knowledge-Based Systems, Málaga, Spain (2008)
9. Schweiger, S.: Lebenszykluskosten optimieren. Gabler Verlag, Wiesbaden (2009)
10. VDI_2221, Methodik zum Entwickeln und Konstruieren technischer Systeme und Produkte. Beuth Verlag, Berlin (1993)
11. Ehrlenspiel, K.: Integrierte Produktentwicklung – Denkabläufe, Methodeneinsatz, Zusammenarbeit. Carl Hanser Verlag, München (2007)
12. Laux, H.: Entscheidungstheorie. Springer, Heidelberg (2005)
13. Goeken, M.: Entwicklung von Data-Warehouse-Systemen. Anforderungmanagement, Modellierung, Implementierung. Universitäts-Verlag I GWV Fachverlage GmbH, Wiesbaden (2006)
14. Abramovici, M., Lindner, A.: Providing product use knowledge for the design of improved product generations. In: CIRP Annals - Manufacturing Technology, Budapest, Hungary (2011)
15. Ruhland, A.: Entscheidungsunterstützung zur Auswahl von Verfahren der Trinkwasseraufbereitung an den Beispielen Arsenentfernung und zentrale Ent-härtung. Dissertation, TU Berlin (2004)
16. Eisenführer, F., Weber, M., Langer, T.: Rationales Entscheiden. Springer, Heidelberg (2010)
17. Lassmann, Wirtschaftsinformatik: Nachschlagewerk für Studium und Praxis. Betriebswirtschaftlicher Verlag Dr.Th. Gabler I GWV Fachverlage GmbH, Wiesbaden (2006)
18. Pahl, G., Beitz, W., Feldhusen, J., Grote, K.: Konstruktionslehre – Grundlagen erfolgreicher Produktentwicklung. Methoden und Anwendungen. Springer, Heidelberg (2006)
19. Strohmeier, S.: Informationssysteme im Personalmanagement: Architektur – Funktionalität – Anwendung. Vieweg+Teubner Verlag IGWV Fachverlage GmbH, Wiesbaden (2008)
20. Dillon, S., Buchanan, J., Corner, J.: A Proposed Framework of Descriptive Decision Making Theories. Icfaian Journal of Management Research 4(2), 65–74 (2005)
21. Hansson, S.: Decision Theory: A Brief Introduction, Department of Philosophy and the History of Technology. Royal Institute of Technology (KTH), Stockholm (1994)
22. Rokach, L., Maimon, O.: Data Mining with decision trees: theory and applications. World Scientific Publishing Co. Pte. Ltd., Singapore (2008)

Machine Fault Diagnosis Using Mutual Information and Informative Wavelet

Reza Tafreshi[1], Farrokh Sassani[2], Hossein Ahmadi[3,4], and Guy Dumont[3]

[1] Mechanical Engineering Program
Texas A&M University at Qatar, Doha, Qatar
rtafreshi@tamu.edu
[2] The Department of Mechanical Engineering
The University of British Columbia, Vancouver, BC, Canada
sassani@mech.ubc.ca
[3] The Department of Electrical and Computer Engineering
The University of British Columbia, Vancouver, BC, Canada
[4] The Department of Electrical and Computer Engineering
University of Tehran, Center of Excellence in Intelligent Signal Processing
{noubari,guyd}@ece.ubc.ca

Abstract. This paper deals with an application of wavelets for feature extraction and classification of machine faults. The statistical approach referred to as *informative wavelet* algorithm is utilized to generate wavelets and subsequent coefficients that are used as feature variables for the classification and diagnosis of machine faults. Informative wavelets are referred to classes of functions generated from elements of a dictionary of orthogonal bases, such as wavelet packet dictionary. Training data are used to construct probability distributions required for the computation of the entropy and mutual information. In our data analysis, we have used machine data acquired from a single cylinder engine under a series of induced faults in a test environment. The objective of the experiment was to evaluate the performance of the informative wavelet algorithm in classifying faults using real-world machine data and to examine the extent to which the results were influenced by different analyzing wavelets chosen for data analysis. The correlation structure of the informative wavelets as well as coefficient matrix are also examined.

Keywords: Fault diagnosis, informative wavelets, wavelet packet analysis.

1 Informative Wavelets, Concept and Approach

Informative wavelets are classes of functions generated from a given analyzing wavelet in a wavelet packet decomposition structure in which for the selection of 'best' wavelets, concepts from information theory, i.e., mutual information [1] and entropy [2,3] are utilized. Entropy is a measure of uncertainty in predicting a given state of a system where a system state refers to different operating conditions such as normal or faulty. Computation of entropy requires calculating state probabilities from training data and supplying them as inputs to the algorithm. An iterative process to identify appropriate informative wavelets is used at each stage, whereby the algorithm selects a wavelet from a dictionary of orthogonal wavelets in a wavelet packet signal decomposition structure, which results in a maximal reduction in entropy. This is equivalent to obtaining maximal reduction in uncertainty of predicting a given system state. In this algorithm, reduction in uncertainty is expressed in terms of mutual information derived from the joint probability distributions of the training data and coefficients. Entropy of a system is defined as:

$$H(S) = -\sum_{i=1}^{M} P(S_i)\log(P(S_i)) \tag{1}$$

where S_1, S_2,..., S_M are the states of the system with probability of occurrences given by $P(S_1)$, $P(S_2)$, ..., $P(S_M)$. Entropy is a measure introduced for the quantification of the information. It can also be considered as a measure of complexity of prediction of the state of the system. The reduction in uncertainty can be regarded as the *quantity* of information about the original system contained in the measurement system, which is referred to as mutual information [4,5,6].

To derive the mathematical definition of mutual information we need to describe the states of the system. Such states can be observed by a measurement system with N possible outcomes $\{T_1,T_2,\ldots,T_N\}$ of a random variable T with a probability distribution $P(T_1)$, $P(T_2)$,..., $P(T_N)$. Mutual information between the states and measurements is defined as the difference between the uncertainty of predicting S before and after the observation of T:

$$J_S(\omega_\gamma) = H(S) - H(S/T) = \sum_{i=1}^{M}\sum_{j=1}^{N} P(S_iT_j)\log\frac{P(S_iT_j)}{P(S_i)P(S_j)} \tag{2}$$

Here $H(S/T)$ and $P(S_iT_j)$ indicate conditional entropy of state S given measurement T and joint probability distribution of $S = S_i$ and $T = T_j$, respectively. ω_γ is the wavelet indexed by the triplet parameter $\gamma = (j, k, m)$, where j, k, m are the indices of scale, oscillation, and translation (time position) in a wavelet packet dictionary. When a given state of a system is independent of the measurements, i.e. $J_S = 0$, a change in the state of the machine will not cause any changes in the probability $P(S_iT_j)$. Then the algorithm selects wavelets that result in a maximal reduction of uncertainty i.e. maximal $J_S(\omega_\gamma)$. In informative algorithm, the measurement system is wavelet. Such wavelets are obtained iteratively where at each stage, the residual signal is considered for further signal expansion. These wavelets are referred to as *informative wavelets*. The iterative selection of the informative wavelets is very much similar to the classical matching pursuit algorithm [7]. Wavelet coefficients are then used as feature variables and as inputs to a neural network classifier for classification [1,5]. Fig. 1 illustrates the main stages of the algorithm. The next section explains different steps of informative wavelet algorithm in more details.

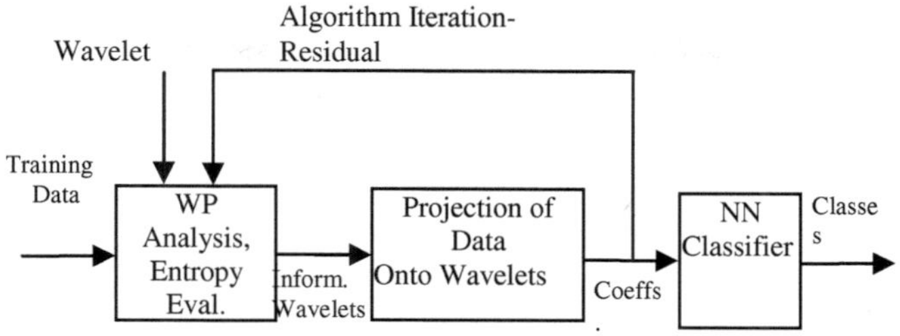

Fig. 1 Block diagram of main stages of informative wavelet algorithm

2 Informative Wavelet Algorithm

The algorithm has two stages: *training* stage and *class recognition* stage. In the training stage, Fig. 2, by using wavelet filters, training data are decomposed into low and high frequencies iteratively to form a wavelet packet (WP) for each training data. Then the collection of these wavelet packet decomposition coefficients is quantized into N fixed and equally-spaced sub-intervals. At this step probability distributions of S, T and joint probability distribution of S and T are obtained. Each wavelet is considered as a measurement system whose output is its decomposition coefficients obtained by projecting data onto the selected wavelet. These wavelet coefficients, which are in fact feature variables, are later fed to a neural network to classify the system state. Using the maximum mutual information ($J_S(\omega_\gamma)$) its corresponding informative wavelet is then selected. In the next step, the corresponding wavelet components are deducted from the residuals of entire training data, much in the same way as in the matching pursuit algorithm. As the informative wavelets are successively selected from these residuals at each iteration, they are less correlated with the ones selected previously.

At the final stage, the coefficients obtained above are used to train the neural network. Once the training is completed the NN weights are attained in order to memorize the main features of different classes. If three classes are considered, these can be, for example, severe fault, mild fault and healthy states.

The input signal along with informative wavelet and neural network weights obtained from the previous stage are inputs to the second or class recognition stage. This stage consists of three steps: projecting the input signal – that we want to identify its class – onto selected informative wavelets, computing their feature vector, and classifying the state of machine (S_2). Fig. 3 shows the flowchart of this stage. This algorithm attempts to match joint state and measurement probability distribution of data with wavelet coefficients, the higher the probability the more the mutual information.

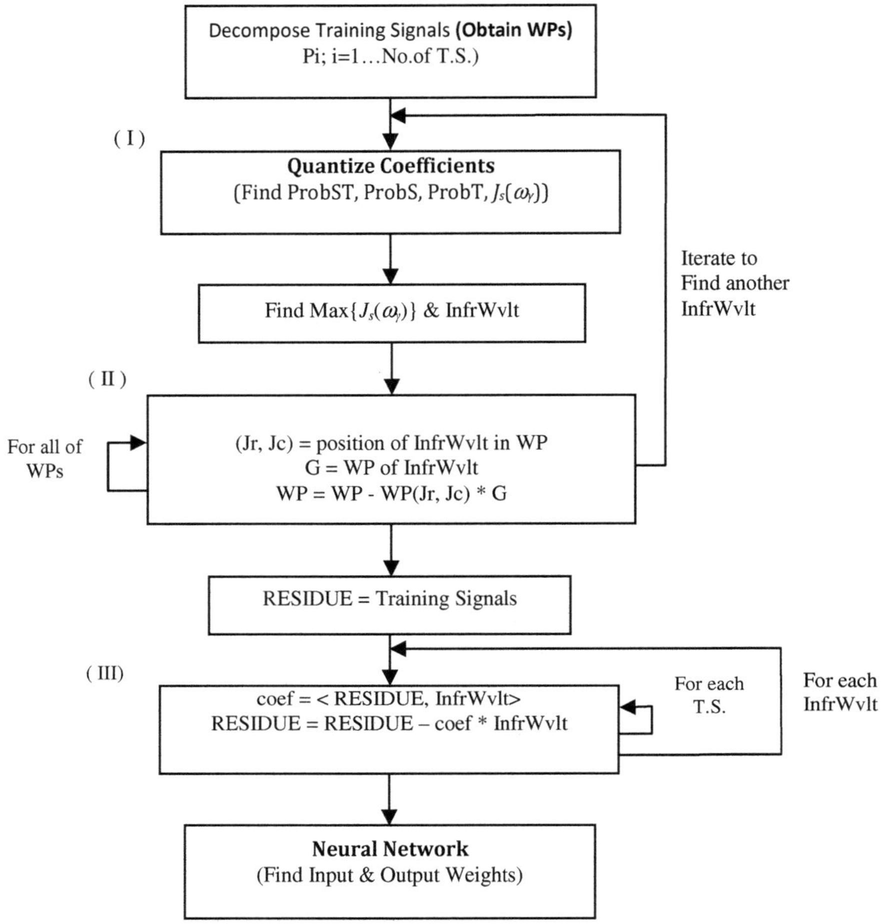

Fig. 2 Informative wavelet algorithm: training stage

The major disadvantage of informative wavelet algorithm may be its computational complexity. The computational time for box (I) is $O(N)$, where N is the total number of training data in all classes. Since the loop of this box must iterate for the whole wavelet packet elements ($n \log_2 n$ times), and for the number of informative wavelets (W), the total cost is $O(W N n \log_2 n)$. This algorithm relies on the evaluation of probability density of training data; consequently, we usually need several training data. An empirical number is about the size of data (n), therefore, the total computational cost is $O(W n^2 \log_2 n)$. If the time for decomposing each training data to wavelet packet coefficients is also added, i.e., $O(n \log_2 n)$, along with other overhead computations, which is not insignificant in this algorithm, the real time cost will approach $O(n^3)$. We note that since probability density function must be evaluated in every iteration, calculation of probability density function is the most time consuming part of the algorithm.

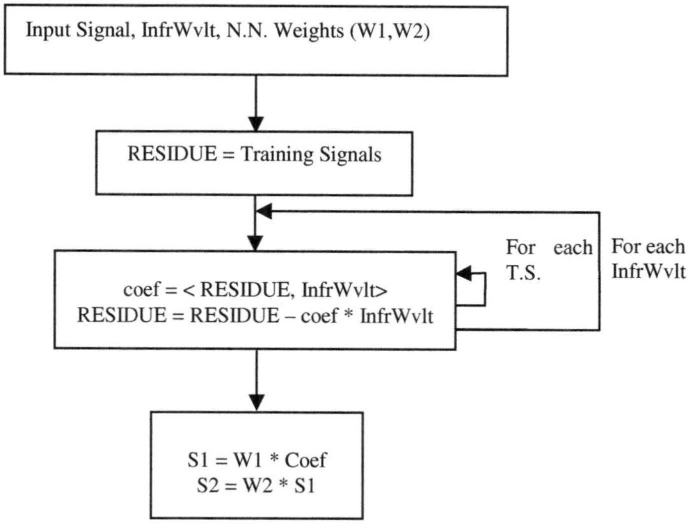

Fig. 3 Informative wavelet algorithm: classification stage

3 Design of Experiments

To evaluate the performance of the algorithm and the accuracy of its classification results we used machine data a single cylinder dual mode unit operating either on diesel fuel or natural gas. Data used here is from diesel mode operation. Acceleration data of the intake valve closing and combustion events from this engine were utilized for data analysis and algorithm testing. Two types of faults, namely intake loose valve and engine knock conditions each with varying intensity levels were considered. Engine knock condition was generated by judicious adjustment of load. Load changes were made in two incremental steps of approximately 15% above nominal load corresponding to 18, 22, 25 HP, respectively.

For loose valve experiments, a set of progressively increasing valve clearances, namely normal, 0.006 in. and 0.012 in. were set on the intake valve. Three categories of data were collected simultaneously: (a) cylinder pressure measured through a connecting tube to the cylinder, (b) block acceleration (vertical vibration) measured at a carefully chosen location on the cylinder head, and (c) engine RPM. Block vibration was actually measured at several places and the best location was found to be at the center of the upper part of the cylinder block which gave reliable signal intensities. Other supplementary data were also collected including engine power, peak cylinder pressure and peak pressure angle. For each test, data from sixteen consecutive cycle runs were acquired.

Fig. 4 shows sample cycle runs of the diesel engine in normal and knock conditions. In this figure, the high amplitude components from left to right correspond to exhaust valve closure, intake valve closure, combustion, exhaust valve opening, and intake valve opening. There were noticeable cycle-to-cycle changes in the signal patterns and intensities even under normal condition, which indicate the complexity and variability of the machine operation. An initial review of data, in which mean values vs. standard deviation of each training data were examined, showed that a certain degree of data clustering and class separation can be found (Fig. 5), though this could not be observed in all of the data sets. Separation of classes was more vivid in training data belonging to valve clearance conditions.

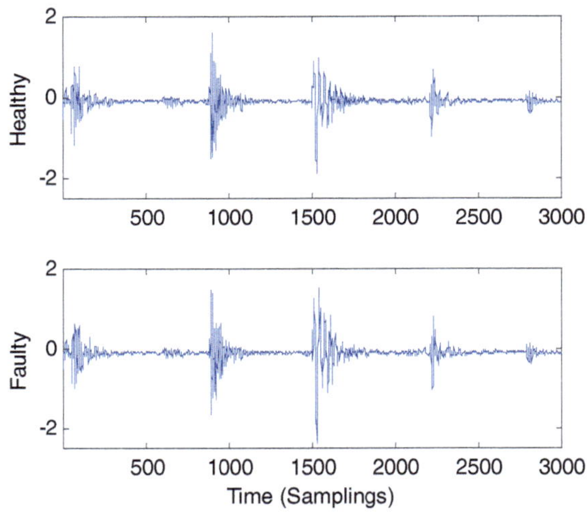

Fig. 4 Sample vibration signal of the single-cylinder diesel engine in normal and knock conditions

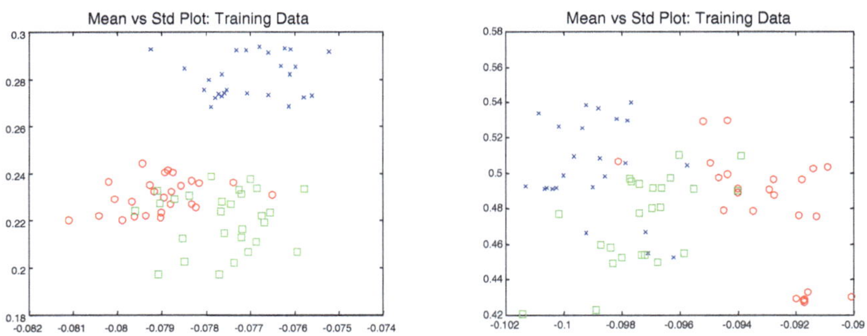

Fig. 5 Training data for valve clearance (first plot) & load change (senod) for three classes of □: normal, x: mild fault, o: severe fault

We note the following in the analysis:

- In informative wavelet algorithm, the "number of informative wavelets" corresponds to the number of feature variables used for the classification. In the absence of any á priori knowledge about a suitable number of feature variables, several values ranging from 1 to 50 were initially considered. At a later stage, the number was confined to a smaller set ranging from 4 to 10.
- Wavelets from orthogonal and biorthogonal wavelet families were used including Daubechies wavelets Db5, Db20, Db40 and Db45 as well as Coif5, Symlet5, Bior3.1, and Bior6.8.
- Multi layer perception backpropagation was used for the neural network classifier. For a three-class data set, five nodes of hidden layer were used in the network.
- We used 30 levels (bins) in quantification of coefficients and training data during construction of the probability distributions.

4 Data Analysis and Classification

As indicated earlier the informative wavelet algorithm is mainly a statistical approach for fault detection and classification in which probability distributions of training data are utilized to generate wavelets during signal expansion. In this algorithm, coefficients of the selected wavelet carry statistical properties that best matched those of the training data.

At the first glance, it may seem that classification results are determined jointly by capturing the statistical properties of the given training data as well as the analyzing wavelet used for data expansion. But our observations using different data and with several analyzing wavelets showed that the former has a higher influence on the classification results. In fact, different analyzing wavelets capture more or less the same amount of statistical information; therefore, the choice of analyzing wavelet does not significantly alter the correlation structure of coefficients, although Coiflet1 wavelet performed marginally better.

Using Coiflet1 we analyzed three load settings (leading to knock) as well as three valve clearance conditions. Mean values vs. standard deviations of the coefficients of training data for three classes as well as histogram of the coefficients were also examined (Fig. 6). Separation of classes in coefficient domain followed a similar pattern as those of training data. For both fault cases, classification errors were below 5%, which were considered to be acceptable. Classification errors for different load changes and knock conditions were influenced to a large extent by the uniformity of the training data in all classes.

4.1 Selected Informative Wavelets

Informative wavelet algorithm is a nonorthogonal signal decomposition in which informative wavelets generated by the algorithm are in general correlated

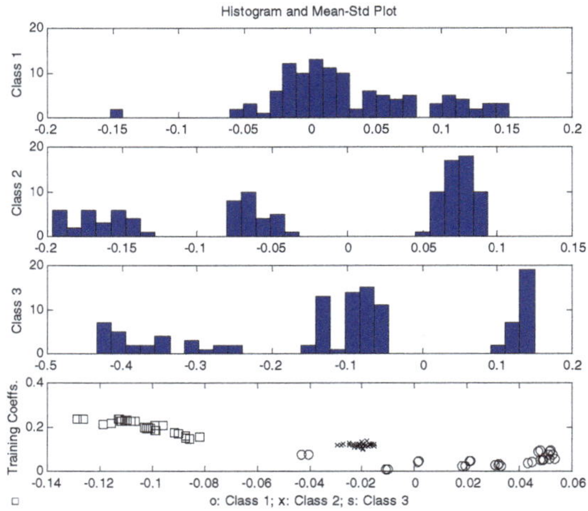

Fig. 6 Histograms of the coefficients as well as mean vs. standard deviations for three classes

with each other and a certain degree of redundancy always exists in signal decomposition. Accordingly, the coefficients generated by projection of data onto informative wavelets follow the same pattern of correlation. Non-orthogonality of signal decomposition is mainly due to the iterative process of selecting informative wavelets where at each stage the residual signal is constructed and used for signal expansion. In our data analysis, we examined deviation from the orthogonality of the informative wavelet for several analyzing wavelets. We examined informative wavelets generated by orthogonal and biorthogonal analyzing wavelets. While informative wavelets in both categories deviated from orthogonality, which was measured by the inner product of the wavelets, the extent of the deviation varied for the two groups. Orthogonal wavelets such as Db family of wavelets or Coiflets, generate informative wavelets with a higher degree of orthogonality as compared with biorthogonal wavelets such as Bior3.1. The same trend can be seen in the correlation structure of coefficient matrix as well.

Correlation structure of coefficient matrix under several analyzing wavelets and for different number of iterations was examined for a given set of data. Differences were observed in correlation of the coefficients for different analyzing wavelets; however, such differences were insignificant to influence the classification results greatly.

4.2 Training Data and Number of Iterations

In our data analysis, a small change in training data resulted in a noticeable change in the informative wavelets selected. For example, a small increase in the number of training data (e.g. a simple repetition of data) caused a different set of informative wavelets to be selected. This could be attributed to the application of matching pursuit type approach in which a small change in the probability distribution of the coefficients leads to changes in mutual information value calculated. Often a small change in the training data caused a change in about half of the informative wavelets.

In the algorithm, the number of informative wavelets (iterations) is chosen á priori as an input. It was observed that increasing the number of iterations in a given data analysis does not alter informative wavelets derived from previous iterations. As a result, there were no changes in the corresponding coefficient values. The additional informative wavelets, selected with larger number of iterations, increased the number of feature variables and thus expanded the dimension of feature space.

In the experiments, mostly 5-10 iterations were used, although higher iterations were also selectively examined. It was observed that an increase in the number of iterations was not always accompanied by an increase in the accuracy of classification results. This could be traced to dilution of information, in which by selection of large number of features unnecessary information is added.

5 Conclusions

In this paper results of an experimental study for an application of informative wavelet algorithm for the classification and diagnosis of machine faults were presented. Several prototype wavelets and different sets of machine data were used. Effectiveness of the algorithm for the classification of two categories of faults namely excess valve clearance and knock conditions each with varying intensity levels were examined. Accuracy of results under different parameters of the algorithm was also studied by employing different analyzing wavelets from both orthogonal and biorthogonal family of wavelets. Some notable results are summarized as follows.

- In majority of the experimental runs, using different analyzing wavelets, satisfactory classification results were obtained when sufficiently large number of training data with adequate uniformity was used. For load changes and knock condition, accuracy of results varied for different training data and different intensity levels of fault conditions.
- Informative wavelets generated by the algorithm varied significantly when small changes were introduced in the number of training data. This was also the case when minor changes were made in training data themselves. While classification results remained almost unaffected under minor changes in the training data, informative wavelets and subsequent coefficient values varied significantly. This was attributed to the particular structure of the algorithm in which minor modifications in the training data are followed by changes in probability distributions which in turn modify mutual information and informative wavelets derived by the algorithm. Changes in the informative wavelets can be amplified by the application of matching pursuit algorithm. In the matching pursuit algorithm, wavelets generated at the later stages are highly sensitive to changes in the wavelets chosen at the early stages.

References

1. Liu, B., Ling, S.F.: On the selection of informative wavelets for machinery diagnosis. Mechanical Systems and Signal Processing 13(1) (1999)
2. Huang, Q., Liu, Y., Liu, H., Cao, L.: A new vibration diagnosis method based on the neural network and wavelet analysis. SAE technical paper series, 2003-01-0363 (2003)
3. Tafreshi, R., Sassani, F., Ahmadi, H., Dumont, G.: An approach for the construction of entropy measure and energy map in machine fault diagnosis. ASME Journal of Vibrations and Acoustics 131(2) (2009)
4. Karmeshu, N.R.P.: Uncertainty, Entropy and Maximum Entropy Principle- and Overview. In: Karmeshu (ed.) Entropy Measures, Maximum Entropy Principle and Emerging Applications. STUDFUZZ, vol. 119, pp. 1–54. Springer, Heidelberg (2003)
5. Ahmadi, H., Dumont, G., Sassani, F., Tafreshi, R.: Performance of informative wavelets for classification and diagnosis of machine faults. International Journal on Wavelets, Multiresolution and Information Processing (IJWMIP) 1(3) (2003)
6. Verron, S., Tiplica, T., Kobi, A.: Fault detection and identification with a new feature selection based on mutual information. Journal of Process Control 18(5), 479–490 (2008)
7. Mallat, S., Zhang, Z.: Matching Pursuit with Time Frequency Dictionaries. IEEE Transactions on Signal Processing 41, 3397–3415 (1993)

Simulation-Based Parameter Identification for Online Condition Monitoring of Spindle Nut Drive

Mahdi Mottahedi[1], Sascha Röck[2], and Alexander Verl[3]

[1] Research Assistant, Institute for Control Engineering of Machine Tools and Manufacturing Units (ISW), Stuttgart University, Germany
[2] Institute of Applied System Dynamics, Aalen University of Applied Sciences, Germany
[3] Institute for Control Engineering of Machine Tools and Manufacturing Units (ISW), Stuttgart University, Germany
Mahdi.Mottahedi@isw.uni-stuttgart.de

Abstract. In this article the development of a method for simulation-based condition monitoring of a spindle nut drive for machine tools will be presented. Thereby, parallel to the operation of the spindle nut drive, an automatic parameter identification of a corresponding simulation model is to be carried out with the aim to identify high-level information like stiffness and damping of the significant components based on the available drive signals. The underlying model for the identification consists of Finite Element (FE) component models and the corresponding component parameters like stiffness and damping of the bearings, spindle nut, etc. Beyond the parameter identification, the characteristics of the components (here stiffness) will be computed by the mentioned model. The identification and the calculation method in this paper is based on finding optimum stiffness parameters which are correspondent to the current state of the system and using a neural network to find the relation between the physical parameters of the system and measurable parameters of the system behavior. The results depict a new diagnostic process which could be also applicable for online condition monitoring of different components.

Keywords: Condition Monitoring, Machine Diagnosis, Parameter Identification, Spindle Nut, Ball Screw Drive.

1 Introduction

As industrial machines and components are becoming more complicated and expensive, the maintenance of their parts is becoming more and more important. The maintenance strategies could be divided into three different categories:

a) The first is Run-to-Break. This strategy is valid for not important components as their failure would not lead to any special losses or stoppage in the production line or human hazards. In this strategy, the probable worn out component and its location is previously known. Hence, the element is left till its breakage. No sign of destruction is considered for such components except poor physical behavior, like enormous vibration, temperature and noises. The out of order part is simply replaced when the failure happens.
b) The other strategy is time-based preventive maintenance strategy. In this method which is applied for more important components, the part will be monitored frequently. This means that in each specific period of time the condition of the part will be analyzed. This analysis is based on specific parameters which could differ from one case to another. Analysis of vibration, acoustic emission, quality of lubricant, temperature, motor current, surface roughness, eddy current etc. are some examples of diagnosis methods to check the machine behaviors in different intervals [1]. Changes of these parameters are a sign of defect emersion in the component. Depending on the method of prognosis and decision making, it could be then decided if the component should be replaced or could continue working.
c) The last strategy, condition-based maintenance, is applied to important components and sophisticated systems, where the location and the kind of failure are not totally clear and the stochastic nature of involved parameters harden the prediction of useful components age. The goal of this strategy is the analysis of the components' behavior during their operation time, determining the current condition of the system as well as predicting the remaining life of the part [2]. This last method has many advantages over the others as it can anticipate the remaining useful life of the components and prevent the down-time cost (almost 80% of down-time is related to defect detection).

Condition monitoring is usually based on vibration diagnosis. In this method, some external accelerometer sensors are used to measure the vibration of the machine in different positions. The diagnosis of these vibrations leads to the description of the system condition [3]. This method of monitoring suffers, however, from some problems as well. The external sensors cost a lot and most of the time they lead to some difficulties while working with the machine. Almost all sensors include noises in the result. These noises make the analysis of different defect modes difficult. In addition, verifying different condition monitoring methods in practice is time-consuming and expensive, as different components with different modes of defects should be analyzed and therefore become artificially defected. Hence, the presented approach is simulation-based condition monitoring. While using this method, different defect modes in different components are modeled numerically and the analysis of specific physical properties of the system leads to diagnosis process. As the influence of defects in different components is analyzed numerically using internal signals, this method has advantages from the cost and time point of view over other methods. In addition, analyzing the components' state in different conditions is faster, deterministic and more precise by using an exact model. There have been studies regarding condition monitoring of spindle nut. Neugebauer [4] did condition-based preventive maintenance of the main spindle by using accelerometers on the outer ring, Acoustic Emission (AE) at housing and considering temperature and eddy current, as well as motor current. The used diagnostic method in his work was Root Mean Square (RMS), Fast Fourier Transformation (FFT) and envelope spectrum. The group understood that the AE is best suited for preventive maintenance. They also concluded that envelope curve can detect damages faster than RMS and even easier than FFT. Sin [5] and his group did their condition monitoring on bearing by putting acceleration sensors on the outer and inner ring. They used RMS, FFT and time-frequency-based techniques. They concluded that time-frequency-based techniques could lead to the best diagnosis. Yan [6] and colleagues worked on modal parameter identification from output - only for the spindle case. They understood modal parameters provide insight into structural changes of the spindle. Zhang [7] and his group did an online condition monitoring based on open system architecture wavelet analysis of spindle vibration and they resolved that wavelet analysis predicts well the frequency of defects compared to theoretical formulae. Saravan [8] and his colleagues did some studies on condition monitoring of spindle bearing. They used vibration, acoustic emission, lubricant analysis and surface roughness by FFT diagnosis method. They found significant peaks at the fault frequencies and also discovered that the vibration level is increased considerably with larger particle sizes. Also based on open system architecture, Li performed a real-time spindle health monitoring [9].

In this paper, however, a new method is presented where the parameters of the system model would be updated online. The way of implementing this method for condition monitoring of a spindle nut drive will be discussed. As a model a Finite Element Model (FEM) including rheological elements (Hookian spring and Newtonian damper) is used. The model includes all important components of a spindle nut drive, namely bearing, coupling, spindle and nut. Using parameter identification the attributes of each component (stiffness and damping) are identified. This identification is performed by using internal signals of the system, namely desk position and motor current. Figure 1 shows an overview of the most important working packages of the project. These are separately: Model construction, model reduction with subsequent parameter identification and development of a prognoses process. In this article it is shown how the parameter identification of the reduced model based on optimization algorithms and neural network could help the condition monitoring process.

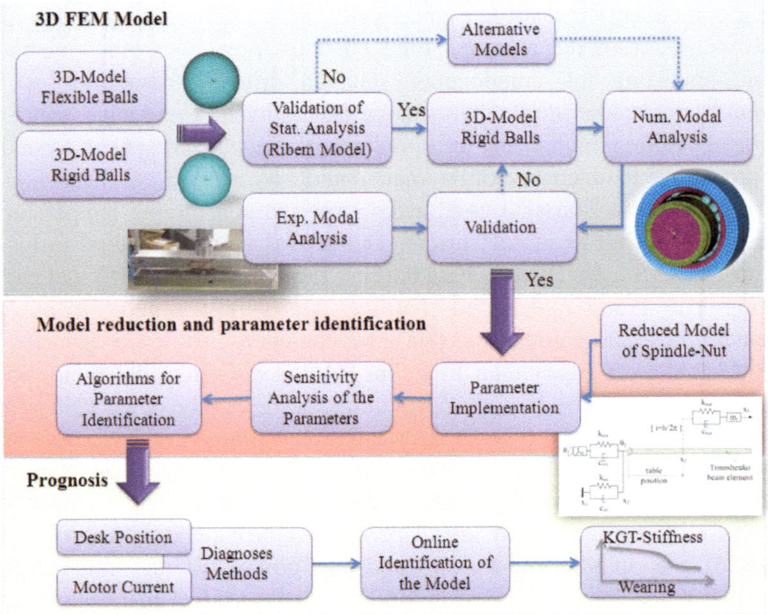

Fig. 1 Overview of the procedure for simulation-based condition monitoring

2 Model Construction

In the first step, a detailed 3D model of the spindle nut drive was generated in order to determine the stiffness of the healthy component. Each separate component, spindle, nut, and balls are afterwards assembled in MESHPARTS. This package of software is working in ANSYS environment and was developed at the ISW [10]. Figure 2 shows an assembled model of the 3D spindle nut [11]. The accuracy of the model was verified by an experimental static and modal analysis. The depicted balks in figure 2 are for leading the force in the axial direction and, therefore, exciting this mode. The parametric model is so constructed that different spindle nut drives could be generated by changing the geometrical inputs. By performing a static analysis, the stiffness of the spindle nut is determined. This stiffness parameter is used later on as an input parameter for the reduced model of the spindle nut.

In the next step, an appropriate reduced model is chosen. As the later diagnosis is based on modal analysis of the system, this model should represent the first two eigenfrequencies of the spindle nut drive exactly [12], [13], [14]. The reduced model depicted in figure 3 consists of a linear spring and a damper for modeling coupling and nut [15] as well as a beam element for modeling the spindle. The parameters of this model should be updated based on the measured frequencies of the real system. In order to perform this identification, two different methods are tested in this paper. As a first method, different identification algorithms are used in order to find optimum stiffness parameters of the model which match the eigenfrequencies of the

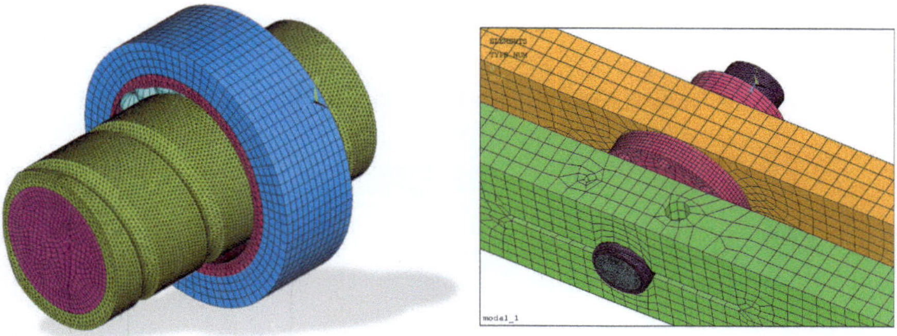

Fig. 2 3-D model of a ball screw and the structure used for modal analysis of the model

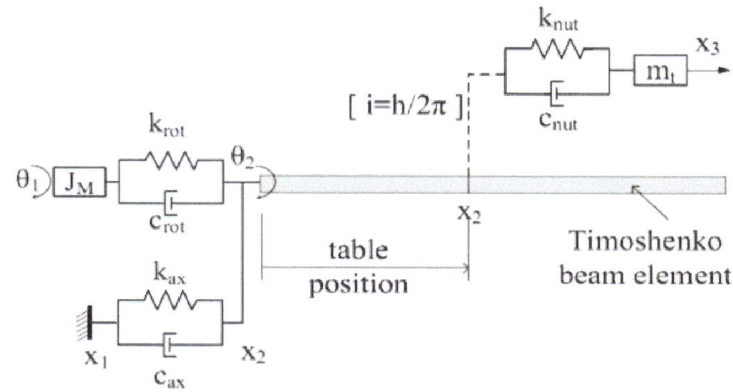

Fig. 3 Reduced model of a spindle nut drive [15]

real system. In the second method, different simulations are performed in order to find the relation between the input and output parameters. These input and output parameters are stiffness and eigenfrequencies of the system. The process of finding the relation between the input and output can be performed by using neural networks. The details of generating the neural network and verifying it are discussed in "Neural Network" (part four) of this article.

3 Model-Based Identification

In order to determine the stiffness parameters, which are relevant to measured frequencies, an identification algorithm can be used. This algorithm, as it has been shown in figure. 4, begins with three initial stiffness coefficients. These variables are bearing, coupling and nut stiffness and are the design variables of the algorithm. By using the reduced model and performing a modal analysis, the relevant eigenfrequencies are calculated. The difference of the calculated eigenfrequencies and measured eigenfrequencies determines the objective function as it has been defined in equation 1. The objective is to minimize this function and determine the relevant stiffness. Error function as it is depicted in

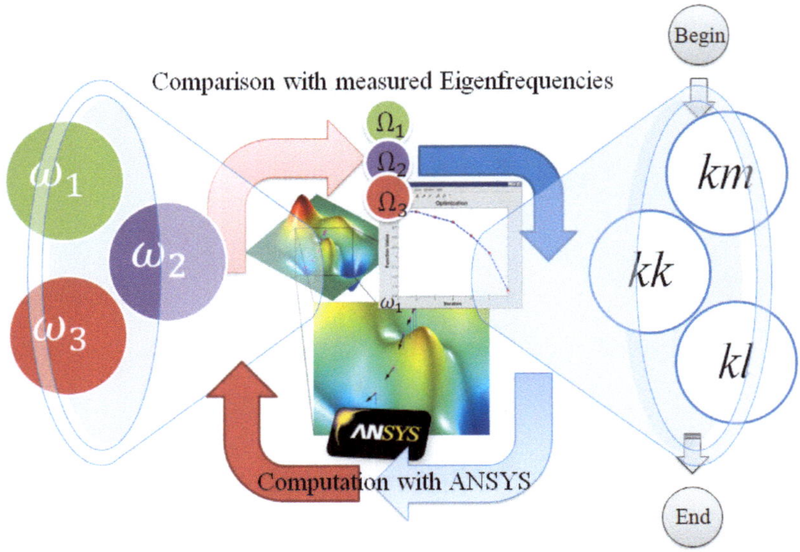

Fig. 4 Identification algorithm used for determination of equivalent stiffness related to measured eigenfrequencies

equation 2 was also used in order to represent the error of each identification algorithm. In equation 1 and 2, ω_1 till ω_3 are computed eigenfrequencies. Ω_1 till Ω_3 are the measured system eigenfrequencies. *KL*, *KM*, *KK* are corresponding stiffness coefficients for the measured eigenfrequencies (reference stiffness), and *kl*, *km*, *kk* are updated stiffnesses of the model in each iteration.

$$\text{Objfun} = (\omega_1 - \Omega_1)^2 + (\omega_2 - \Omega_2)^2 + (\omega_3 - \Omega_3)^2 \tag{1}$$

$$\text{Error\_fun} = \left[\frac{|kl - KL|}{KL} + \frac{|km - KM|}{KM} + \frac{|kk - KK|}{KK} \right] / 3 \tag{2}$$

While testing different identification algorithms, the best one is determined and the effect of using 2 or 3 eigenfrequencies of the system on ascertaining the stiffness of different components is analyzed. In another try, the influence of performing a modal analysis in different desk positions on identification and robustness of the method was tested.

Different methods were examined while using APDL coding in ANSYS. These methods were based on subproblem approximation, first order method, random design, sweep generation, factorial evaluation and gradient evaluation. The results as they have been shown in figure 5 and table 1 reveal that the best method of identification in this case is the first order method, which uses the gradient of the objective function in order to minimize it. Hence, in the rest of the analysis this method is used. In the first row of table 1 the reference values has been written. So the goal of the identification is to reach the measured eigenfrequencies by changing stiffness values. The variables' space was shown in the second row and the start variables in the third row of table 1.

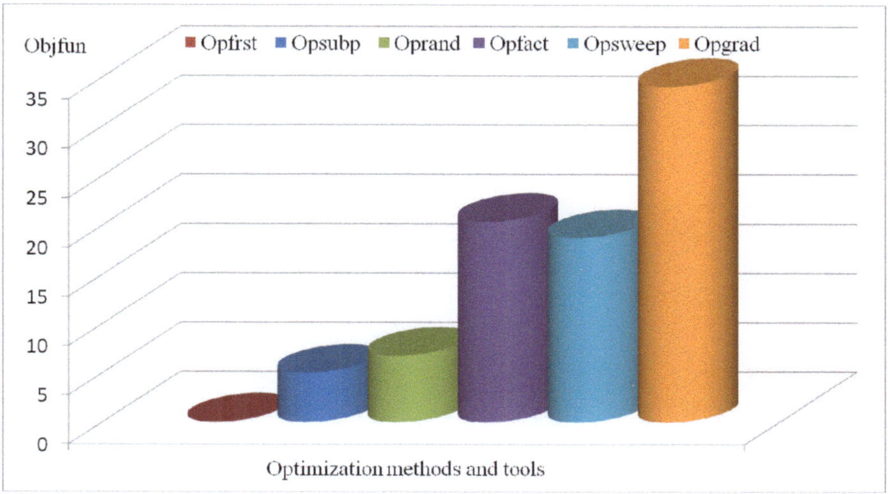

Fig. 5 Comparison of different identification algorithms based on objective function defined in equation 1

Table 1 Calculated stiffness parameters for different identification algorithms and their comparison with reference values

KL:270e6	*KM*:550e6	*KK*:170e3	Ω_1:70.95	Ω_2:382.81	Ω_3:866.15
Identification methods	Variables space	*kl*:200e6-400e6	*km*:400e6-600e6	*kk*:100e3-200e3	
	Start values:	*kl*:250e6	*km*:570e6	*kk*:140e3	
		kl e6	*km* e6	*kk* e3	OBJFUN
1	First order	265	552	171	0.04
2	Subproblem	259	561	173	5.01
3	random	264	558	158	6.75
4	factorial	200	600	200	20.37
5	Sweep generation	250	570	200	18.74
6	Gradient evaluation	250	570	140	34.08

The effect of using different eigenfrequencies in different desk positions is shown in figure 6 and table 2. The identification algorithms begins with start values and changes the stiffness parameters in variable space shown in the first row of table 2 till reaching the measured eigenfrequencies. The error is defined by comparison of computed stiffness (*km*, *kl*, *kk*) and reference values (*KM*, *KL*, *KK*). The identification is firstly based on the first two eigenfrequencies and then by means of the first three eigenfrequencies in different table positions.

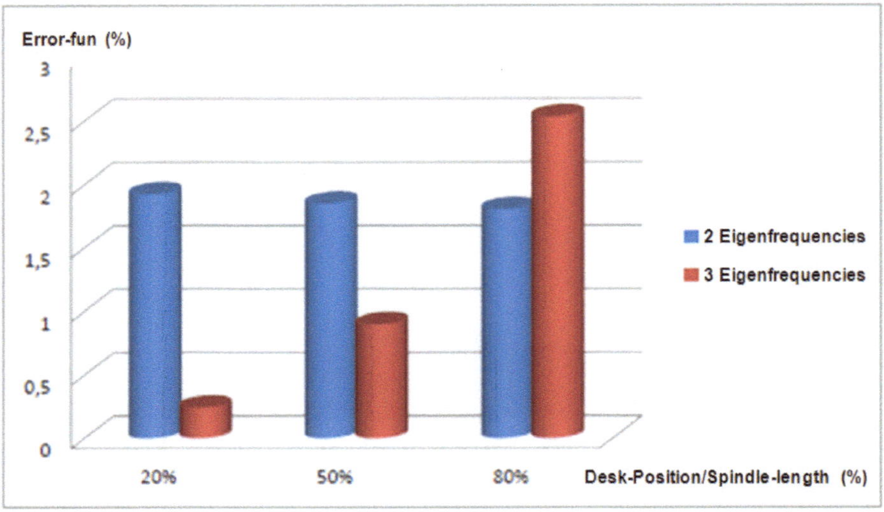

Fig. 6 Error function based on equation 2 for different table positions and by using 2 or 3 eigenfrequencies and first order identification algorithm

Table 2 Comparison of error function for different table positions by using 2 or different eigenfrequencies for performing identification based on first order method

	Variables Space	*kl*:200e6-400e6	*km*:400e6-600e6	*kk*:100e3-200e3			Start values	*kl*:250e6	*km*:570e6	*kk*:140e3
Using first order method and first second or third eigenfrequencies for identification **Table position=20%**										
	KL	*KM*	*KK*	Ω_1	Ω_2	Ω_3	*kl* e6	*km* e6	*kk* e6	Error in %
1	270e6	550e6	170e3	82.50	380.98	-	265	571	170	1.9
2	270e6	550e6	170e3	82.50	380.98	657.92	268	550	170	0.2
Using first order method and first second or third eigenfrequencies for identification **Table position=50%**										
	KL	*KM*	*KK*	Ω_1	Ω_2	Ω_3	*kl* e6	*km* e6	*kk* e6	Error in %
3	270e6	550e6	170e3	70.95	382.81	-	265	569	170	1.9
4	270e6	550e6	170e3	70.95	382.81	866.15	265	552	171	0.9
Using first order method and first second or third eigenfrequencies for identification **Table position=80%**										
	KL	*KM*	*KK*	Ω_1	Ω_2	Ω_3	*kl* e6	*km* e6	*kk* e6	Error in %
5	270e6	550e6	170e3	63.29	383.73	-	266	569	170	1.8
6	270e6	550e6	170e3	63.29	383.73	901.73	260	566	171	2.5

As it could be seen from figure 6 and table 2, diagnosis based on three eigenfrequencies leads to more precise results; however, using first two eigenfrequencies results also in a low amount of errors. And as more table positions are used, more exact results would be achieved.

4 Neural Network

In order to determine the relation between stiffness and eigenfrequencies of the system, neural network algorithms were also examined. This approach was previously attempted in Silva's work for the case of cutting process and it provided promising results for feasible applications of the method [16]. The goal of this paper is training of a network in which by implementing the eigenfrequencies the equivalent stiffness would be achieved. The advantage of this method over the model-based identification method is faster calculation of stiffness and hence its application in real-time identification. In this paper, the results of different algorithms are compared and the exactness of the networks, while first eigenfrequencies and then stiffness as input are chosen and then compared. The process of training a neural network begins with the generation of a data bank. In the case of this project, 189 series of eigenfrequencies and equivalent stiffness were given to the code. This range of data covers the stiffness changes of each component up to 50%, which ensures a sufficient scale of input for feasible applications of the method. The neural network generator uses 70% of these data as input to train and generate the network (training step). 15% of the data is used to evaluate the accuracy of the network (validation step). If the error of the network output was more than a specific value, the training step is repeated again and the weights are adjusted, until the accuracy of the network output is acceptable. The network is then saved and 15% of the rest of data is used to test the network (test step). This step has no effect on the generation of the network, but it gives only a feedback how accurate the network is, in case new values would be employed. Implemented algorithms are shown schematically in figure 7. The relation between the input and output could be determined by means of hidden neurons. In figure 7 only one hidden layer was depicted; however, depending on the algorithm, more layers could be used. The figure also shows how the weight parameters between the neurons are changed in a loop in order to get the least difference between the output and targets. The mean square error as defined in equation 3 is used for demonstrating the accuracy of the network in cases of stiffness and eigenfrequencies as output of the network. In this equation km^{ii}, kl^{ii} , kk^{ii} and ω^{ii}_i are calculated stiffness and eigenfrequencies in *ii* th series respectively. KM^{ii} , KL^{ii} , KK^{ii} and Ω^{ii}_i are the reference stiffness and eigenfrequency in the data bank. Different algorithms were tested and the results can be seen in figure 8. The first method "Bayesian Regularization Method" minimizes a linear combination of squared errors and weights. The second method "Conjugate Gradient with Powell/Beale Restarts" adjusts the weights in the steepest descending direction. The third method was the "Fletcher-Powell Conjugate Gradient". This method generates conjugate directions using only a one-dimensional search at each iteration. The fourth method, "Resilient Back-Propagation", takes into account only the sign of the partial derivative over all patterns (not the magnitude), and acts independently on each "weight". The fifth method is "Scaled Conjugate Gradient back-propagation (SCG)" and the last method (the sixth) was "Levenberg-Marquardt Backpropagation". In this method, validation vectors are used to stop the training early if the network performance on the validation vectors fails to improve or remains the same for a maximum number of failures in a row. Test vectors are used as a further check. One example of network performance by using "Levenberg-Marquardt Backpropogation" can be seen in table 3 and the definition of error in this case is shown in equation 4. Figure 8 shows that the best neural network algorithm for both cases of getting stiffness from eigenfrequencies and eigenfrequencies from stiffness is the "Levenberg-Marquardt Backpropogation". The low amount of error in table 3 and the short training time (almost 1 hour on a quad-core system) depicts that the neural network is appropriate for determining the corresponding values of stiffness and eigenfrequencies.

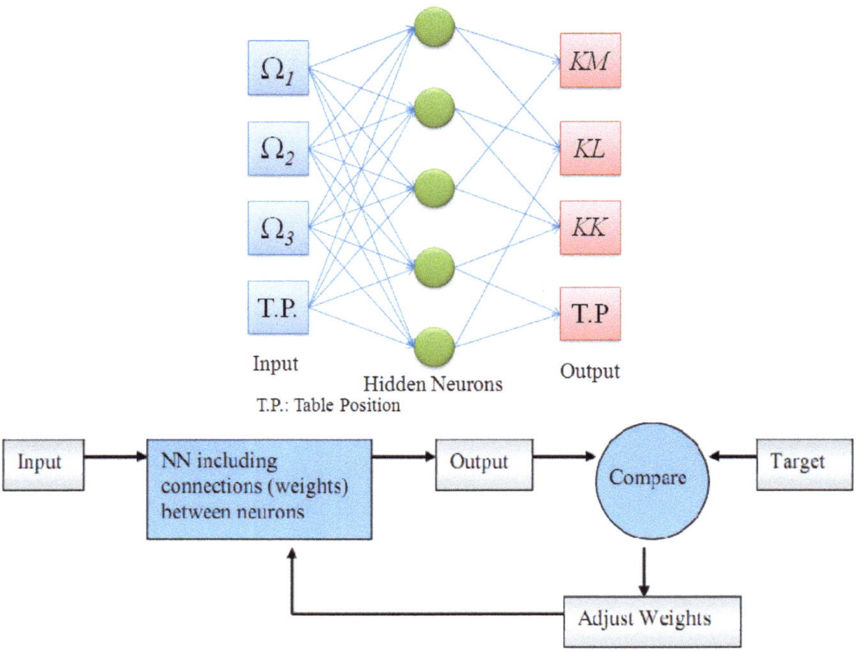

Fig. 7 Schematic algorithm of utilized neural network

$$\begin{aligned} &\text{Error } (\Omega \rightarrow K) = \frac{1}{n}\sum_{ii=1}^{n}\left[(\text{km}^{ii}\text{-KM}^{ii})^2 + (\text{kl}^{ii}\text{-KL}^{ii})^2 + (\text{kk}^{ii}\text{-KK}^{ii})^2\right], \; n=189 \\ &\text{Error } (K \rightarrow \Omega) = \frac{1}{n}\sum_{ii=1}^{n}\sum_{i=1}^{3}(\omega_i^{\text{ii}} - \Omega_i^{\text{ii}})^2, \; n=189 \end{aligned} \tag{3}$$

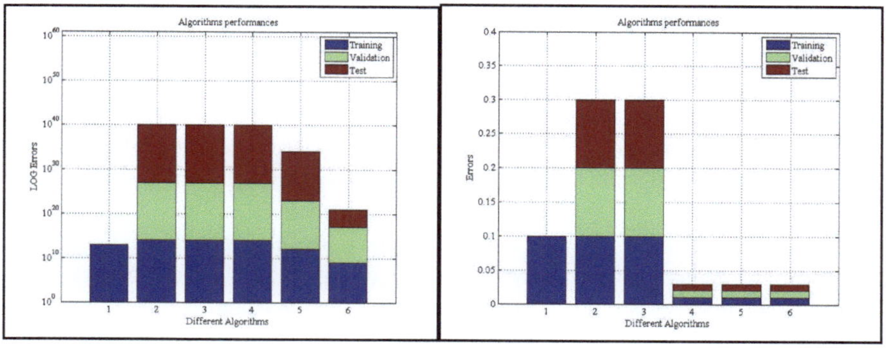

Ω →K (Log-Linear) K → Ω (Linear-Linear)

Fig. 8 Comparison of different neural network algorithms based on the error function defined in equation 3

Table 3 Comparison of Error defined in formula 4 based on "Levenberg-Marquardt Backpropogation" algorithm for finding the stiffness out of eigenfrequencies and vice versa

Output (Input: Ω_1, Ω_2, Ω_3, T.P.)	Reference Value	Network Result	Error %
KL (N/m)	270 e6	272.44 e6	0.91
KM (N/m)	550 e6	544.41 e6	1.02
KK (N/m)	0.170 e6	0. 172 e6	1.18
T.P. (%)	50	51.47	2.94
Output (Input: *KL*, *KM*, *KK*, T.P.)	Reference Value	Network Result	Error %
Ω_1 (Hz)	68.19	68.04	0.22
Ω_2 (Hz)	377.85	377.67	0.04
Ω_3 (Hz)	809.56	809.44	0.01
T.P. (%)	50.00	49.97	0.06

$$\text{Error} = \frac{\text{Network Result} - \text{Reference Value}}{\text{Reference Value}} \tag{4}$$

5 Conclusion

In this paper, the idea of simulation-based condition monitoring was presented. The procedure for anticipating the condition of a system based on series of previous tests on its numerical model was demonstrated. For certifying the purpose, a spindle drive system was chosen. A reduced model of a ball screw drive, which can precisely represent the first two eigenfrequencies of the ball screw drive, was introduced. A parametric 3D model of a spindle nut drive for determining the stiffness of a healthy spindle nut drive (reference stiffness) was also exhibited. As the prognosis step is based on the comparison of stiffness in different components, different identification and neural network methods were tested on the reduced model in order to determine the relationship between the eigenfrequencies and stiffness parameters of the system. The best algorithm of each method was determined and it was shown that these methods could be helpful in determining the stiffness of different spindle nut drive components based on the first two measured eigenfrequencies of the system. The implementation of these methods could be, hence, useful in performing condition monitoring of other systems. However, in case of the spindle nut drive, they could be improved by performing a frequency response analysis and using not only the poles but the whole frequency response for parameter identification of the system. Although the method was tested numerically, it is necessary to verify the results in practice as well, which could be the task of further studies.

Acknowledgments. The authors would like to thank the German Research Foundation (DFG) and the companies TRUMPF, CADFEM, ISG for financial support of the project within the Cluster of Excellence in Simulation Technology (EXC 310/1) at the University of Stuttgart.

References

1. Jantunen, E.: A summary of methods applied to tool condition monitoring in drilling. Journal of Machine Tools & Manufacture 42(2), 997–1010 (2002)
2. Zhang, L., Yan, R., Gao, R.X., Lee, K.: Design of a Real-time Spindle Health Monitoring and Diagnosis System Based on Open Systems Architecture. In: International Smart Machining Systems Conference, France (2007)
3. Schopp, M.: Sensorbasierte Zustandsdiagnose und –prognose von Kugelgewindetrieben. Ph.D. desertation, Institute for production technique (wbk), Karlsruher University (2009)
4. Neugebauer, R., Fischer, J., Praedicow, M.: Condition-based preventive maintenance of main spindles. German Academic Society for Production Engineering Journal (WGP) (September 2010), doi:10.1007/s11740-010-0272-z
5. Sin, M.L., Soong, W.L., Ertugrul, N.: Induction machine on-line condition monitoring and fault diagnosis - a survey. In: Power Engineering Conference (AUPEC 2003), pp. 1–6. CDROM, Christchurch (2003)
6. Yan, R., Gao, R., Li, Z., Lee, K.B.: Modal Parameter Identification from Output-only Measurement Data: Application to Operating Spindle Condition Monitoring. In: ICFDM 2008, Tianjin, China (September 2008)
7. Zhang, L., Yan, R., Gao, R.X., Lee, K.: Design of a Real-time Spidnle Health Monitoring and Diagnosis System Based on Open Systems Architecture. In: International Smart Machining Systems Conference (2007)
8. Saravanan, S., Yadava, G.S., Rao, P.V.: Condition monitoring studies on spindle bearing of a lathe. Journal of Advance Manufacturing Technology (2005), doi:10.1007/s00170-004-2449-0

9. Zhang, L., Yan, R., Gao, R.X., Lee, K.: Design of a Real-time Spindle Health Monitoring and Diagnosis System Based on Open Systems Architecture. In: International Smart Machining Systems Conference, France (2007)
10. Dadalau, A., Verl, A.: Bottom-Up Component Oriented FE-Modelling of Machine Tools. In: ACUM 2011, Stuttgart (2011)
11. Mottahedi, M., Dadalau, A., Röck, S., Verl, A.: Simulation Based Condition Monitoring of Roll Bearing. In: ACUM 2011, Stuttgart (2011)
12. Kamalzadeh, A., Erkorkmaz, K.: Compensation of Axial Vibrations in Ball Screw Drives. Journal of Manufacturing Technology 56(1), 373–378 (2007)
13. Varanasi, K.K., Nayfeh, S.A.: The Dynamics of Lead-Screw Drives: Low-Order Modeling and Experiments. Journal of Dynamic Systems, Measurement, and Control 126(2), 388–397 (2004), doi:10.1115/1.1771690
14. Yan, R., Gao, R.X., Zhang, L., Lee, K.B.: Modal Parameter Identification from Output-only Measurement Data: Application to Operating Spindle Condition Monitoring. In: International Conference on Frontiers of Design and Manufacturing, Tianjin, China (September 2008)
15. Frey, S., Dadalau, A., Verl, A.: Expedient Modeling of Ball Screw Feed Drives. Production Engineering 6(2), 205–211 (2011), doi:10.1007/s11740-012-0371-0
16. Silva, R.G.: Condition Monitoring of the Cutting Process Using a Self-organizing Spiking Neural Network Map. Journal of Intelligent Manufacturing 21(6) (2010), doi:10.1007/s10845-009-0258-x

On Designing a Unified Ontology for Holonic Manufacturing Networks

Giouvanni Désiré Jules, Mozafar Saadat, and Nan Li

School of Mechanical Engineering, University of Birmingham, United Kingdom
{gdj039,m.saadat}@bham.ac.uk, nxl968@adf.bham.ac.uk

Abstract. Small and medium enterprise (SME) manufacturers are generally better off being part of groups of integrated companies which collectively add value to an end-product. An SME is limited in two ways: by its resources and by its knowledge. In contrast, a well-created manufacturing network should have the necessary competencies in resources and knowledge it needs. While an individually owned ontology inherits the heterogeneous nature of the SMEs, the manufacturing network system integrator needs a universal knowledge base and ontology; an ontology that the SMEs would understand and 'willingly' contribute information to. This paper presents the manufacturing system ontology with a foundational framework from the Product Resource Order Staff Architecture (PROSA). With multi-agent system environment in mind, the ontology is designed for agent interpretability. The exchange and processing of production, production execution and process information need to be automated as far as possible. This paper intends to present a reusable and scalable ontology in Ontology Web Language (OWL). The paper highlights the concepts and slots that constitute the ontology and a knowledge base with a set of rules that allows selection of resources for the manufacturing of a product. The proposed ontology is finally appraised against a set of criteria and compared with a number of existing ontologies for manufacturing networks.

Keywords: Ontology, Knowledge Base, Expert System, Manufacturing System, Manufacturing Network, PROSA.

1 Introduction

Small and medium manufacturing enterprises share a unique set of characteristics that make them suitable for participation in networks. They have lean structures, oriented to a high-tech market segment, adaptable to the changes in the market segment and strive in subcontracting relations [1]. Due to their lean structure, SMEs have very few core competencies. Also being at the receiving end of outsourcing and subcontracting, they behave like independent network nodes. They also follow a set of decision making rules which are often limited to the scope of the SMEs' activities; rules in the form of relations, recommendations, directives, strategies and heuristics [2]. In contrast, a manufacturing network is designed to have all the competencies in resources and knowledge it needs, to fulfill a temporary market demand i.e. job shop production. There is a general agreement across the research field of industrial production, entrepreneurship and economics that SMEs add more value as part of a network than on their own [1-4]. SMEs within the network are held together by a sense of reliability, responsibility and commitment; in other words, trust binds a network. An empirical study was carried out to determine what makes manufacturing networks successful [5]. The study showed that, in order of importance, reliability, commitment on behalf of the network, capability and information technology are critical success factors for long term survival. A network of SMEs working together, for the first time, has to be closely monitored. During this incubation period, the network is coordinated by a system integrator. After many successful deliveries and when the reliability of the network converges towards maturity, trust is established and the intervention of the system integrator would become less critical. Trust is also a function of the length of the collaboration [6]. Referring back to the incubation stage of the network, in order to perform its role, the system integrator needs a centralized source of organized information consisting of process, capacity, performance of nodes (or SMEs), inter-node transport and a set of coordination rules [1]. The manufacturing network system integrator would benefit from a unified ontology that the SMEs would contribute information to. The ontology is one of two prerequisites for constructing an expert system. The second is the knowledge base which holds decision making rules.

Determination of the content of ontology has been the subject of much research. Manufacturing strategists take into account order information, product structures, routing data, resource information and production feedback data among others, to determine the manufacturing processes to be used. Subsequently, the manufacturing cost and leadtime of a product are derived from the manufacturing processes used [7]. In a study the need for data such as the routing, bill of materials, state of the resources, availability of resources, production schedule, priority of order and inferred permission is highlighted, when investigating the online simulation in a holonic manufacturing system [8]. In an evaluation of the reliability of network plans in cell manufacturing systems, various equations, from variables mean time between failure and mean flow time have been derived [9, 10]. An extensive case study was carried out to identify what attributes have had significant influence in the long-term success of manufacturing SMEs [11]. A strategy focusing on company orientation, price determination, production experience, product life cycle and quality control have been identified as the top five attributes for long-term survival. Moreover, it is suggested that pre-process, in-process and post-process inspection are common attributes of successful manufacturers [12].

Product resource order staff architecture (PROSA) implements the concept of autonomous co-operating agents to manufacturing systems. Agent is a computer science term and the term 'holon' is its counter-part in the physical world. Holon is something that is simultaneously a part of another whole and a self-contained whole to its subordinated parts [13]. PROSA provides three basic holons of type product, resource, order and an ad-hoc holon of type 'staff'. It is suggested that from these four types of holons only, a holonic manufacturing system (HMS) can be built. The result is a reconfigurable system, with a high degree of self-similarity, scalability and compatibility [14]. The development and application of the holonic concept in manufacturing has been widely reviewed [15]. However, the most complete holonic system so far, has been developed for the ADACOR project where a multi-agent system and a rule-based engine were utilized [16]. Based on literature review, it is understandable why manufacturing networks are likely to consist of SMEs. The paper, therefore, proposes the ontology for the domain of manufacturing network that implements the principles of PROSA. The scope of the ontology is limited to the type of information that the system integrator needs in order to carry out its function. Section 2 presents the research methodology, while section 3 explains the structure of the ontology in terms of concepts and slots. Section 4 highlights the rules used to select resources for the manufacture of a product. Section 5 explains how the multi-agent system, the ontology and the rule-based engines work together. In section 6 the proposed ontology is appraised against a set of criteria and compared with existing ontologies for manufacturing networks.

2 Research Methodology

There are several methodologies for building ontologies that have been developed over the past 15 years. This paper uses an adaptation of the Uschold and King's method [17].

The road map, shown in Figure 1, depicts the development phases of the ontology for the unification of manufacturing system's knowledge for manufacturing networks.

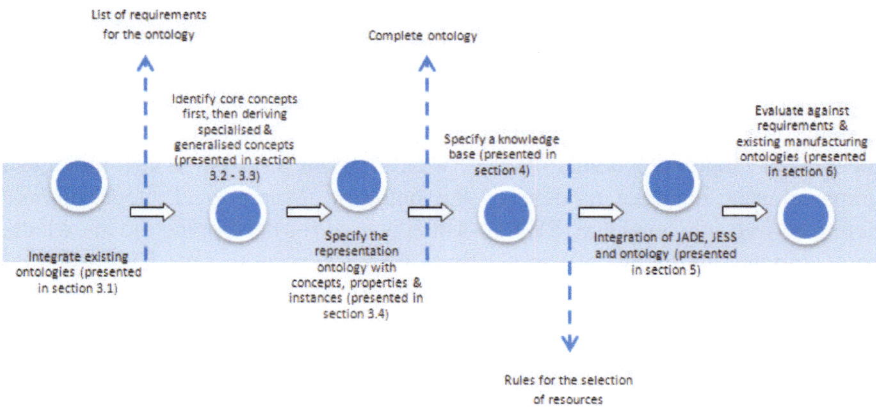

Fig. 1 Proposed methodology for ontology development

As the first step, the non-functional requirements of the ontology have been established by considering a number of fundamental points [18]. The requirements together with their explanations are given below:

S1 – Versatility of ontology to support the logistic, technical and control aspect of domain: This means that the ontology is designed to be reused by all agents involved in the system. An individual agent uses the part of the ontology that is important to its functions and ignores the irrelevant parts.

S2 – Ease of defining rules from ontology, for the knowledge-base: This means that the ontology is designed using a consistent methodology, the naming convention used is as close to the terms used in the domain of interest and the ontology is consistent with the right constraints in place. These allow the rules to be generated intuitively and to model accurately the decision-making process.

S3 – Appropriateness of ontology as communication tool for interacting agents: This means that the ontology can be used by agents to transmit objects encoded in XML or string format over a distributed network and the objects would be recognized by all agents using the same ontology.

S4 – Industrial accessibility of the data the ontology is designed to store: This means that the ontology has been designed for data that user can transfer over the network that is proprietary to the network.

S5 – Relevance of data for investigating reliability of manufacturing networks: This means that the ontology has to recognize quality, cost and delivery (QCD) data that are used to investigate reliability and process capabilities.

S6 – Accuracy of ontology to model the structures of data, used in the coordination of manufacturing networks for job shop production: This means that the ontology should model the data circulating in manufacturing networks and not the data used on manufacturers' shopfloors.

S7 – Ease of extending scope of the ontology by integrating specialized ontologies: This means that if the ontology needs to be specialized, for instance on the technical aspect of the product design, the ontology should have an extension point to integrate the specialized ontology.

To achieve these requirements, the tools available to the research community are investigated i.e. development platform, evaluation tools, extensions with inference engines and ontology generators for multi-agent systems. A number of available ontology tools that are in use today include Ontolingua Server, WebOnto, Protégé, WebODE, OntoEdit, OntoStudio, KAON, Observer, MnM, COHSE and UBOT AeroDAML. In this research the Protégé platform was chosen.

Protégé platform is an ontology-editor and a knowledge-base framework system. Protégé has relevant advantages over the other platforms. It supports two methods of modeling a domain, one of which is Protégé-OWL. It also provides a graphical user interface to develop the ontology. Moreover, Protégé supports Semantic web rule language (SWRL) which is used for developing the knowledge base. Protégé also allows the translation of SWRL rules to Java Expert System Shell (JESS) rules. JESS and SWRL will be explored in more details in Section 4.

3 Development of Ontology through an Industrial Case Study

The construction of the ontology is divided into four sections. The knowledge captured in the ontology is heavily based on literature on small and medium manufacturing enterprises, manufacturing systems in general, and holonic manufacturing systems in particular.

The appropriateness of the information to the system integrator of a manufacturing network is judged on the basis of a case study that was carried with a company acting as a system integrator to a manufacturing network. Gruppo Fabricazione Meccanica (GFM) Srl is a private company located in the province of Bergamo, Italy. GFM is a system integrator for a network of more than 30 manufacturing companies and over 500 specialized suppliers which collectively provide hundreds of processing capabilities. The company manages the production of parts and assembly equipment for gas turbines, steam turbines and electrical generators [19]. The company subcontracts orders to manufacturers based on their capability. GFM has control over the selection of manufacturers and the logistics surrounding the product i.e. collection and delivery of raw material, semi-finished and finished products, and would regularly monitor the progress of its orders to ensure that the logistics is not disturbed. The company also performs the quality inspection on the semi-finished products prior to their delivery to the next manufacturer, or on the finished product prior to its final delivery to the customer. However, GFM has no control over the manufacture of the products, which are independently managed by the manufacturers unless the manufacturers are under-performing. Thus, the system integrator acts as a coordinator and in order to perform its role, it would require the right information, which is modeled by the ontology proposed in the following sections. The ontology also captures the required information to evaluate the probability of the logistics failing during the makespan of the product i.e. the reliability of the manufacturing network.

3.1 Integrating Existing Ontologies

One of the key benefits of ontology is the opportunity to merge it with existing ontologies. Using existing ontologies not only saves time and effort but gives structure that is required for compatibility with particular applications. For instance, in the case of this paper, the ontology imports 'OWLSimpleJADEAbstractOntology.owl', 'swrla.owl' and 'sqwrl.owl' ontologies. The former allows our ontology to be compiled using Bean Generator tool which generates a FIPA compliant java-based ontology for the multi-agent platform JADE. The 'swrla' and 'sqwrl' ontologies allow the use of semantic web rule language (SWRL) to create the knowledge base for our ontology.

3.2 Identification of the Abstract Concepts

The proposed ontology is built using the structure of the existing ontologies. In line with the methodology given in Figure 1, the sub-classes of the class *'Beangenerator:Concept'* are first identified. Using a middle-out strategy [17], the abstract concepts are initially identified and shown in Figure 2. The purpose of the concept is described as follows. *Products* capture the production details such as bill of material, network plan and list of capable resources. *Resources* represent capability and historical records of operation, performance, order and product. *Orders* capture the logistical aspect of production such as fixed due time, quantity and contracted resources. *Operation* captures the type of operation and technical description of operation. *Beangenerator:AID* stands for Agent ID and gives agent its name and location e-addresses in the network. *Beangenerator:AgentAction* captures the type of actions performed by agents to change its internal and environment states. *ValuePartition* is the additional information used to refine the concepts and to indicate the state of the concepts.

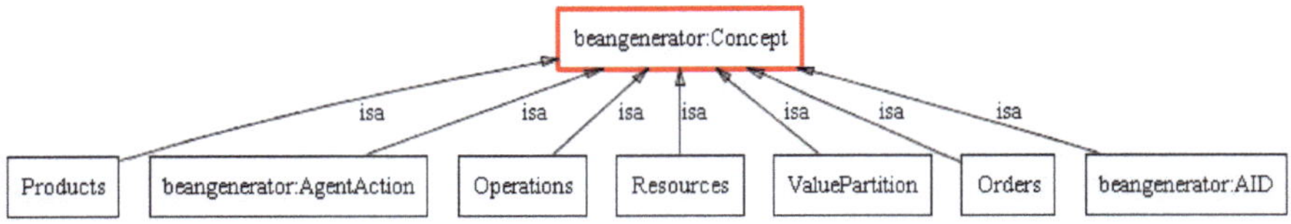

Fig. 2 Main concepts of the ontology

3.3 Identification of the Specialized Concepts

The *'Beangenerator:Concept'* are specialized into more specific concepts. The ontology must maintain a good balance between its usability and reusability. The scope of the ontology is also limited to the information that the system integrator needs, to project-manage manufacturing networks. For example, it may be tempting to specialize the concept 'Resources' into 'manufacturer', 'inspector', 'haulier', 'warehouse' and 'packager'. However, apart from their difference in the services they provide, they all have the same property types such as 'name', 'product history', 'order history', 'operation history' and 'performance history'. Figures 3 – 8 show the proposed taxonomy for a manufacturing network.

Assembly, Subassembly, Component and *RawMaterial* are specializations of the *Products* concept as shown in Figure 3a. *BoughtStockOrder* and *MakeToOrderOrder* are specializations of *Orders* as shown in Figure 3b. *BoughtStockOrder* contains the order name, quantity, arrival time and due time. *MakeToOrderOrder* contains the order name, quantity, price, due time and a checklist for delivery on time, quality and external assistance required.

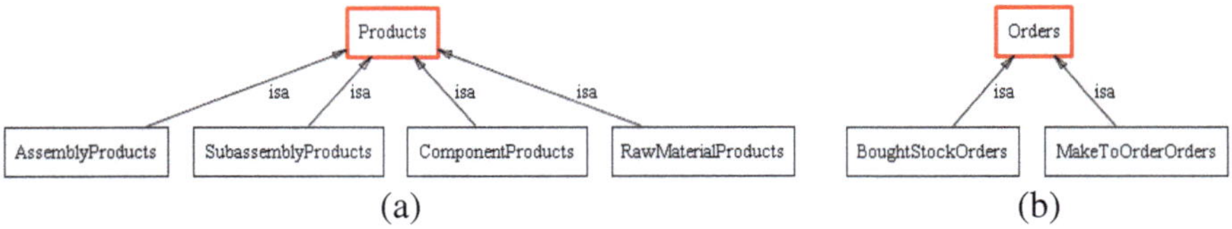

Fig. 3 Products and orders concepts

ProductHolonAID, ResourceHolonAID, OrderHolonAID and *StaffHolonAID* are the specializations of *Beangenerator:AID* as illustrated in Figure 4a. *ProductHolonAID* consists of the product managed and the actions for managing the product. *OrderHolonAID* consists of the order managed and the actions for managing the order. *ResourceHolonAID* consists of the resource managed and the actions for managing the resource. It can consist of other *ResourceHolonAID*. *StaffHolonAID* consists of adhoc holons for the scheduling and sequencing of order. *Beangenerator:AgentAction* specializes into *OrderHolonAction, ResourceHolonAction and ProductHolonAction* as Figure 4b shows.

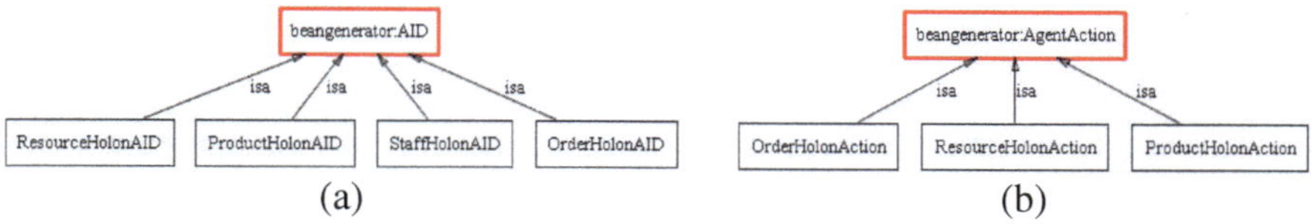

Fig. 4 AID and AgentAction concepts

ProductHolonAction further specializes into *SetupNetworkPlan* and *RepairNetworkPlan* as shown in Figure 5a. *SetupNetworksPlan* creates many alternative network plans for a product and finds potential resources to form the networks. A network plan is equivalent to a process plan. *RepairNetworkPlan* finds an alternative network plan to a faulty network. *StartWork, StopWork* and *UnderRepair* are specializations of *ResourceHolonAction* as shown in Figure 5b. *StartWork* indicates the the resource has started processing an order. *StopWork* indicates that the resource is idle. *UnderRepair* indicates that the resource is affected by a breakdown. *OrderHolonAction* is specialized into *AllocateOrder, HandleDeadlock, MonitorProgress, PenaliseResource, RewardResource* and *UnallocateOrder* as illustrated by Figure 5c. *AllocateOrder* contracts a resource with a product via an order agreement. *HandleDeadlock* resolves the conflicts between order holons needing the same resource. *MonitorProgress* monitors the tardiness, progress and status of an order. *PenalizeResource* penalizes the resource for breaching order agreement. *RewardResource* rewards the resource for a well delivered order. *UnallocateOrder* voids the contract with a resource.

Beangenerator:Predicate specializes into *Deadlock, FaultyNetworkPlan, OrderAllocation, OrderPriority, OrderProgress, OrderStatus, Performance, NetworkPlan, ResourceStatus, ScheduledStartTime* and *ScheduledFinishTime* as shown by Figure 6. *Deadlock indicates* the conflicting orders and the target resource. *FaultyNetworkPlan* indicates the faulty resources affecting the network. *OrderAllocation* shows the list of potential resources for the order. *OrderPriority* indicates the priority assigned to order. *OrderProgress* indicates the percentage of order completion. *OrderStatus* indicates that an order has started, finished, has been accepted or has been rejected. *Performance* indicates the resource-operation reliability, usage and the mean time between failures. *NetworkPlan* shows the sequence of operations and the list of resources having at least one operation required by a product. *ResourceStatus* indicates the work state of the resource. *ScheduledStartTime* shows the start time of the order and *ScheduledFinishTime* shows the finish time of the order.

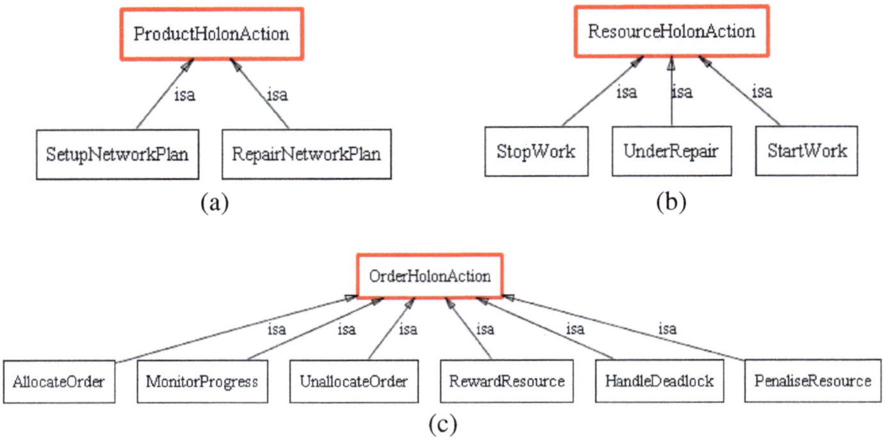

Fig. 5 ProductHolonAction, ResourceHolonAction and OrderHolonAction concepts

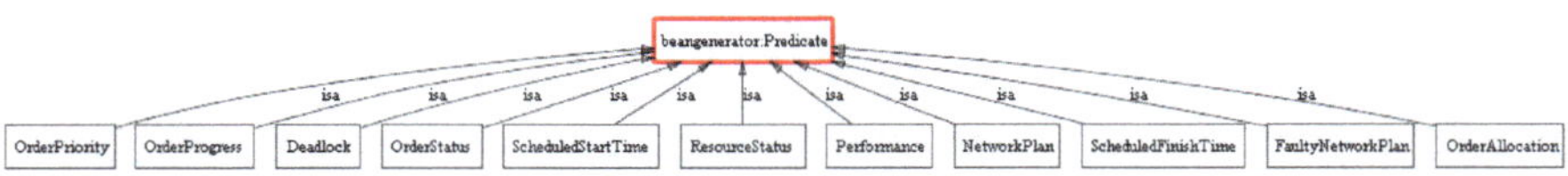

Fig. 6 Predicate concept

3.4 *Identification of Slots of Concepts*

A slot is an attribute which defines the characteristics of a concept. The property of a slot is called a facet. A facet represents the cardinality of a slot, the type of a slot and default values [20]. Table 1 presents all the slots that define the concepts described in section two of this paper. In the context of a multi-agent system, the slots with the dynamic data will be monitored at regular intervals by the agents. Slots having static data will be monitored by the agents only when a static data has been modified, which would be a rare occurrence.

Table 1 All the slots that define the concepts of the ontology

Slot names			
Slots that contain static data			
canBeContractedWith	hasOrderHistory	onOperation	hasDueTime
hasActions	hasPerformanceHistory	requiresOperations	hasName
hasContractWith	hasProduct	hasActionID	hasNotRequiredExternalSupervision
hasInputComponent	hasProductHistory	hasArrivalTime	hasPassedQuality
hasInputRawMaterial	hasProductSpecification	hasBeenDelivered	hasPriceRatio
hasInputSubassembly	hasResource	hasBestLeadtime	hasQuantity
hasOperationHistory	hasSubordinates	hasBestPrice	
hasOrder	isType	hasDescription	
Slots that belong to subclasses of beangenerator:Predicate are slots that hold dynamic data			
approvedResourceHolons	hasWorkStatus	hasOrderProgress	hasStateDuration
conflictingOrderHolons	resourceHolonsForNetworkPlan	hasPriority	hasUsageFrequency
faultyResourceHolons	hasFinishTime	hasReliabilityScore	hasTardines
hasOrderStatus	hasMeanTimeBetweenFailure	hasStartTime	

4 Development of Knowledge Base

The ontology enables the representation of concepts and their slots. However in order to develop an expert system, the ability for decision making needs to be implemented in the form of a knowledge base. Semantic web rule language (SWRL) is an expressive OWL-based rule language [21] that is used to define the relationship between individual concepts. An inference engine such as java expert system shell (JESS) [22] interprets the relationship and carries out the decisions made.

Below are examples of two sets of rules that enable the selection of suitable service providers and manufacturers for products *p*.

(Rule 1) ResourceHolonAID (?rh)∧hasProductHistory(?rh, ? p)∧hasProduct(?ph, ?p)∧ProductHolonAID(?ph)
→canBeContractedWith(?ph, ?rh)

(Rule 2a) ProductHolonAID(?ph)∧ProductHolonAction(?networkplan)→hasActions(?ph, ?networkplan)

(Rule 2b) ProductHolonAID(?ph)∧hasProduct(?ph, ?p)∧hasActions(?ph, ?action)∧requiresOperations(?p, ?operation)∧
swrlx:makeOWLThing(?networkplan, ?ph)→NetworkPlan(?networkplan)∧problemSolvingAction(?networkplan, ?action)

(Rule 2c) ProductHolonAID(?ph)∧hasProduct(?ph, ?p)∧requiresOperations(?p, ?operation)∧hasOperationHistory(?rh, ?operation)∧
ResourceHolonAID(?rh)∧hasnetworkplan(?ph, ?networkplan)∧NetworkPlan(?networkplan)
→resourceHolonsForNetworkPlan(?networkplan, ?rh)

(Rule 2d) NetworkPlan(?networkplan)∧swrlx:makeOWLThing(?rh, ?networkplan)→ResourceHolonAID(?rh)∧
hasmegaresourceholon(?networkplan, ?rh)

(Rule 2e) NetworkPlan(?networkplan)∧hasmegaresourceholon(?networkplan, ?megaresourceholon)∧
ResourceHolonAID(?megaresourceholon)∧resourceHolonsForProcessPlan(?networkplan, ?rh)∧hasnetworkplan(?p, ?networkplan)∧
Products(?p)→hasSubordinates(?megaresourceholon, ?rh)∧hasProductHistory(?rh, ?p)

Rule 1 establishes a relationship between the product history of the resource holons *rh* and product *p* of the product holons *ph*. In other words, the rule matches product holons with resource holons which have worked on the same products before.

Rules 2a-2e are for the scenario where a product has never been produced before and has no process plan. The rules are followed sequentially in response to the 'networkplan' action of the product holons. The inference engine queries for all the resource holons that have the operation capability to satisfy all the operation requirements of product holons. The rule enables the rule-based engine to form networks of resource holons that collectively provide the required operation capability. It must be noted that resource holons can consist of other resource holons by definition. The *'hasProductHistory'* slot of the new resource holons is then updated. Finally by re-using rule 1, these resource holons potentially can be directly contracted with the product holon, when the next order is placed.

5 Integration of JADE, JESS and Ontology

JESS is an instrument that can be used to add artificial intelligence to multi-agent systems that was built using JADE. JADE is a JAVA based software agent middleware and it provides an environment and the services that the agents need, to work. In contrast to other development framework such as JACK®, JADE does not provide the tools for developing intelligence in its agents while JESS can be used to implement a rule-based type of intelligence in individual agents. The ontology is essential to enable accurate and effective communication in the multi-agent system and for JESS to work.

JADE provides the communication method which enables decentralized agents to transmit data objects. The ontology plays a vital role in communication. The sender agents transmit the data objects in an XML-based language and the receiver agents convert the XML-based messages back into data objects. The XML-schema that is used to convert an object into an XML-based message and vice versa is stored in the ontology. This communication method removes the need for the serialization of data objects resulting in a faster transmission performance. Moreover, XML-based messages have a low memory utilization footprint. Also, the communication method is effective even when the agents are distributed on devices with different operating systems as long as the device has a Java Virtual Machine (JVM).

Once the data objects are received, the agents need to use the data and make decisions. This intelligence can be implemented in many ways but JESS provides a slightly faster and more insightful method as shown previously in section 4. The advantage of using JADE, JESS and the ontology together is that JESS can be configured to receive, process and transmit the XML-based messages without conversion. The XML-based messages are only converted into objects to take user inputs and to display information to the user.

The multi-agent system is being developed to assist the users during the coordination of a manufacturing network. The ontology models the type of data that are important for the coordination of the network of manufacturing shop floors. JESS is used to model the decision making process taking place during the formation and operation of a manufacturing network.

6 Evaluation against Requirements

The proposed ontology has been compared to a number of existing ontologies in the literature with respect to the requirements defined in section two of the paper, as shown in Table 2. S1 to S7 represent the non-functional requirements of the required ontology.

Ontology 1 has been developed to represent the manufacturing resources of a shop floor producing electronic connectors. Here, the ontology does not provide important information such the schedule of raw material, and finished product delivery. It also does not show the bill of material. Moreover, no history of resource performance is available. Ontology 2 is well designed but the logistics, technical and control information is intermixed. A decoupled ontology is preferred to facilitate its use by heterogeneous agents and, and also for maintenance. Moreover, the ontology has no history for order tardiness, resource breakdown, quality failure, etc. Ontology 3 is clearly decoupled into customer, product, manufacturer, transport. The top level ontology is very reusable, but it does not contain history of performance. Also, the top level ontology is very basic while the domain level is too subjective to be reusable for the domain of job shop production. Ontology 4 is well designed and accurately describes a manufacturing plant. However, the ontology represents the domain of mass production. Also the naming convention used for the slots of the concepts, is not appropriate. See Table 1 for examples of the correct naming convention. Ontology 5 is good but acts as a bridge between those with different syntax. Thus it is not designed to contain relevant data for the coordination of a manufacturing network. MASON is the most comprehensive ontology for manufacturing in literature. It is also freely available online in OWL format. The downside here is that it is too specialized for shop floor applications. The system integrator cannot use this ontology to coordinate a manufacturing network. Moreover, the ontology demands information that the system integrator does not have access to since much of the information is owned by the manufacturers.

Literature review reveals that the availability of reusable ontologies in the field of manufacturing is fairly limited, but this is likely to improve significantly due to increasing availability of development tools. Future work will involve the comprehensive development of the knowledge base using artificial intelligence. This will be achieved through a further case study with GFM Srl. The proposed ontology and the emerging knowledge base will be used in tandem with a multi-agent system. This will facilitate the investigation of the effects that rules, relations, recommendations, directives, strategies and heuristics have on the reliability of manufacturing networks.

Table 2 Evaluation of the proposed ontology and the existing ontologies with respect to requirements

Manufacturing ontologies	S1	S2	S3	S4	S5	S6	S7	Comments
Proposed ontology	+	+	+	+	+	+	+	Reducible to essentially order, product, resource domain
Ontology 1 [23]	-	+	+	+	-	+	+	Ontology is limited to resource domain
Ontology 2 [24]	-	+	+	+	-	+	+	Ontology shows no control aspect of resource domain
Ontology 3 [25]	+	+	+	+	-	-	+	Ontology is functionally sound but concepts used are inaccurate
Ontology 4 [26]	+	-	-	+	-	+	-	Naming convention, for relation between concepts, is complex
Ontology 5 [27]	+	+	+	+	-	-	+	Ontology acting as a mediator between dissimilar ontologies
MASON [28]	+	+	+	-	-	-	+	Ontology is very specialized for in-house production

+ Satisfies the requirements **- Does not satisfy the requirements**

7 Conclusion

In this paper, a type of knowledge that a unified ontology should capture has been developed. The proposed ontology was designed based on the principles of product resource order staff architecture (PROSA). Then, a knowledge base was presented with examples of rules for the selection of resources for manufacturing a product. The proposed ontology was evaluated against a set of non-functional requirements and compared with existing manufacturing ontologies. The proposed ontology has met all the requirements that are relevant to the scope of its future use. Its strong foundation from PROSA allows scalability without compromising compatibility, whilst the system integrator can use it to request information from its manufacturers and vice versa. Furthermore, the ontology is uniquely designed for network coordination and its reliability evaluation during the makespan of products.

Acknowledgments. The authors wish to thank GFM Srl (Italy) for their kind contribution in providing the case study for this research.

References

1. Mezgár, I., Kovács, G.L., Paganelli, P.: Co-operative production planning for small- and medium-sized enterprises. International Journal of Production Economics 64(1-3), 37–48 (2000)
2. Durkin, J.: Expert Systems - Design and Development. Prentice Hall International Inc., NJ (1994)

3. Galbraith, C.S., Rodriguez, C.L., DeNoble, A.F.: SME Competitive Strategy and Location Behavior: An Exploratory Study of High-Technology Manufacturing. Journal of Small Business Management 46(2), 183–202 (2008)
4. Noori, H., Lee, W.B.: Dispersed network manufacturing: adapting SMEs to compete on the global scale. Journal of Manufacturing Technology Management 17(8), 1022–1041 (2006)
5. Sherer, S.A.: Critical success factors for manufacturing networks as perceived by network coordinators. Journal of Small Business Management 41(4), 325–345 (2003)
6. Sharyn Smith, S.H.: Role of trust in SME business network relationships. In: 1997 USASBE/ICSB World Conference, San Francisco, California (1997)
7. Halevi, G., Wang, K.: Knowledge based manufacturing system (KBMS). Journal of Intelligent Manufacturing 18(4), 467–474 (2007)
8. Cardin, O., Castagna, P.: Using online simulation in Holonic manufacturing systems. Engineering Applications of Artificial Intelligence 22(7), 1025–1033 (2009)
9. Das, K., Lashkari, R.S., Sengupta, S.: Reliability consideration in the design and analysis of cellular manufacturing systems. International Journal of Production Economics 105(1), 243–262 (2007)
10. Seifoddini, H., Djassemi, M.: The effect of reliability consideration on the application of quality index. Computers & Industrial Engineering 40(1-2), 65–77 (2001)
11. Kim, K.S., Knotts, T.L., Jones, S.C.: Characterizing viability of small manufacturing enterprises (SME) in the market. Expert Systems with Applications 34(1), 128–134 (2008)
12. Abdul-Aziz, Z., Chan, J.F.L., Metcalfe, A.V.: Quality practices in the manufacturing industry in the UK and Malaysia. Total Quality Management 11(8), 1053–1064 (2000)
13. Koestler, A.: The act of creation. Picador, London (1964)
14. Van Brussel, H., et al.: Reference architecture for holonic manufacturing systems: PROSA. Computers in Industry 37(3), 255–274 (1998)
15. Babiceanu, R., Chen, F.: Development and Applications of Holonic Manufacturing Systems: A Survey. Journal of Intelligent Manufacturing 17(1), 111–131 (2006)
16. Paulo, L.: Agent-based distributed manufacturing control: A state-of-the-art survey. Engineering Applications of Artificial Intelligence 22(7), 979–991 (2009)
17. Uschold, M., et al.: The Enterprise Ontology. Knowl. Eng. Rev. 13(1), 31–89 (1998)
18. Schalkoff, R.J.: Intelligent System: Principles, paradigms and pragmatics. Jones and Bartlett Publishers, Boston (2011)
19. Jules, G., Saadat, M., Owliya, M.: A holonic systems approach to the formation of manufacturing networks. In: 2010 IEEE 9th International Conference on Cybernetic Intelligent Systems (CIS), Reading, UK (2010)
20. Asuncion Gomez-Perez, M.F.-L., Corcho, O.: Ontological Engineering with examples form the areas of Knowledge Management, e-commerce and the Semantic Web. Springer, London (2004)
21. Ian Horrocks, P.F.P.-S., Boley, H., Tabet, S., Grosof, B., Dean, M.: SWRL: A Semantic Web Rule Language combining OWL and RuleML (2004)
22. Friedman-Hill, E.J.: Jess, The java Expert System Shell (1998)
23. Lin, L.F., et al.: Developing manufacturing ontologies for knowledge reuse in distributed manufacturing environment. International Journal of Production Research 49(2), 343–359 (2011)

24. Jiang, Y., Peng, G., Liu, W.: Research on ontology-based integration of product knowledge for collaborative manufacturing. The International Journal of Advanced Manufacturing Technology 49(9), 1209–1221 (2010)
25. Yan, J., et al.: Ontology of collaborative manufacturing: Alignment of service-oriented framework with service-dominant logic. Expert Systems with Applications 37(3), 2222–2231 (2010)
26. Giret, A., Botti, V.: Engineering Holonic Manufacturing Systems. Computers in Industry 60(6), 428–440 (2009)
27. Lin, H.K., Harding, J.A.: A manufacturing system engineering ontology model on the semantic web for inter-enterprise collaboration. Computers in Industry 58(5), 428–437 (2007)
28. Lemaignan, S., et al.: MASON: A Proposal for An Ontology of Manufacturing Domain. In: IEEE Workshop, DIS 2006, pp. 195–200 (2006)

Application Specific Process Development for MEMS Design and Fabrication

Rainer Brück[1] and Thilo Schmidt[2]

[1] Universität Siegen, Lehrstuhl Mikrosystementwurf, 57068 Siegen, Germany
[2] ELMOS Semiconductor AG, Heinrich-Hertz-Straße 1, 44227 Dortmund, Germany
rainer.brueck@uni-siegen.de, thilo.schmidt@elmos.com

Abstract. With MEMS (Micro electro-mechanical systems) entering fast moving consumer markets, the need for a more efficient design approach becomes apparent. One of the biggest challenges in this context is that virtually every MEMS-product requires its own specifically designed and optimized manufacturing technology. The only currently feasible solution to the problems arising from this so-called "MEMS-law" seems to be the extensive modularization and reuse of existing manufacturing technologies.

PDES (Process Development Execution Systems) provide a framework to handle MEMS manufacturing technologies. In this article a new visual approach to process modeling built on top of PDES along with a simulation interface that allows setting up and performing virtual experiments and optimization is presented. The new approach supports the device engineer in selecting an appropriate manufacturing technology based on a set of device-cross-sections. For this purpose dedicated software tools have been developed that are able to analyze the cross-sections and map the analysis results to specific technologies. The results are used to synthesize abstract process-templates that form the basis for the development of new application specific fabrication processes. The new approach is particularly suited for fabless MEMS companies that are in need to develop application specific manufacturing processes.

Keywords: Micro Electro Mechanical Systems (MEMS), MEMS Design, microfabrication, process synthesis, knowledge acquisition and management, trading zone.

1 Introduction

In the last few years the focus of micro and nano technology (MNT) has moved from the automotive market to the consumer market [1]. Accelerometers, MEMS microphones, micro mirrors and many other MEMS products provide the core functionality for many innovative consumer applications.

This recent development goes along with severe consequences on the product development process. New efficient product development strategies are necessary to cope with the requirements of new application areas, shorter development cycles and stronger competition. In the field of micro electronics that has seen a similar development in the past, this was accompanied by the emergence of new business models like the fabless design house or the pure-play foundry. There has been a clear trend towards cooperative and flexible product development strategies. The recently increasing importance of MEMS products calls for a similar approach. Fig. 1 gives an overview of some business models that have emerged along the MNT value chain. According to [3] this development has been accelerated by the global economic crisis of the last years.

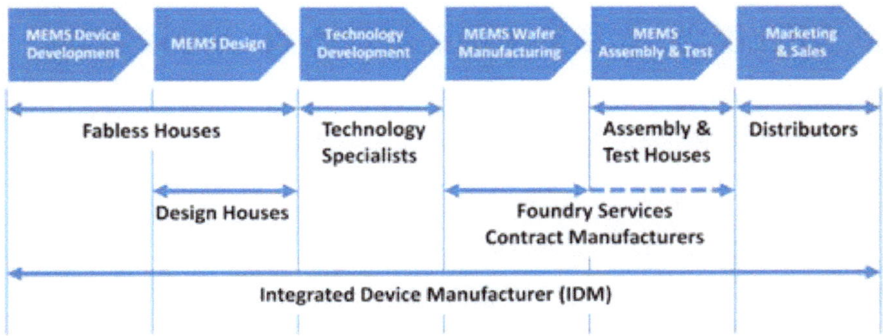

Fig. 1 Current business models along the MNT value chain

A technical challenge distinguishing MNT from micro electronics is the strong interdependency of system design and fabrication technology. For the MNT that goes along with a missing common technology platform like CMOS for microelectronics [2]. This important finding has become widely known as the so-called MEMS-Law ("One Product, one Process") [3]. It requires a much stronger cooperation of system design and fabrication technology than it is known from micro electronics. In this field design and fabrication have been made nearly completely independent by sophisticated abstraction mechanisms. Fig 2 illustrates this important difference between MEMS and micro electronics: In micro electronics the knowledge transfer between the system design phases and the fabrication technology is accomplished by employing design rules as a simple unidirectional abstraction mechanism. The constraints of the fabrication technology are expressed as geometric rules for mask layouts. This is only possible because the underlying fabrication process is relatively static. For the MEMS flow this knowledge transfer is represented by a more complex task called "technology management" that operates bidirectionally between fabrication and design and that influences not only the mask layout but all stages of product design, even the very early abstract ones. This article is on technology management and how it can be supported by dedicated tools.

The technology management task can be characterized by two critical aspects that are concerned with the interface between the design and fabrication phases of MEMS development – on the one hand a comprehensive TCAD simulation

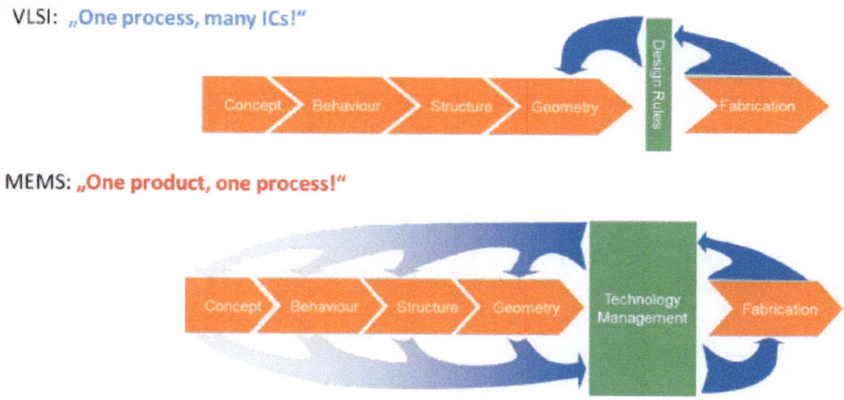

Fig. 2 Dependencies of design and fabrication in micro electronics and MEMS

interface and on the other hand an effective procedure to synthesize process sequences from appropriate representations of the product under development.

This part of technology management brings forward a cooperative product development strategy where design and fabrication are performed by different partners. Therefore it is an important step towards establishing a fabless-foundry business model in the MEMS area. The software supported process synthesis procedure is in the focus of this article: It is part of the research performed by the authors in cooperation with various companies and research institutes in this area [4], [5]: The goal of the research was to assist a cooperative distributed product development strategy tailored towards the needs of MNT industry and supporting this strategy with design automation tools.

Section 2 of this article gives a short introduction into how a fabless-foundry design strategy for MEMS products might be organized. Section 3 gives a comprehensive introduction into the new process selection and synthesis procedure and tools. The other direction of the knowledge flow is supported by a TCAD simulation interface. However, from a scientific point-of-view this is the less innovative part of the procedure and it will hence not be presented in this article. For a detailed discussion on this aspect see e.g. [6]. Section 4 finally shows the potential for further research work in this field.

2 The Fabless-Foundry Model in MNT

Business models in the MNT industry will more and more make use of product development cooperations between various companies along the value chain. Each of these companies provides specific knowledge and design services to the product development. In particular, a model where one development partner concentrates on the system design (fabless design house) whereas the other one acts as a technology provider focusing on MEMS production (pure-play foundry) turns out to be very promising. Three types of cooperation seem feasible for such a kind of business models:

1. The technology partner provides several predefined fabrication processes. The development partner has to adapt its system design to one of these processes. This is the approach that we typically see in the micro electronics industry.
2. The technology partner offers various generic fabrication technologies. In close cooperation with the development partner both system design and fabrication processes are adapted to each other.
3. The technology partner designs a specific fabrication process suited to the requirements given by the development partner and specifically dedicated towards the particular product under design.

The first approach is by far the fastest and least expensive one. However, because of the MEMS-Law it is only applicable in rare cases of products with moderate design requirements. The third approach is the most flexible one; however, it is the one that is also the most expensive with regard to time and cost. Particularly in the consumer area it is therefor probably not economically applicable. The second approach promises a viable compromise between flexibility and cost. It is hence the most probable scenario for cooperative MNT product development projects. Because of the strong interdependency between fabrication technology and system design the development partner must select a specific technology partner at a fairly early stage of the design project. To accomplish a decent technology selection, either the development partner needs detailed knowledge about all possible fabrication processes of all possible technology partners, or he will have to give detailed information about the product idea to all possible fabrication partners. Despite the complication and effort of interpreting and analyzing the possible partners` specifications this also might cause severe IP issues on both sides. Neither system designers nor technology providers are interested in disclosing details of their future products respectively fabrication processes to possible competitors.

To efficiently support a fabless-foundry business model in the MNT field it would hence be necessary to

1. have a pre-selection of potentially compatible fabrication processes for a given product,
2. give away information about fabrication processes that is required to perform product design in a selective and abstract manner (e. g. as customer-specific process design kits (PDK)).

These two steps will have to be accomplished without disclosing more knowledge than absolutely necessary. Making use of PDES is one way to support this scenario. However, in PDES, the necessary interfacing procedures are not automated as a start and hence require considerable time and knowledge. Process synthesis as presented in section 3 is one promising approach to at least achieve a partial automation of this interface between design and fabrication.

3 The Process Selection and Synthesis Procedure

3.1 The Cross-Section Drawing as a Trading-Zone

A major issue when trying to establish a link between system design and fabrication technology arises from the fact that this is an inherently interdisciplinary task.

Knowledge from various engineering disciplines is required including electrical engineering and mechanical engineering (for designing the desired functionality) and physics and chemistry (for adapting fabrication technologies) and computer science (for providing the appropriate design tools).

Fig. 3 shows a representation of the knowledge domains involved in MEMS design. It has been derived from the well-known DIKW model (Data – Information – Knowledge – Wisdom) introduced in [7] and [8]. The model shows 3 separate knowledge pyramids that are connected on the wisdom level. It indicates that a substantial understanding of all three partially disjoint domains is a prerequisite to a successful MEMS design project. Appropriate communication mechanisms are required to close the gaps between the lower – unconnected parts – of the knowledge pyramids on the D, I and K levels.

A useful concept for bridging these gaps is the notion of "trading zones" that has been introduced by Galison [9]. According to Galison the concept of a trading zone can be described as follows

"Two groups can agree on rules of exchange even if they ascribe utterly different significance to the objects being exchanged; they may even disagree on the meaning of the exchange process itself. Nonetheless, the trading partners can hammer out a local coordination, despite vast global differences. In an even more sophisticated way, cultures in interaction frequently establish contact languages, systems of discourse that can vary from the most function-specific jargons, through semi-specific pidgins, to full-edged creoles rich enough to support activities as complex as poetry and metalinguistic reflection" (cited from [9]).

A trading zone, in other words, is a communication mechanism, where the two communicating partners need not necessarily agree on a common understanding. Nevertheless a trading zone transports knowledge to both involved partners` benefit.

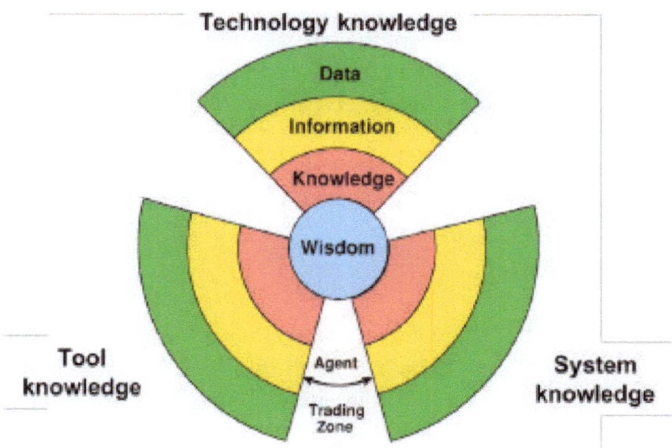

Fig. 3 Knowledge categories for MEMS design and fabrication

As pointed out in [6] for silicon-based micro technologies cross-section drawings may act as such trading zones between MEMS system designers and technology providers. System designers use this representation to sketch the basic functionality of a MEMS system without really being aware of the technological implications of the various layers they draw. Process engineers on the other hand make use of cross-section drawings in order to create a set of processing steps that are required to generate the specified sequence of layers. During this process they are usually not aware of the functionality of the device represented by the cross-section drawing (at least there is no need for them to be). A typical example of such a cross-section drawing of a MEMS device is shown in Fig. 4.

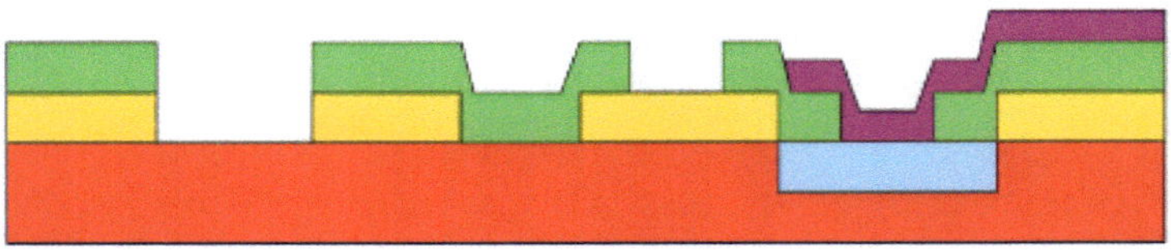

Fig. 4 Example of a cross-section drawing

The usual approach followed in design projects is to mutually exchange cross-section drawings or to create them in common face-to-face meetings of design and process engineers. The effort involved in this procedure is fairly high and has the disadvantage to require considerable additional engineering effort to be spent on both sides. The approach presented in this article aims to increase the efficiency of this process by making use of appropriate software tool support.

A first step to improve communication between the involved engineers is a cross-section editor. It offers a set of drawing tools especially tailored to the effects of thin-film fabrication technologies and supports the definition of nongeometric constraints (like resonance frequencies, etc.) [10], [11]. Fig. 5 shows the current implementation of the cross-section editor that has been realized in the authors` institute. The next step is to offer tools that assist in an automated translation of a cross-section drawing given by a MEMS designer into a process step skeleton that is oriented towards the technology portfolio of a given technology provider. In the following sections the first practically usable prototype of such a system will be presented. For details on the approach see [6].

3.2 The Layer Model

The cross-section drawing is a relatively natural knowledge representation, easy to understand and to handle for engineers, however, relatively hard to handle for an algorithmic procedure. What makes the situation even worse is the fact that in general more than one cross-section drawing is required to denote all relevant structures in a MEMS design.

Therefore in our approach cross-section drawings are first transformed into a so called layer model. The layer model is an abstract model that is based on the assumption that the final fabrication is derived from a silicon micro machining technology. In silicon micro-machining devices are produced by iteratively applying layer generating and layer modifying process steps.

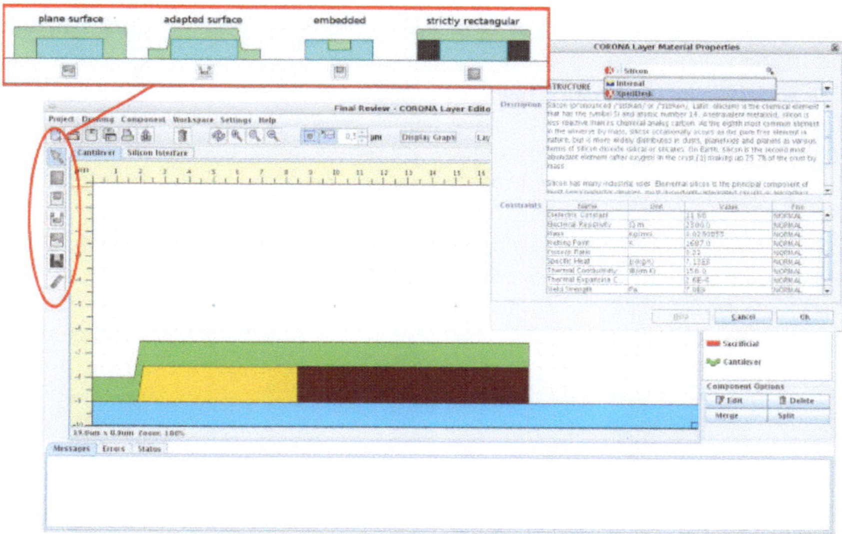

Fig. 5 Cross-section editor

To derive a layer model from a cross-section drawing all geometric elements with similar properties (thickness, shape, material …) are combined into layers. Assuming a limited set of basic shapes, shape modifications can be extracted that result in the final realistic shape [12]. The upper-left part of Fig. 6 shows a cross-section drawing along with the respective layer model including layer modifications.

In the end the layer model consists of a set of layers and modifications, the dependencies among which can be modeled as a directed graph. A welcome side-effect of the layer model is that it usually cannot be traced back to the cross-section drawing that has been used as a starting point. Therefore it can be distributed without disclosing the original product idea.

Furthermore the layer model can also be regarded as an abstract description of a fabrication technology. From this point of view the layer model can represent the capabilities and combination potential of a technology – without disclosing the underlying process recipes. A technology layer model is generated in much the same way as a design layer model. The process engineers draw sample cross-sections for the technology that are then used to derive the layer model. The only difference lies in the fact that in a technology layer model it is possible to label certain layers and modifications as optional. Furthermore it can be defined that certain layers may be generated several times and that some of the modification steps can be performed more than once. In Fig. 6 a design layer model and a technology layer model are shown. The above mentioned differences can clearly be seen.

3.3 Technology Mapping

If a design layer model and a technology layer model are given, it is possible to determine whether both layer models are compatible. The mapping algorithm that is used to determine compatibility proceeds by trying to map elements and dependencies of both layer models to each other. As a result a third combined layer model is generated.

On the right-hand side of Figure 6 a design and a technology layer model are shown along with the combined layer model derived by the mapping algorithm. For clarity reasons only the layer objects are shown in the figure. It can be noticed that the technology layer model uses two more layers than implied by the cross-section drawing. The nitride and passivation layers do not have any correspondence in the design layer model. The nitride layer is labeled optional and can hence be omitted. The passivation layer, however, is a mandatory part of the technology layer model and must be taken into account when deriving the combined layer model. In practice that means that passivation has to take place, even though it is at this point not part of the design.

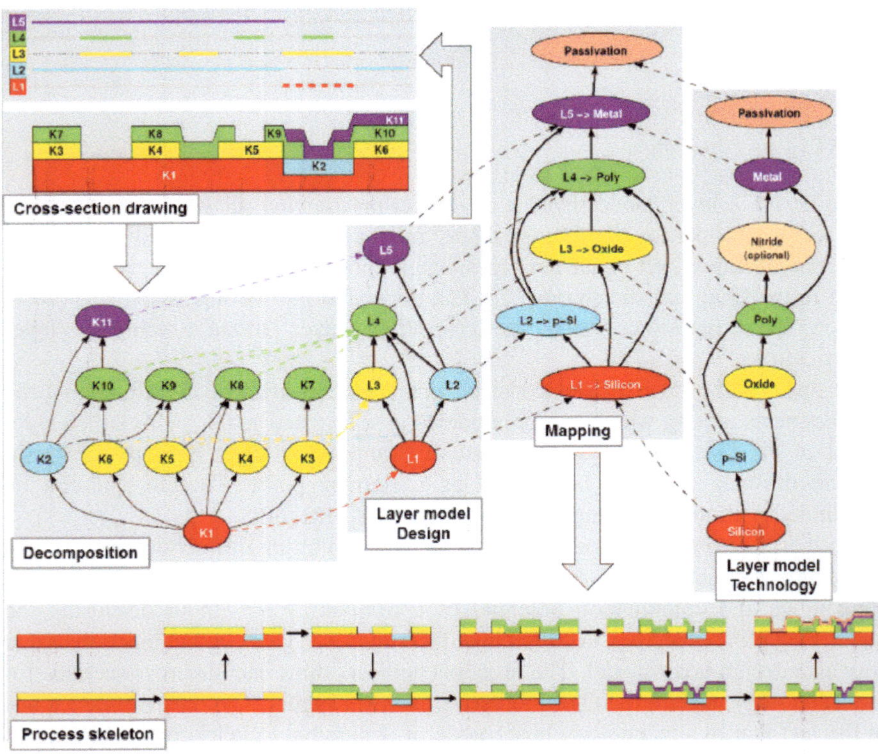

Fig. 6 The process selection and synthesis procedure

A technology is compatible to a design layer model, if each element of the design layer model can be mapped to an element of the technology layer model. In practice that means that a design with a given cross-section can be realized making use of the fabrication process that is abstractly represented by the technology layer model.

3.4 Process Sequence Derivation

Based on the common layer model the technology provider can now assemble an appropriate fabrication process sequence. A knowledge-based process skeleton generator assists in this process by making proposals of an appropriate process skeleton that includes layer deposition, lithography and layer modification process steps. The process skeleton generator makes use of the specific geometric properties extracted from the cross-section drawing to deduce process optimizations like eg. self-adjustment.

The process skeleton can then be used to generate a dedicated real process recipe. For this step it makes use of PDES systems like e.g. the XperiDesk system [13]. During this procedure the process skeletons are extended by concrete process recipes of the chosen technology. The PDES system includes a process consistency checking module that makes sure that the generated process step sequence is technologically feasible. The interface to TCAD simulations tools permits fast evaluations of the process capabilities and can hence show whether the mapping process is technologically valid [14]. If the design partner decides to use the proposed fabrication process the PDES provides further tools to validate and characterize the process making use of real experiments and prototypes [13].

4 Conclusions

In this article a software supported methodology has been presented that allows customer-specific MEMS fabrication processes to be made usable in a structured and efficient manner. In particular appropriate IP protection has been implemented so that no undesired knowledge transfer will occur between the involved parties during the process generation procedure. With this methodology a considerable increase in the efficiency of fabless-foundry-models for the MEMS industry can be achieved.

Along with the use of PDES systems like XperiDesk this methodology introduces the possibility of efficient reuse and advancement of existing fabrication process IP.

Making use of the selective export capabilities that are available in PDES systems a step towards the generation of customer-specific process design kits can be achieved. This is the first step to a complete integration of process and product design for MEMS and hence towards the automated synthesis of MEMS devices.

References

1. Senturia, S.: Perspectives on MEMS Past and Future: the Tortuous Pathway from Bright Ideas to Real Products. In: Digest Tech. Papers Transducers 2003 Conference, Boston, USA, June 8-12, pp. 10–15 (2003)
2. van Heeren, H.: Appearance of a Moore's law in mems? Trends affecting the MNT supply chain. In: Proc. of SPIE: MEMS, MOEMS, and Micromachining II, vol. 6186(1) (2006)
3. Yole Developpement: Status of the MEMS Industry. Yole Developpement (2010)
4. Schmidt, T., Hahn, K., Mielke, M., Brück, R., Ortloff, D., Popp, J.: Distributed and Collaborative Product Engineering for MEMS. International Journal of Microelectronics and Computer Science 1(3) (2010)
5. Schröpfer, G., Lorenz, G., Rouvillois, S., Breit, S.: Novel 3d modeling methods for virtual fabrication and eda compatible design of mems via parametric libraries. J. of Micromechanics and Microengineering 20(6) (2010)
6. Schmidt, T.: Technologiemanagement und anwendungsspezifische Prozessentwicklung in der Mikrosystemtechnik. Doctoral thesis, University of Siegen (2011)
7. Ackoff, R.L.: From data to wisdom. Journal of Applied Systems Analysis 16, S. 3–S. 9 (1989)
8. Bellinger, G., Castro, D., Mills, A.: Data, Information, Knowledge, and Wisdom. Version (2004), `http://www.systemsthinking.org/dikw/dikw.htm` (last inspected: December 18, 2011)
9. Galison, P.: Image and logic: a material culture of microphysics. University of Chicago Press (1997)
10. Hahn, K., Wagener, A., Popp, J., Brück, R.: Process Management and Design for MEMS and Microelectronics Technologies. In: Proc. of SPIE: Microelectronics: Design, Technology, and Packaging, vol. 5274 (2003)
11. Schmidt, T., Mielke, M., Hahn, K., Brück, R., Ortloff, D., Popp, J.: A visual approach on MEMS process modeling using device cross-sections. In: Proc. of Microtech. (2010)
12. Schmidt, T., Hahn, K., Brück, R.: A Knowledge Based Approach for MEMS Fabrication Process Design Automation. In: Proc. of IEMT (2008)
13. Ortloff, D., Popp, J., Schmidt, T., Brück, R.: Process development support environment: a tool suite to engineer manufacturing sequences. Int. J. Computer Mater. Sci. and Surf. Eng. 2009 (IJCMSSE 2009) 2, 312–334 (2009)
14. Ortloff, D., Popp, J., Laughlin, E., Greiner, K.: Efficient virtual manufacturing for MNT. In: Proc. of COMS 2010 (2010)

Industrialization of Customized AI Techniques: A Long Way to Success!

Ralf Montino and Christian Weber

Elmos IT, Dortmund, Germany
ralf.montino@elmos.eu

Abstract. Implementing and integrating a complex artificial intelligence (AI) powered system into industrial production is a long way to go. Even if there is already a working system in the lab, there is a number of hurdles between a successful prove of concept in a scientific environment and the acceptance of a system by the engineers which should use it. A result oriented company will not invest resources to fuel interesting solutions but only to increase the performance, quality and/or efficiency of products and processes. The return on investment has to be clear and fast. In consequence the justification of not only of an implementation but of a working integration process is a must have for a return on investment.

The good news is: There are still relevant open questions in existing production companies with a need and a potential for approaches out of the variety of intelligent solutions. A solid fraction of these questions has a real chance for a successful approach with customized AI techniques. Proofing, planning and pre-customizing an applicable technique in an early project stage, combined with a conscious integration process, is the path towards a successful application.

This paper captures what is important to bring an AI powered solution into application from the industrial point of view. It shows a well walked path for a successful implementation within the semi-conductor industry for a combination of feature selection and neural networks which are supporting the root cause identification within the complex environment of the production line and presents a novel approach to solve production control issues with AI techniques.

Keywords: AI Techniques, Feature Selection, Artificial Neuronal Networks, Cost Savings, Industrial Application.

1 Introduction

Today a state of the art production environment is controlled closely by a variety of computer systems. Together with the computerization of the production equipment itself this emphasizes the capturing of product, equipment and production data. This together fuels aspects as traceability, failure analysis, failure detection and more compromising aspects as prediction and state reports on a daily or short term base. As markets and industrial orientations are fluidly and constantly changing, they promote a customized and dynamic definition of quality. This could be today a robustness of the product in a variety of environments while tomorrow the desired quality could be a steady throughput and low production cost for another segment of the market.

Emerging from the new dynamic the need to capture data at every graspable part of a product is becoming ubiquitous and is still increasing. This is especially the case in the semiconductor industry where a high technological demand meets strict and changing market requirements. The question arises how to accompany the flow of data with extraction and analysis tools in this changing environments.

Through this paper a fusion of known AI techniques in a new combination with sophisticated application-near structures and interfaces is proposed to create a common AI powered framework which is specialized to grasp the data and analysis connected to the harsh production environment and still general enough to switch the world view on demand to sub-worlds or even complete new fields of application.

2 High Potential for Flexible and Adaptive AI Solutions

Market regions with high quality demands have their zenith in the concept of “Zero Defect”, meaning that absolutely no malfunction is being accepted by the customer. Here the production and development does not work without controlling systems based on and facilitated by a groundwork of captured data. This data does not only capture the control parameters for today’s systems but also parameters, sets and systems of the next development cycles of the actual processes and production targets. Thus they could already include parts of the potential “tomorrow”. This development together with a steadily present uncertainty concerning market changes and situations, announces the critical need for a general flexibility. A flexibility which starts with the planning of new products and production facilities but goes down to the creation and integration of fitting and customizable tools and which captures the need for a flexible and dynamic behavior regarding data and its analysis.

On top, due to the dependencies in complex high technology products, there are still unwanted effects to capture which are known but not fully understood and thus could not be avoided yet. Every disturbance in the production flow or direct quality loss at the products will cost time and money and could even trigger the failure of a whole production line, or -even worse- generate a quality problem at the customer. There are various efforts to increase the scope of control and the discovery of new control potentials. Any kind of intelligent approach which could grasp and extract effects and dependencies would be a great improvement in front of the established tools and methods (i.e. statistical process control, statistical equipment control, multivariate analysis).

The use and integration of AI methods is promising and already proved to provide excellent solutions in delimited applications, including long and complex production flows and concepts and in the direction of extracting insights out of huge data sets. Words like “adaptive” or “dynamic” are mostly pointing into the direction of the pure data but not towards situation and application adaptivity when used in connection with AI powered solutions. If a company wants to make use of an AI solutions it needs a reconsideration of the costs and methods of an implementation.

An industrial implementation has to surpass a cost gap for AI methods out of its complexity of integration and an AI solution comes with the pressure to last adequately longer or bring better results than a regular solution to pay off. As a result an implementation should be able to adopt itself to situation changes which could render it else wise inapplicable after a period of use.

The hurdles for long term use and integration of AI powered solutions are high and lead to the reconsidering if to implement an AI solution in house or in cooperation with an institution or to simply buy an already working software package.

2.1 *Make or Buy*

Typically companies try to buy instead of investing into development. For an off the shelf product the investment is known before the purchase decision and most products even come with a success story and/or a test installation to evaluate results beforehand. There are already software suites available, which are talking about using AI techniques for information classification and process control [1, 2]. As well as there is a broad landscape of AI techniques working below the surface of already common applications when it comes to data and gaining insights out of data with the help of AI subfields like machine learning [4].

The downside of commercial software is that they are in most cases complete black boxes to the users and even to the company's engineers. These packages are developed as single-sell products which address one specific topic or use and could prove to be unusable if the situation and the environment changes. Or the company has to pay for an adoption with additional extensions through add-ons or customization contracts.

In the end the real downside is not the black box character itself, though there are good concepts to build custom software arrays out of black box components [3], but the gap between the availability of complex AI methods and techniques and the missing knowledge to judge them for buying as a product. This aside, software companies filed a lot of patents regarding the usage of AI technologies for data analysis purposes [5] and have their own interests to not allow deep insights into their software functionality. But, the use of AI techniques is finally one –maybe relevant- piece to solve a complex puzzle. Without ideas what is working beyond the curtain, a big portion of the possible advantage might be lost.

This renders the development of in-house solutions an interesting opportunity. It provides the flexibility to arrange the application and application area right to the spot of current and future interests. On top a designed software solution is able to match and address more requirements than a predefined software package and could include the insights and usage strategies of the application experts.

But even a well planned in-house project could lead to costs similar to commercial software. In this regards it would be an improvement to be able to reuse parts or components of other packages and/or designing the project in a way that it is reusable at least regarding potential black box aspects [3]. There are holes and bottlenecks to pass and even an insightful planning will not prevent all of them. To reduce the overall "costs" in an abstract manner a system should hence strengthen its return on investment.

Beside the general and desired quality improvement of the connected production processes, the return on investment could also be increased through reusing the software in different application areas. The goal for the project described by this paper is to model a kind of this software as an adaptive framework which then could be apply to different fields of industrial application through generalizing its structure and interfaces.

2.2 *Custom vs. Standard*

Important for developing a reusable framework solution is to keep a steady eye on the complexity of the components in general but especially on the complexity of use. Our target was to keep the complexity of use of the target AI framework solution to a level between a full custom solution and a standard product.

A full custom solution captures the needs of a laboratory environment. Here it is important to have full and direct access to all available parameters and interfaces to have a potential link to every possible entity within the available solution space. The direct downside of such an approach is the complexity of the resulting interface which renders it impossible to have a natural understanding of the software without investing a huge amount of time. A standard solution is quite the opposite, with easy to use interfaces but without extended possibilities to access algorithmic parameters or to adapt to different environments.

These two directions define the space for a trade of solution which presents a mix of customization and usability. While being more application specific is no direct harm, the challenge is not to cross the "red line" for adaptivity. Crossing the line to left hand side, the application will not be able to help solving the problem (Fig. 1). A to complex access to the framework will keep it unused independent of its potential.

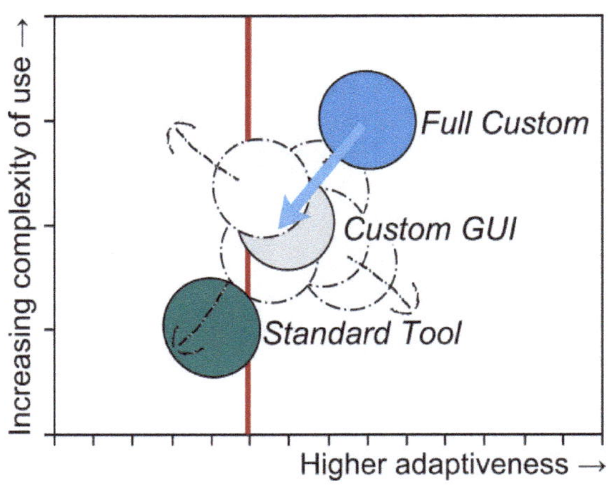

Fig. 1 Solution space of a framework between adaptivity and complexity of use

2.3 Integrating and Connecting AI Techniques

For a trade-off between grasping the application and adapting to new areas, a system has to include a certain level of generalization [10]. For regular software solutions the path is well walked but if this experience should be bridged to AI powered systems the environment is different and requires additional attention.

OpenCog [13] is one of the systems which went ahead and created a generalized and adaptive solution to integrate a variety of different but cooperating AI techniques in one working framework. It rivals in this manner with the well-known RapidMiner framework [11] which includes popular and proved extensions like WEKA [12]. What renders OpenCog unique among them is the general concern towards a global structure which enables the cooperation of different methods in one framework. While all this frameworks are still being too complex and "heavy" to power our desired solution, OpenCog defines a good basic model for grasping AI frameworks [9] like shown in Fig. 2.

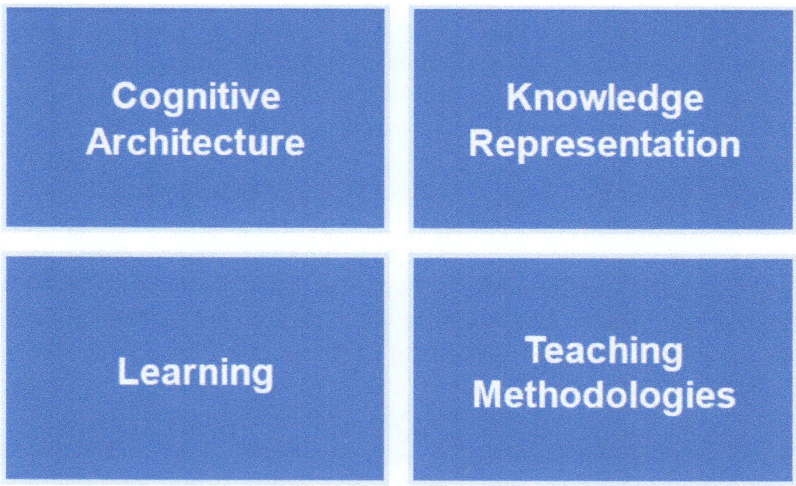

Fig. 2 AI Framework key aspects [9]

These aspects are directly connected to AI methods and their understanding represents the view of an ideal world regarding AI integration in this environment. For a solution developed for product near environments the framework has to mix and exchange the AI aspects with the human factor and direct system requirements. In the following these aspects are interpreted for the industrial AI framework approach [9]:

Cognitive Architecture: The overall design and connection model of the components which include AI techniques as well as modules which connect the real world and the human user.

Knowledge Representation: For an AI system the representation captures the knowledge included and processed by the components. For an industrial application framework the system also has to store flat information like the component states of triggered methods and information about how the data were fed, what were the sources and how it were transformed. Thus it increases the traceability of every process to a level that clearly connects the results to the inputs and the human understanding. For this human understanding it is essential to be able to generate representations which are human readable like rule sets and thus expand the direct insight into the data, results and reasoning through visualization components.

Learning: An AI framework should at its best facilitate methods of learning in a way that it could learn and encode new knowledge and, as the highest goal, could even adjust parts of its code towards a change of situation. While this level of situation awareness is fitting the general wish of being dynamic and adaptive like for the first part (i.e. Artificial Neuronal Networks (ANN)) which stores process knowledge), the later goal of changing the framework on code level collides with the strict requirement in an industrial environment of traceability and transparency about what happened at which time.

Teaching Methodologies: These grasp every technique which enables and assists connecting to new knowledge. As it addresses more sophistical methods in the pure AI context, it captures in the industrial context the data interfaces, data- and as well feedback- streams which as well include user feedback which for modifying the direction of the next framework implementation cycle.

Within the context of the semiconductor industry and in a strong scientific cooperation with the University of Siegen, a multifunctional AI core component was developed and proven in a different project frame [7]. Thus it was available for the new and generalized AI framework solution and is providing, beside options and interfaces, one kind of return on investment. Around this core an AI component composition is developed and evaluated to bridge the needs and potentials of an industrial customized AI framework. This framework will face and handle huge amounts of data like they are daily occurring within the wafer production and will support and be used by the engineers who cope with decisions based on the data mined from the production.

The preselected, available AI core component is a solution for knowledge discovery from huge amounts of data. The component is a combination of a feature selection (FS) functionality for identification of relevant parameters, and an ANN for learning and verification of assumptions gathered by the FS. While being promising to cope with the data analysis, it proved to be not available on the market in this combination and in a form fitting to the application needs. Moreover, even scientific publications of this kind of combination could hardly be found [6].

To satisfy the need for more options of adaption to further increase the return on investment and as well the quality improvement, the working concept has to be pushed further into the shape and possibilities of a framework. The result is the new system AIFind – Adaptive Intelligent Framework for interrelation discovery.

3 AIFind Industrialization

To create the adaptive framework, the existing AI core component and its modules had to be extended to bridge the gaps towards the framework key concepts, gathered in the previous section. The first results and prove of the core component concept and implementation were summarized and presented in [8]. A basic overview of the algorithmic core component is shown in Fig. 3 below.

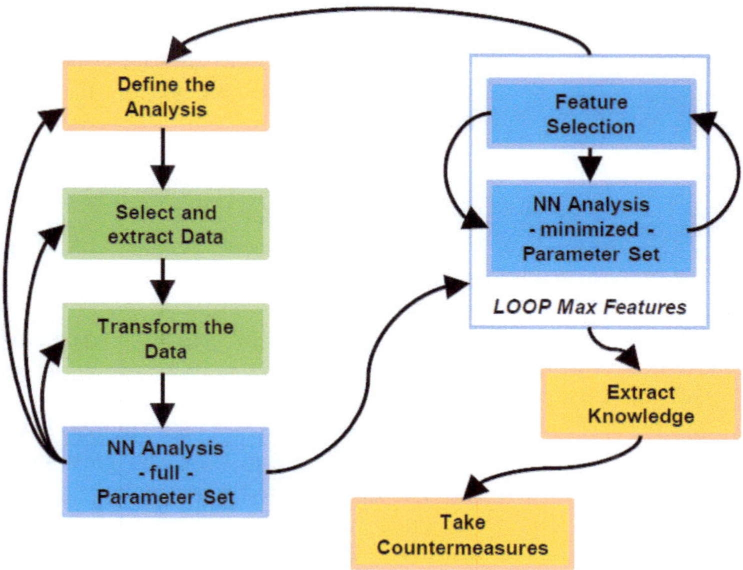

Fig. 3 System overview [8]

For a successful deployment of an AI solution in the industrial environment, archiving results as in [8] is necessary but not sufficient. For a good complexity of usage a full customized lab system like the core component with a command line interface has high hurdles for a direct use by engineers in their daily work. Further there are still lacks regarding capturing data about the ongoing software use which are needed for traceability and as a feedback for the further development.

Table 1 captures the framework goals regarding the aspects of section 2.3, which are already fulfilled by the core component, to derive what is left to bridge for the targeted framework solution.

There are two major sections in Table 1 which need improvements and additional implementation. On one hand data and performance driven requirements to divide the components in a way that enables them to operate independently of shared hardware platforms and integrate data from different and

Table 1 Contents and gaps of the existing core component

AI framework key aspects	Core component's contents and gaps
Cognitive Architecture	+ extraction, transformation and loading is divided from the FS and ANN component + all components are using one standardized data exchange format as a common data interface - missing general process structures to capture framework runs as connected analysis cases - missing final user interface concept.
Knowledge Representation	+ basic component states are saved via reloadable files which allows access to the captured information + extracted/transformed data collections are saved into specialized data base structures from which they are parsed to files + calls to the components generates result reports + the ANN interface exports reports with human readable rules - missing union of process steps as one standardized case - missing option to store analysis runs over time - no visualizations except pure numbers in reports
Learning	+ the ANN is able to learn and grow over time fitting to the presented data - no direct option to repeat previous analysis with modified parameters
Teaching Methodologies	+ the data extraction and transformation module is able to adopt new data sources + toolbox of normalizations to fit to even new data types - no user and use driven interfaces. - only active gathered feedback is available

distributed resources. On the other hand there is still a strong need for an easy use of the framework. A user should be able to see what is happening, have visualization options to judge the result and preliminary steps and have a number of options and ways to grasp the processes in parts or as a whole. Internal algorithm parameters should be visible and changeable but not necessarily have to be changed for an analysis case and should come with standards. They should improve but not block the overall results.

In the following, additional requirements from the industrial view are captured together with the derived changes for the fundamental upgrade of the AI core components to an AI framework.

3.1 Meeting System Requirements

There is a range of system requirements out of the detected gaps, as well as based on application specific industrial and product motivated adaptations. So, besides the "standard requirements" of the software industrialization the new framework has to reach additional goals:

(a) Making the software robust and stable for a steady trust into the system analysis.
(b) Handle a bigger load of requests, also in parallel by several users.
(c) Moving the software to the data center and connecting it to productive databases.
(d) Setting up log files, backups and usage rights.
(e) Setting up system monitoring / adequate error handling and a saving mechanism for analysis process runs.

3.2 The AIFind Composition and Architecture

To cope with these goals the new AIFind system leverages an extended and improved architecture concept, which is presented in Fig. 4. Components are now divided and could be called from and located on completely different platforms. This enables a set of handlers which are triggered by the user interface through choosing the type of job and which take and process the triggered jobs based on their resources.

To fuel the traceability, reusability and performance of connectable data sets, the architecture utilizes core parts of data warehouse concepts. The data extraction is extended to a fully controllable Extraction, Transformation and Load (ETL) process and is fusing and transmitting data for an analysis from the productive data base to a data warehouse solution. The framework components could access and re-access the resulting data in a pre-transformed and normalized format.

An iteration of the framework is shown by the numbers in Fig. 4. An analysis triggered and parameterized by a user (1) will put a data extraction job on a central but openly accessible job list (2). A controlling daemon will grasp then the parameters and trigger the extraction process to store the data into the data warehouse (3). Potential errors and messages will be fed back by the daemon towards the user interface to report and allow the job modification. From the data warehouse a result file is parsed and placed into a data pool (4), from where it is fed, together with the analysis parameters, to the ANN/FS component (5). Results will be stored and returned with potential messages as a report to the interface directly or via mail to give a status and the opportunity to re-modify the parameters and grasp the results. Out of the data pool the, potentially time intensively created, data sets are reused without the need to retransform.

Via this structure it is not only possible to have a basic kind of load balancing but also to have the parts of the analysis processed as jobs while the interface is closed and inactive. The GUI is acting independent of the underlying processes.

A user could create a set of test analysis to see whether they are following the right direction and then place the promising ones with extended data and analysis sets into the queue to have them calculated over night to return to the result reports the next morning.

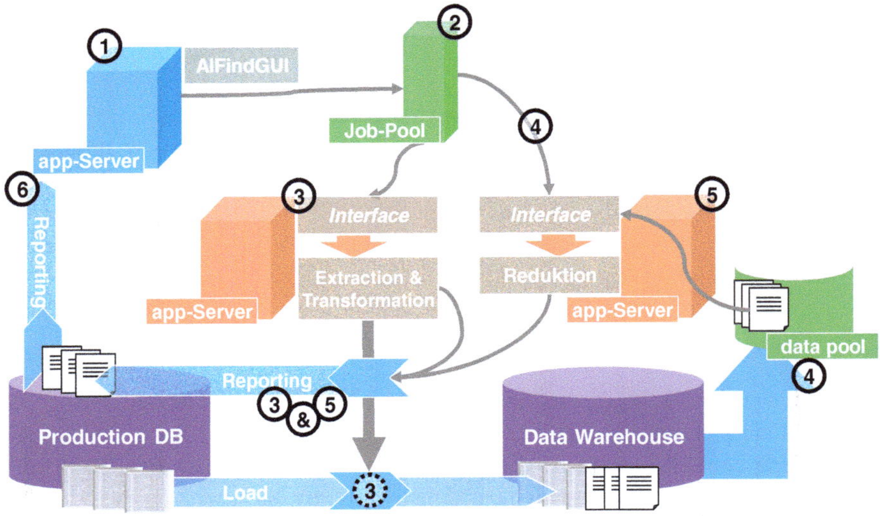

Fig. 4 The architecture of AIFind

3.3 Meeting User Interface Requirements

One mayor and most important aspect of the new AIFind system framework extension is the introduction of a capable, customized and easy to use interface which is based on the needs of the users. The target was to add a user friendly interface to the system and reduce the customization necessity for analysis jobs and parameters within the framework as far as possible. A starter has to be able to directly make use of the system while specialists should still be able to customize the system for even better or more specialized results to archive an early but lasting acceptance by the customers.

To achieve this target, an easy to adapt central accessible web based GUI solution were developed, which embeds naturally into the intranet service landscape. Besides the main task of reducing the complexity of the user interface for a high acknowledgment, this GUI has to fulfill further requirements:

(a) Make the system available at any place in the corporate network without the necessity of software installation
(b) Show the status of the system and the tasks of the users
(c) Automate the data collection from and to the data warehouse

(d) Keep the results and the corresponding input parameters together and fuse them to named cases
(e) Make the AI powered analysis and their results available for a greater number of people

A user will start the processing with an overview page as desktop to start from (Figure 5). Here the currently running and historical sessions are visible in one, central view. A session is the synonym for an analysis case. It is composed out of data definition and the analysis definition itself. Both definitions will be filled and converted into results by the underlying processing modules and a ready set is indicated by a “complete” button which leads to the fitting result page. The user could return to have a steady impression of the sessions and as well could have a reduced look on his cases or the cases of other users by applying filters to the list. Through ”create new session“ the user triggers the session definition page.

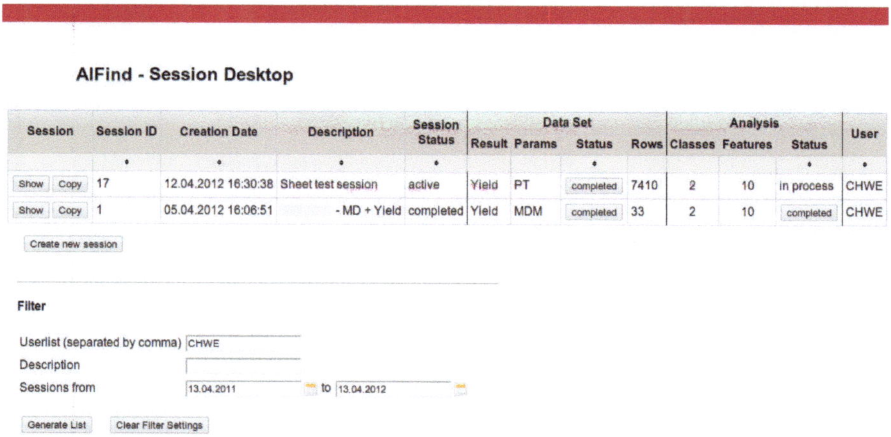

Fig. 5 Grasp the system states on one unified session list overview

Within the session definition dialog (Figure 6) the user defines the two basic blocks for the case analysis - the data definition and the analysis sources and parameters. While being able to be specific about the data sets and the analysis parameters a starter has just to give a minimum of information to get the first promising results. An important aspect is here the presented look and feel. A user notes his ideas virtually on a piece of paper and could save it, share it with other users and come back to change the parameters as often as he likes. Finally pushing the confirmation button puts the paper into the job box for processing where the sub-processes will fetch it and send a confirmation when the process is ready, which proves to be easy and natural to use.

The system reuses the metaphor by providing the option to copy or recycle parts of previous sessions and fetches the desired pre-used paper and/or the complete finished data set and attaches it to a new session.

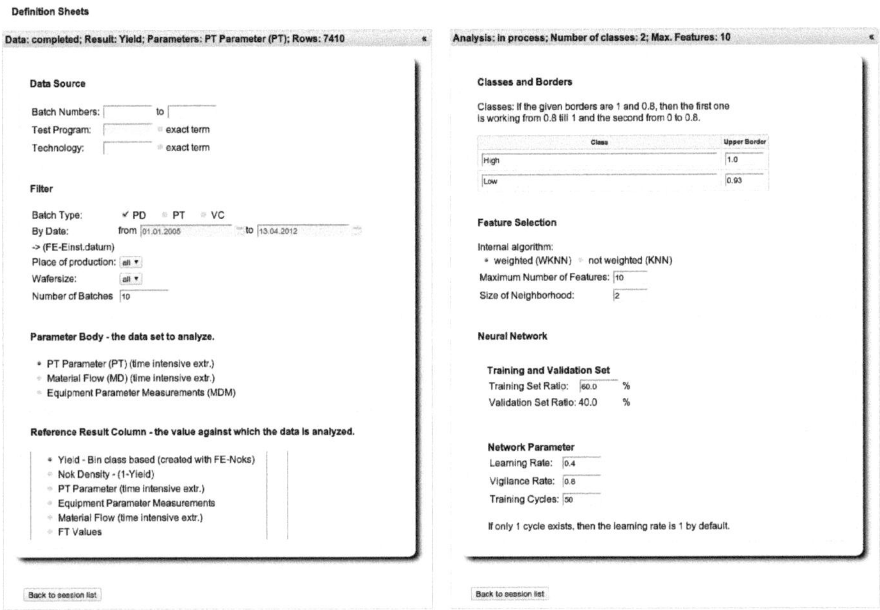

Fig. 6 "Define on a paper" concept for naturally creating the analyzing case

By the time the defined data set is extracted, normalized and prepared, the data overview becomes available through the session desktop (Figure 7). Here the user could grasp an overview of the value distribution of every parameter involved in the session and analysis. Since the values are normalized to the interval [0, 1] for the ANN the view calculates back to original parameter values in a comparison table. In this view the user could form the decision how to define classes for the problem case.

Finally the analysis result view becomes available (Figure 8). It captures the results of the FS created and ANN evaluated feature reduction. As the AI algorithm is targeting to build smaller sets of data, grasping an entropy equal to the full data set, the view lists the calculated feature sets. Supported by the overview table and class diagram the user chooses the most promising reduction sets and triggers to visualize the quality of the ANN proof evaluation, together with a textual representation of the collected features and a summary what changed in comparison with the sets before.

Data Result Overview:

	SetId	Result Column Name	Rows	Columns
Show result	1212	YIELD	7410	152

Parameter Overview List:

	Parameter Name	Original Number of Occurrence
Show	PT//11810//\	1379
Show	PT//11811//\	591
Show	PT//5187//\	1629
Show	PT//5251//\	1729
Show	PT//5252//\	1729
Show	PT//5253//\	1729
Show	PT//5254//\	1729
Show	PT//5255//\	1729
Show	PT//5256//\	1729
Show	PT//5257//\	1729

Data Normalization Overview of value

Shown Norm Value	Original Value
0.68	58.737449999999995
0.67	58.42574333333333
0.66	58.114036666666664
0.65	57.80233
0.64	57.49062333333333
0.63	57.178916666666666
0.62	56.86721
0.61	56.555503333333334
0.6	56.24379666666667
0.59	55.93209

Result Column Histogram

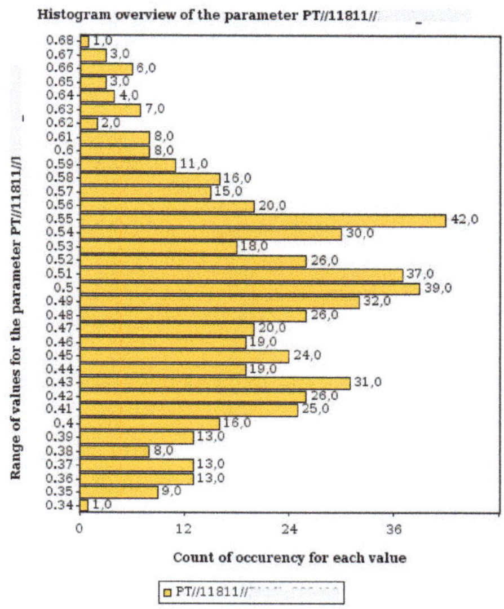

Fig. 7 Data result set normalization and visualization overview

What renders the view appealing is the combination of different easy to read and cooperating visualizations, including seamlessly the information about internal algorithmic parameters which blends with analysis related result parameters.

4 Results

The implementation of the system described in the previous sections of this paper is proven to be successful based on the laboratory tests and the usage and feedback of the user group. The system is being accepted by the engineers and used frequently for a variety of questions.

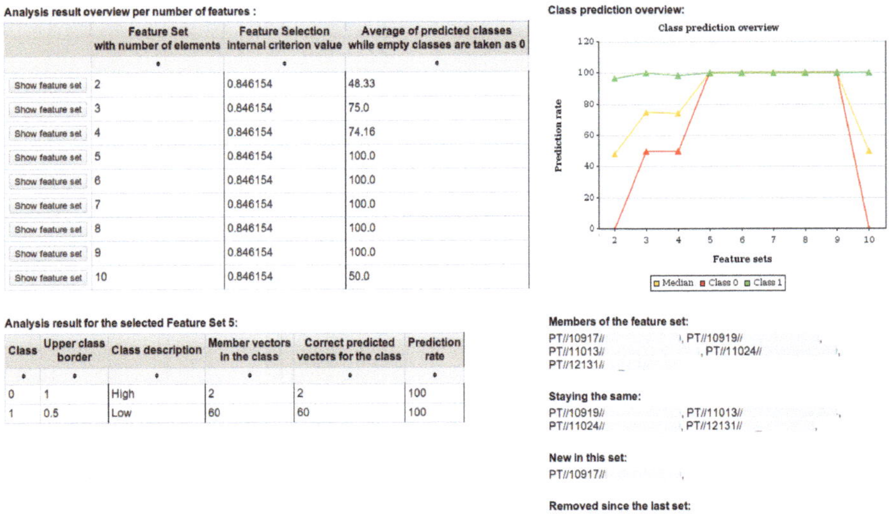

Fig. 8 Analysis result overview page together with visualization strategies

The main point for this user acceptance was the "easy to use GUI", which renders it natural for the engineer to "ask" the system for a hint, even if the user may not fully understand what is going on "behind the curtain". An electrical engineer, a chemist or a physicist can boost his competence in statistical data analysis and process knowledge by simply using this tool, available in the intranet.

In the beginning, generating the command line tool to prove the core concept was only half of the way to go. But the integration of the user as well as the preparation and integration of data sources, which are available in the company, took a lot more effort than estimated.

In a successful cooperation between science and industry a completely new tool for data mining was developed and integrated as a component and thus achieved a short way from theory to practice and a partial return on investment through software reuse. Today the new framework presents a multipurpose problem solver which will become a central tool in the analysis chain of the company and complete the return on investment through environmental adaptation and integration.

The usability of the system (Fig. 9) is, as targeted, above a typical commercial data mining tool without failing the aimed adaptiveness (in this case, the data mining tools available do not even solve the problem).

The result is an AI powered customized solution which fits better and is easier to use than a piece of commercial software. Hence, if there is a blank area on the solution map, it might be worth to invest the effort.

5 Outlook

Besides continuously providing further data sources and functionality to the user (vertical enlargement) we are planning to apply the system also concerning further areas. The general attempt is a global one, looking on all data belonging to

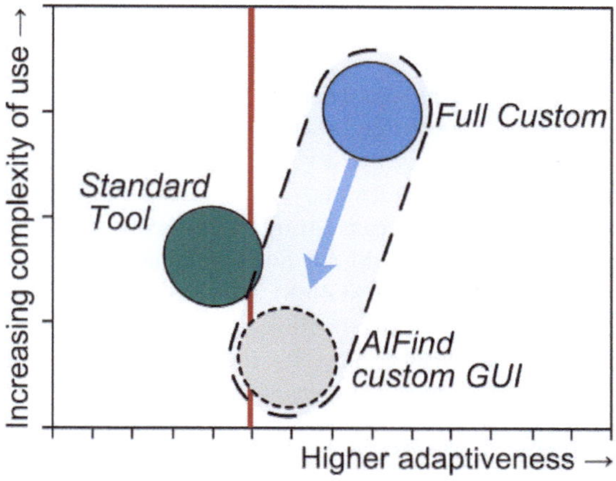

Fig. 9 Reduction of complexity achieved

the product analyzed. Without any input about the interrelation of this data but the general target, the framework is detecting the root cause for unwanted characteristics of products. As described, the AIFind system shows good results by identifying interrelated sets of parameters, which are responsible for the deviance, out of several hundreds available.

The next step is planning to use the system in a more limited environment, only exploring one or a small number of process steps in a semiconductor fabrication (fab). Today, semiconductor fabs typically use control software for Advanced Process Control (APC). The task of these packages is the control of customizable parameters in recipes, describing the process and variables of a production machine. Typically derived or measured results of other process steps are taken into account. For -high volume, high throughput- production sites with a small number of different products, this approach works fine and needs always to be monitored and adapted by specialists.

For flexible and small production sites this approach does not work well. If all products are treated equally, the statistical basis for the control algorithms is fair, but if the product data is not that homogeneous, due to the different behaviour of different products the results are not as good as desired. Thus if the decision is taken to set up a control loop for every different product, the number of data available is often not big enough. Even worse, the statistical weight of the individual measurements is weighted with the age of the data. By experience the reduction of significance by adding a kind of "shelf life" for data, has turned out to increase the quality of the results, but it additionally reduces the amount of data for the control loop. As a result the success of Run to Run (R2R) and Feed Forward solutions is limited.

The planning for the horizontal enlargement of the usage of AIFind inside the company is to develop and set up a standard APC control algorithm frame and increases the relevance of the input parameters for the algorithms inside by AIFind. This should help to find products with similar behavior regarding a specific process step and will enlarge the statistical basis for the control processes and could thus overcome the current limitations.

References

1. Applied Materials E3 Automation Software, http://www.appliedmaterials.com/services-software/library/e3
2. Rudolph Technologies Genesis Software, http://www.rudolphtech.com/AnalysisProduct_Genesis.aspx
3. Ravichandran, T., Rothenberger, M.A.: Software reuse strategies and component markets. Communications of the ACM 46(8), 109–114 (2003)
4. Seagarn, T.: Programming Collective Intelligence. O'Reilly (2007)
5. Shanmugasundram, A.P., Schwarm, A.T., Prabhu, G.B.: Feedback control of a chemical mechanical polishing device providing manipulation of removal rate profiles, US Patent 7160739 (2007)
6. Montino, R.: Crystal Ball: Die Gewinnung von verwertbarer Information aus Datenobjekten mit unscharfem Zusammenhang, Südwestdeutscher Verlag für Hochschulschriften, 102ff (2010)
7. Montino, R., Bensch, M., Bogdan, M., Schröder, M., Rosenstiel, W., Czerner, P., Soberger, G., Linke, P., Schmidt, R.: Neural network algorithms for generalised online tool control in medium size semiconductor fabs. In: 6th European Advanced Equipment Control/Advanced Process Control (AEC/APC) Conference, Dublin, Ireland, April 6-8 (2005)
8. Montino, R.: Combined Usage of different knowledge generating processes in the high tech industry. In: 3rd International Conference on Integrated Systems Design and Technology 2009, Anchorage, Alaska (2009)
9. Hart, D., Goertzel, B.: A Software Framework for Integrative Artificial General Intelligence. In: Wang, P., Goertzel, B., Franklin, S. (eds.) Proceedings of the 2008 Conference on Artificial General Intelligence 2008: Proceedings of the First AGI Conference, pp. 468–472. IOS Press, Amsterdam (2008)
10. Cagan, J., Grossmann, I.E., Hooker, J.: A conceptual framework for combining artificial intelligence and optimization in engineering design. In: Research in Engineering Design, vol. 9, pp. 20–34. Springer, London (1997)
11. Rapid-I RapidMiner Software, http://rapid-i.com
12. Machine Learning Project at the University of Waikato Weka Software, http://www.cs.waikato.ac.nz/ml/weka/
13. OpenCog Foundations OpenCog Software, http://opencog.org
14. Leon, F., Atanasiu, G.M.: Integrating artificial intelligence into organizational intelligence. In: The 9th European Conference on Knowledge Management, Academic Conferences Limited, pp. 417–424 (2008)
15. Brooks, R.: Intelligence without representation. In: Mind Design Two, pp. 395–420. MIT (1997)
16. Lieberman, H.: User Interface Goals, AI Opportunities. In: AI Magazine, Association for the Advancement of Artificial Intelligence (AAAI). Special Issue on Usability of AI Systems, vol. 30(4) (2010)
17. Malinowski, E., Zimányi, E.: Advanced Data Warehouse Design from convential to Spatial and Temporal Applications. In: Data-Centric Systems and Applications. Springer (2008)

Modeling the Diffusion Process for Developing Optical Waveguides for PC-Board Integration

Thomas Kühler and Elmar Griese

University of Siegen, Theoretical Electrical Engineering and Photonics, Hölderlinstraße 3, D-57068 Siegen/Germany
{thomas.kuehler,elmar.griese}@uni-siegen.de

Abstract. Increasing demands for high bandwidth electronic systems lead to increasing needs for new interconnection concepts and technologies. A promising concept is to extend the established electrical interconnection technology by optical interconnections on system, module and component level. As printed circuit boards belong today as well as in future to the most important interconnect devices, the realization of electrical printed circuit boards with integrated optical interconnects is a very important interdisciplinary R&D task. This paper addresses the ion-exchange technology for realizing optical layers with integrated multimode waveguides. The ion-exchange process requires an appropriate simulation model which allows to analyze and to optimize all process parameters. Based on the principle diffusion characteristics an approach for modeling and simulation of the ion-exchange process is presented. The results are used to derive a model for calculating the process parameters necessary for obtaining waveguides with desired optical characteristics.

Keywords: Optical interconnects, EOCB, ion-exchange, multimode waveguides, electrical optical interconnection technology.

1 Introduction

The bandwidth used in broadband technologies rises every year. For example, in 2011 the total amount of data transferred was about 300 Exabytes (1 Exabyte = 10^{18} Bytes), ten times of the value of 2006 [1]. The long-distance data rate using optical fibers is up to 100 Gbit/s per channel. The real data rate per fiber is much higher taking into account established multiplex technologies like Dense Wavelength Division Multiplex (DWDM). This of course results in very high bandwidth demands within the nodes. Several new concepts are under investigation to enhance significantly the throughput in these nodes. One promising approach is to realize the high data-rate interconnects within electronic systems by optical interconnects [7]. On printed circuit board level this has to done by the integration of optical layers, containing optical waveguides, into electrical pc boards as depicted in figure 1.

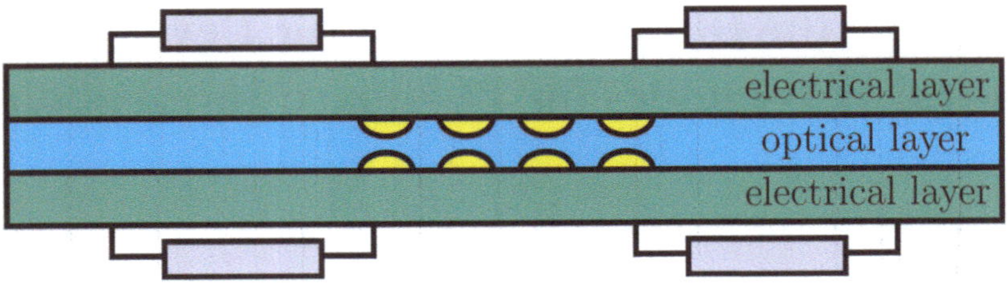

Fig. 1 Concept of an electrical printed circuit board with an integrated optical layer

During the last years several methods for realizing printed circuit boards with integrated optical layers have been developed. The main challenge of all approaches is to combine the advantages of optical and electrical interconnects in combination with the highest possible compatibility to the established electrical interconnection technology [8]. The main compatibility requirements are mechanical tolerances to be in the range on PC-board technology and resistance to temperature and pressure during the PC-board lamination process. The required performance and the compatibility can only be obtained if the expert knowledge of electrical and/or electronic engineers, physicists, chemists, and material scientists can be merged.

2 Manufacturing of Optical Layers with Optical Waveguides for PC-Board Integration

A couple of approaches for the fabrication of optical layers are known. All of them have in common, that the optical layer is embedded between conventional electrical layers. In the following subsections, three technology approaches are described very roughly. The processes are of course much more complicated taking into account the compatibility requirements mentioned above. Moreover, the necessary materials have to be developed and optimized in parallel, in order to obtain the necessary optical, temperature and pressure characteristics.

2.1 Photolithography

Using photolithography technology, UV-curing polymers are applied to shape optical waveguides with rectangular cross sections. As shown in figure 2, the carrier material (substrate), mostly glass or FR4, is coated with a polymer with a low refraction index. On top of that, a second layer of polymer with a higher refraction index is added. For forming the waveguide core, a mask based UV-curing is used. The cores are covered with the lower reflection index polymer. After that step, the carrier material is placed on top.

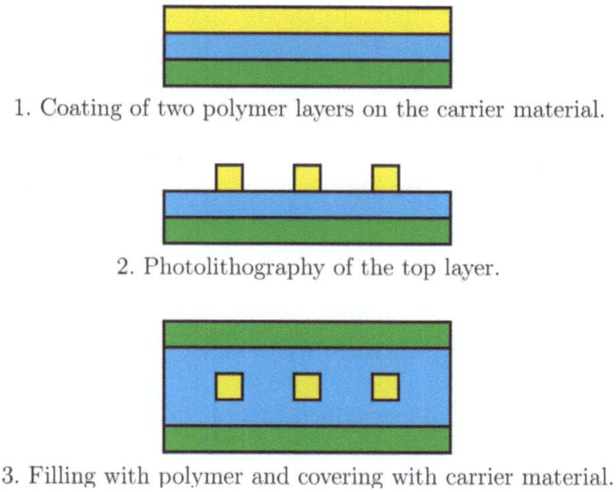

Fig. 2 Polymer optical waveguides realized by photolithography

2.2 Waveguide Manufacturing by Laser Ablation

Another method to manufacture optical waveguides is the use of a Laser beam for excavating grooves into a glass sheet as depicted in figure 3. The grooves get

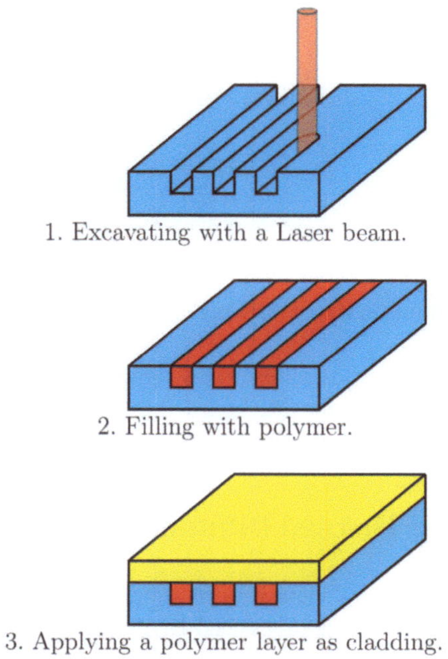

Fig. 3 Polymer-glass waveguides realized by laser ablation

filled by a polymer with a higher refractive index as the glass material. An additional polymer layer with a refractive index similar to the glass is coated on top. Using this technology is has to be taken care of the surface quality of the grooves. In case of roughness high waveguide attenuation is unavoidable.

2.3 Ion-Exchange Waveguides

A new and very promising approach for fabricating integrated optical waveguides is the ion-exchange technology. While the diffusion of ions into glass is known since the middle ages for the coloration of windows [10], the first waveguides fabricated by diffusion were presented by Izawa [2] in 1972. Figure 4 shows the principle steps of the process. Firstly, a mask is fitted on a glass sheet. This is placed in a salt melt containing ions which are able to diffuse into the glass at high temperature. The next step is the removal of the mask and a short inverse diffusion for burying the ions below the surface. In this step some of the previously exchanged ions are removed from the glass sheet again.

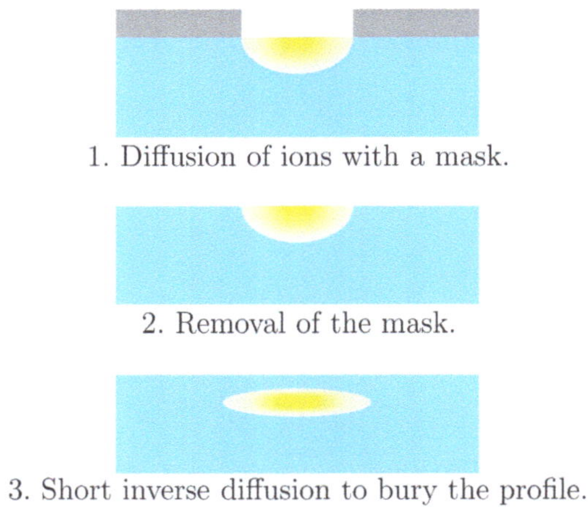

Fig. 4 Ion-exchange multimode waveguides in thin glass

Besides the simplified above described approaches some more methods to fabricate optical waveguides for PC-board integration and are known. A good overview with advantages and disadvantages is given in [3].

3 Modeling of the Thermal Diffusion Process

From systems design point of view it is of significant importance to get detailed information about the index profile caused by the diffusion of ions into the glass sheet. Therefore, the ion density inside the material has to be known. The principle diffusion process is illustrated in figure 5.

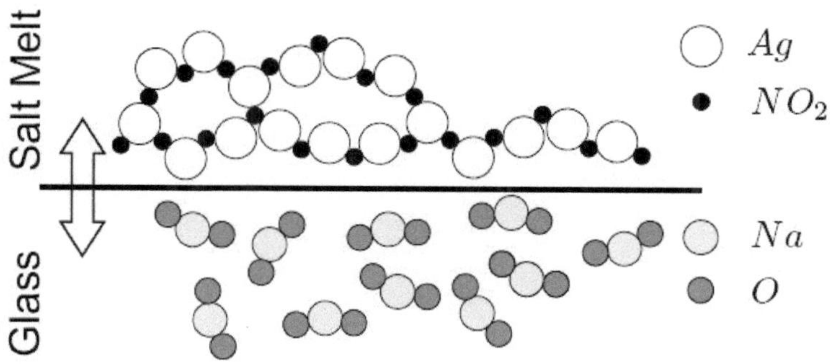

Fig. 5 Schematic one-dimensional diffusion process of silver ions in silicate glass

Considering a planar, one-dimensional diffusion and assuming that the glass sheet is located in an $AgNO_3$-melt, the silver ions migrate at high temperature out of the melt into the glass substrate. Several other ions can be used for the process, but silver ions have the benefit of causing a high increase of the refractive index and the melt is nontoxic. In exchange, sodium ions diffuse out of the glass. This process can be described by Fick's law for time-variant diffusion:

$$\frac{\partial c}{\partial t} = \frac{\partial}{\partial x}\left(D\frac{\partial c}{\partial x}\right), \tag{1}$$

where D is the diffusion coefficient and c the ion concentration. In this one-dimensional case the solution of (1) is given by

$$c(x,t) = \frac{c_2 - c_1}{2}\left[1 + \operatorname{erf}\left(\frac{x}{2\sqrt{Dt}}\right)\right] + c_1 \tag{2}$$

where erf is the error function and $c_1 = c(x<0, t=0)$ and $c_2 = c(x>0, t=0)$ are the initial conditions. In figure 6 the computed normalized ion concentration as a function of the diffusion depth x with the diffusion time t as a parameter is illustrated. The corresponding simulation parameters are given in table 1.

For modeling a graded-index waveguide obtained by using an aluminum mask, the two-dimensional diffusion process has to be investigated. These results in the fact, that equation (1) cannot be solved analytically and a numerical solution is necessary. Figure 7 shows the simulated silver ion density for a buried graded-index waveguide in comparison to the measured ion density.

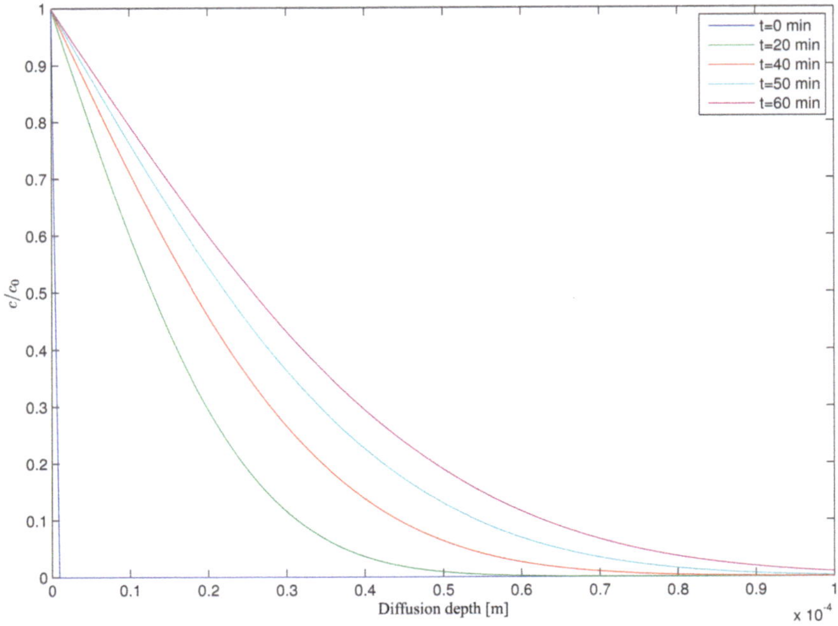

Fig. 6 Normalized ion-concentration versus diffusion depth

Table 1 Parameters for the diffusion process

Parameter	Value
Diffusion coefficient	$0.26 \cdot 10^{-14}\ m/s^2$
Glass thickness	$100\ \mu m$
Melt concentration	$0.02\ mol/m^3$
Temperature	*350 °C*

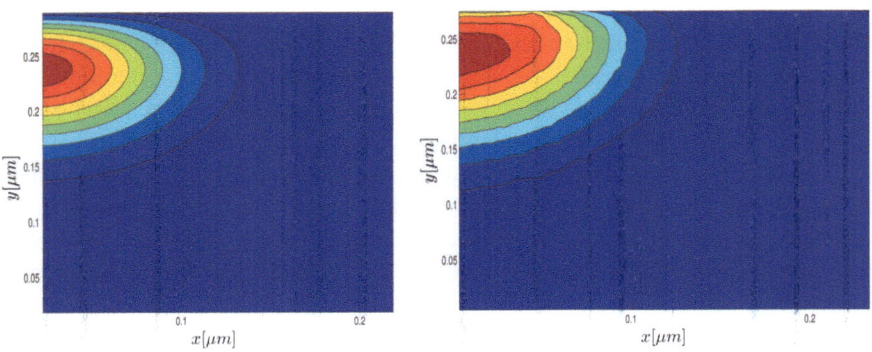

Fig. 7 Simulated (left) and measured (right) ion density of a graded-index waveguide

As the ion density cannot be measured directly the comparison of simulation and measurement results can only be done indirectly. Therefore, the computed ion density has to be transformed to the refractive index profile which can be measured using RNF (refracted near-field) equipment. The transformation can be done applying the Gladstone-Dale relation [5]

$$n_s = 1 + \frac{r_s}{v_s}, \tag{3}$$

where the specific volume v_s and the Gladstone-Dale refractive parameter r_s are material specific constants. For a standard borosilicate glass (Schott, D263T), the refractive index results to $n_s = 1.522$. In [5], the index change is given by

$$\Delta n = 0.072 \cdot c\,. \tag{4}$$

With the known silver-ion concentration, a location-dependent refractive index is obtained:

$$n(x, y) = n_s + 0.072 \cdot c(x, y)\,. \tag{5}$$

Using these relations, the diffusion process can be described very accurately.

4 Modeling of the Field-Driven Diffusion Process

For realizing optical layers providing the same or more degrees of freedom as electrical layers, three-dimensional optical structures, like lenses for coupling, tapers and splitters, or optical vias, which are connections between the opposite glass layers, are necessary. Such structures cannot be achieved by thermal diffusion processes. The idea is to process the basic refractive index profile by thermal diffusion. In a following process, an appropriate electrical field is applied in order to bury the silver ions deeper with a desired spatial distribution into the glass sheet. The electrical field causes both the silver and sodium ions to move in the direction of the potential gradient. The concentration profile can be described with [6] and erfc as the complementary error function by

$$C_B(x,t) = \frac{C_{B0}}{2}\left[\operatorname{erfc}\left(\frac{x - \mu E t}{2\sqrt{Dt}}\right) + e^{\frac{\mu E x}{D}} \operatorname{erfc}\left(\frac{x + \mu E t}{2\sqrt{Dt}}\right)\right]. \tag{6}$$

The initial ion density is defined by C_{B0}. The constant μ describes the mobility of the slower moving silver ions in the glass sheet. It is connected to the diffusion constant by the Einstein relation

$$D = \frac{\mu k T}{q}, \tag{7}$$

where q is the electrical charge of an ion and k the Boltzmann's constant. For large values of the electrical field, the second summand of equation (6) turns to zero and equation (6) simplifies to

$$C_B(x,t) = \frac{C_{B0}}{2}\,\mathrm{erfc}\left(\frac{x-\mu E t}{2\sqrt{Dt}}\right). \tag{8}$$

Figure 8 shows the calculated ion concentration as a function of the diffusion depth. The ion mobility is given with $\mu=6.24\cdot10^{-8}\ m^2/(Vs)$ and the diffusion constant with $D=0.26\cdot10^{-8}\ m/s^2$. The electrical field strength is *100 V/m*. The direction of the electrical field is normal to the glass surface.

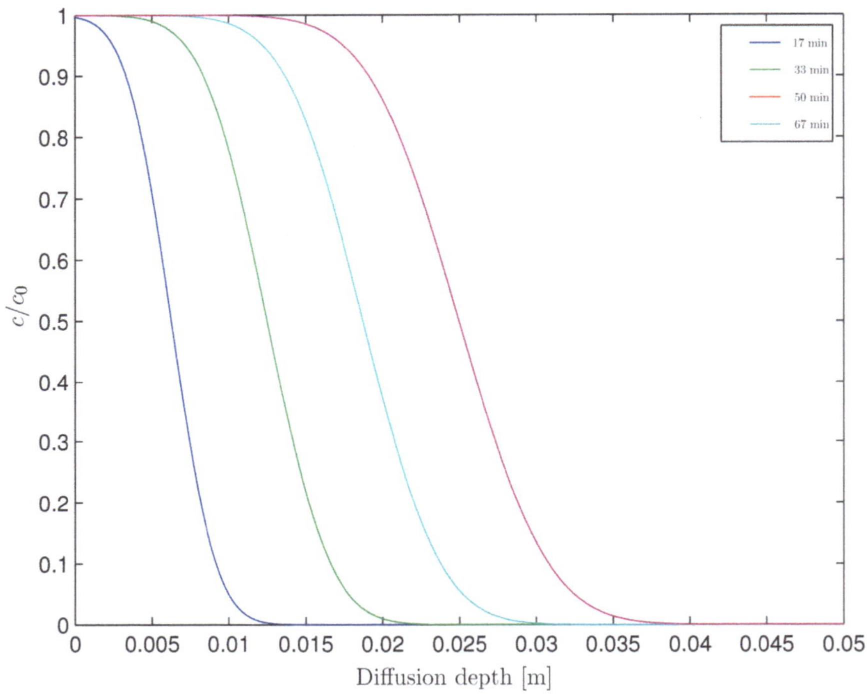

Fig. 8 Normalized ion density versus diffusion depth

5 Deriving Process Parameters

From manufacturers point of view the derivation of the process parameters from the desired position and shape of the refractive index profile is of great interest. For that reason, the diffusion process must be investigated in order to obtain the dependencies between the different parameters like diffusion time, geometrical mask parameters and concentration of the salt melt. The two-dimensional refractive index profile can be reconstructed with elementary functions by using a few characteristic data like the position x_{max} and value n_{max} of the maximum of the refractive index, the height and width of the index profile. By means of this values, the dependency of the process parameters can by analyzed with the aid of the method shown in section 3.

Figure 9 shows the dependency of the maximum value of the refractive index and its position from the mask width and the temperature.

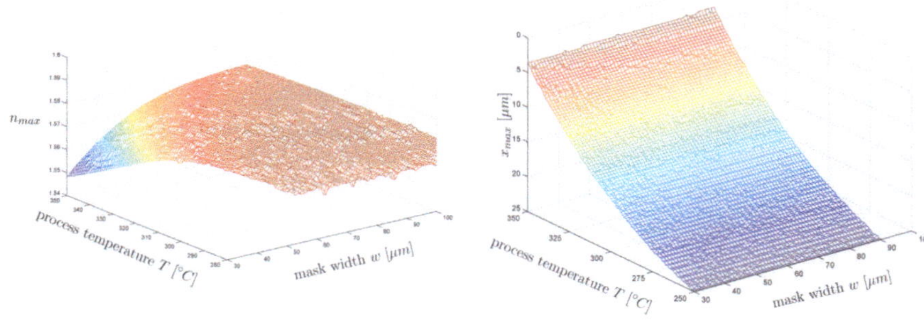

Fig. 9 Position and maximum value of the refractive index in dependency of temperature and mask width

The maximum value of the refractive index is nearly constant for a wide range of the mask width and the process temperature. Only for small mask openings and low temperatures a variation can be observed. The position of the maximum value is nearly independent from the mask width and shows approximately an exponential behavior for changes of the process temperature.

The ratio between the periods of diffusing ions into and out of the glass layer has great influence on the maximum value of the refractive index. Figure 10 displays on the left side the influence of the diffusion time for three values of the back diffusion time. On the right side, the back diffusion time is varied for three fixed values of the diffusion time. The maximum value of the refractive index changes nearly linear for an increasing diffusion time. For increasing back diffusion time with constant diffusion time on the right, n_{max} decays nearly exponentially.

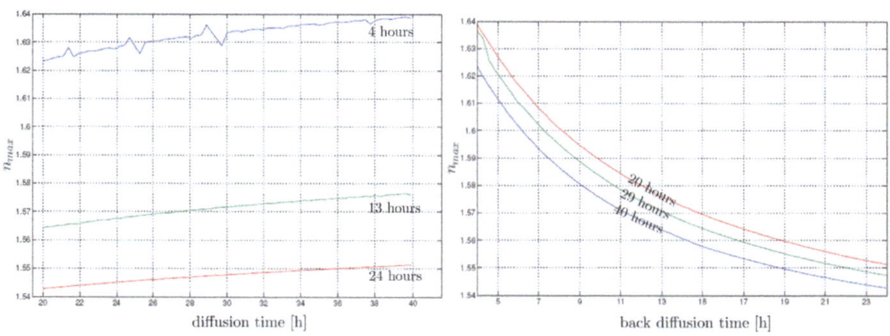

Fig. 10 Maximum value of the refractive index in dependency of diffusion and back diffusion time

With these correlations, first determinations of process parameters can be made, when the position and value of the maximum of the refractive index are given.

6 Conclusions

The demonstrated approach to develop a model to calculate the process parameters for given optical characteristics is very promising. The modeling of the thermal and field-driven ion exchange process is a challenging task but it leads to very good results. The refractive index profile can be calculated the simulated ion concentration with the aid of the Gladstone-Dale relation and is in very good agreement with measurement results. Detailed information regarding the methodology used is given in [11].

The method derived for describing the diffusion process is used to calculate the dependencies between the parameters. The results are examined with regard to correlations. Most dependencies can be approximately interpolated with analytical functions. The first promising results are able to derive process parameters for given optical waveguide characteristics. Further work will concentrate on extending the approach for being able to derive all process parameters for three-dimensional optical structures.

References

1. Photonics21: Lighting the way ahead – second strategic research – Agenda in photonics. European Technology Platform Photonics21 (2010)
2. Izawa, T., Nakagome, H.: Optical waveguide formed by electrically induced migration of ions in glass plates. Applied Physics Letters 21, 584 (1972)
3. Hunsperger, R.G.: Integrated Optics. Theory and Technology. Springer (2001)
4. Forrest, K., Pagano, S.-J., Viehmann, W.: Channel Waveguides in Glass via Silver-Sodium Field-Assisted Ion Exchange. Journal of Lightwave Technology LT-4(2), 140–150 (1986)
5. Lilienhof, H.J., Voges, E., Ritter, D., Pantschew, B.: Field-Induced Index Profiles of Multimode Ion-Exchanged Strip Waveguides. IEEE Journal of Quantum Electronics QE-18(11), 1877–1883 (1982)
6. Ramaswamy, R.V., Najafi, S.I.: Planar, Buried, Ion-Exchanged Glass Wave-Guides - Diffusion Characteristics. IEEE Journal of Quantum Electronics 22(6), 883–891 (1986)
7. Griese, E.: Reducing EMC problems through an electrical/optical interconnection technology. IEEE Transactions on Electromagnetic Compatibility 41(4), 502–509 (1999)
8. Griese, E.: A high performance hybrid electrical-optical interconnection technology for high-speed electronic systems. IEEE Transactions on Advanced Packaging 24(3), 375–383 (2001)

9. Griese, E., Krabe, D., Strake, E.: Electrical-optical printed circuit boards: Technology – Design – Modeling. In: Grabinski, H. (ed.) Interconnects in VLSI Design, pp. 221–236. Kluwer Publisher, Boston (2000)
10. Honkanen, S., West, B.R.: Recent advances in ion exchanged glass waveguides and devices. Physical Chemical Glass Science and Technology B 47, 110–120 (2006)
11. Kühler, T.: Modellierung der Ausbreitungseigenschaften und des Herstellungsprozesses von Gradientenindexwellenleitern in Dünnglasfolien. Shaker-Verlag (2012)

Control and Energy Management of a Cascade Heating System by Fuzzy Logic Control Embedded into a LONWORKS®- LOCAL Operating Network- System

Reza T. Daryani and Alexander Rebel

Cologne University of Applied Sciences
reza.talebi-daryani@fh-koeln.de

Abstract. Generation and consumption of heat power for domestic demand should consider economical and ecological aspects. Fuzzy Logic provides, by evaluation of the thermal behavior of the heating system, a powerful rule base for decision making in order to guarantee optimal operation of the heating system. Analysis of the dynamically thermal behavior of the Building and the heating system was necessary, in order to use measurement information as input variables for different Fuzzy Controllers.

The whole system consists of three different Fuzzy Controllers with the following functions: Fuzzy PID- Controller for a supply temperature Control loop, a Fuzzy Controller for optimal evaluation of heat power demand, and a Fuzzy Controller for the operation of a Cascade Heat Center with high efficiency and lowest contaminated exhaust emission.

1 The Objective Project

The objective of the project was to combine the advantages of control and energy management for cascade heating systems, using fuzzy control with the advantages of LOCAL OPEARTING NETWORKS TECHOLOGY (LONWORKS® technology). [1]

1.1 Germany's Share in Final Energy Consumption as the First Driving Force for the Project Goals

Figure 1 shows the major energy consumption sources in Germany.

As we can see from the figure 1, the major energy consuming sources in Germany and also in many countries are the heating systems for buildings and for Process industries. We can see from the Figure 1 that the Energy Consumption for heating is about 52 % from the total Energy consumption for Germany. The State

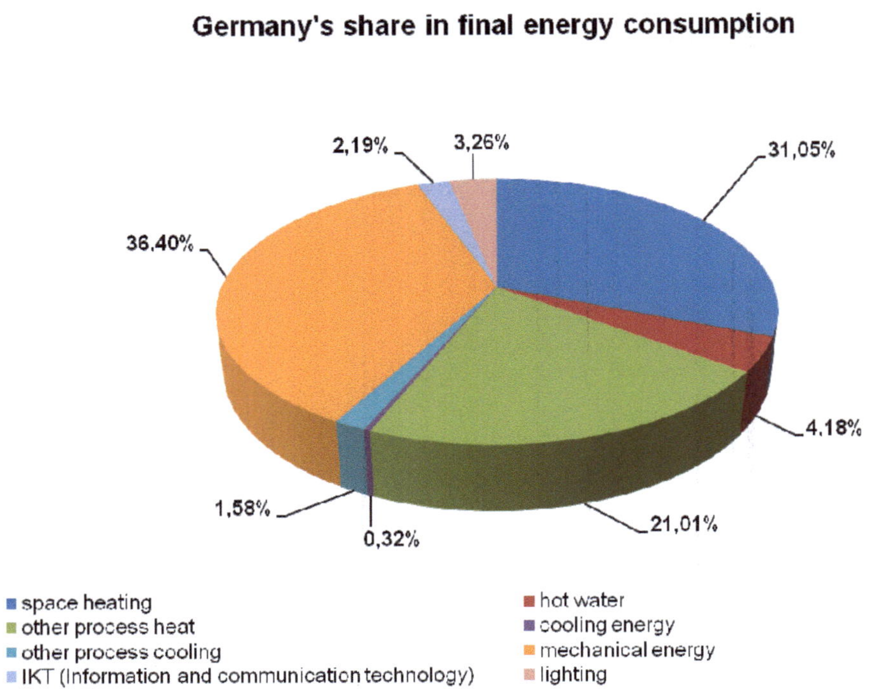

Fig. 1 Germany's share in final energy consumption [2]

of the art of the heating technology is, that most of the building are very old and do not have efficient operating heating systems. These systems are operating in an inefficient way, and produce more energy then, are demanded.

The focus of this project was to optimize the system behavior for a demand oriented energy production from environmental point of view, and to improve the quality of the control loop for Supply temperature.

1.2 Using of Open Network Systems for Building Automation

As the **second driving force** for this project, is using the Open Network Systems for System Integration.

For Building Automation, a powerful automation system should offer open information exchange based on ISO/ OSI- model for different automation units within complex for different aspects/ 3 /. Using intelligent control and management technologies converts the building technologies into smart buildings for efficient system operation and energy management from the environmental point of view.

2 State of the Switching Strategies for Start Stop Control and Optimally Operation of the System

The goal of the developed strategy is to achieve an operation mode for both heaters in an optimal partial load utilization range and a low start stop frequency in order to reduce environmental pollution. Figure 2 shows the impact Emission characteristic of a heating system for contaminated exhaust gas emission.

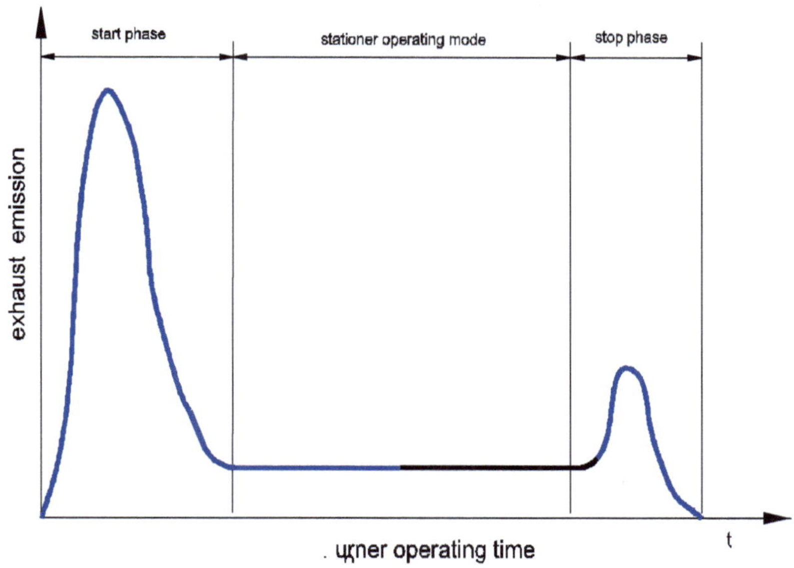

Fig. 2 Emission characteristic of a heating system for contaminated exhaust gas emission [4]

One will also seek to prevent overshooting of the supply temperature through improved supply temperature control which may well lead to switching off of the heater and therefore unnecessary cycles. The autonomous control operation of the heaters is used in this case, that is every heater is fitted with its own temperature sensor.

Figure 3+4 show different switching strategy for FLC based cascade heating system. None of the strategies fulfill the requirements form the optimally operation of the system from economical and environmental point of view.

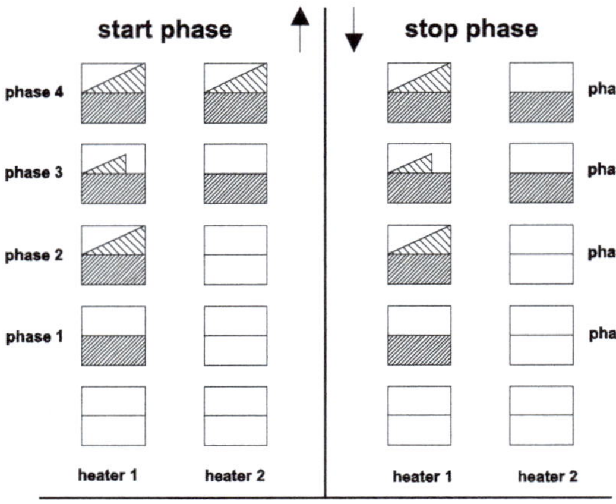

Fig. 3 Serial switching Strategy for 2 heaters cascade

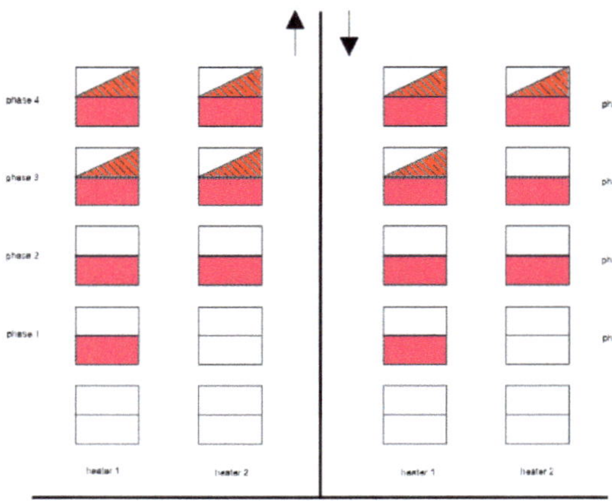

Fig. 4 Parallel switching strategy 2 heaters cascade

2.1 New Switching Strategy for Optimal Operation of the Cascade Heating System

As we can see from figure 5, the second (following) heater is switched on if the burner modulation of the first leading heater is so high that it can be covered by the base load of the second heater.

This avoids the leading heater having to switch into full load mode, which increases the thermal efficiency of the heater. In order to avoid switching, the second heater should not be directly switched on if there is exceeding of the sum of the base loads. It is therefore necessary to implement a hysteresis for.

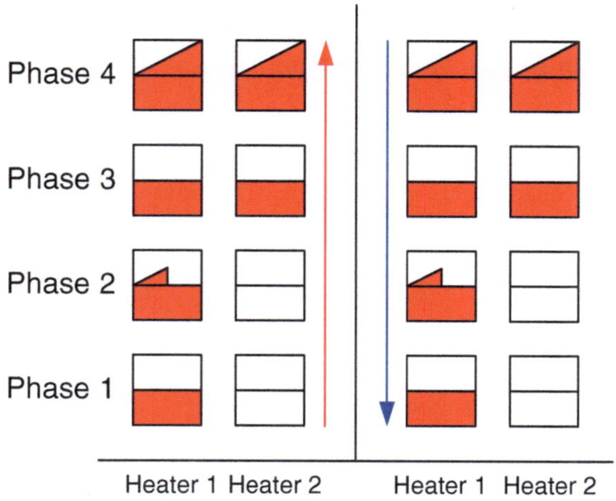

Fig. 5 Optimally switching strategy for 2 heaters cascade

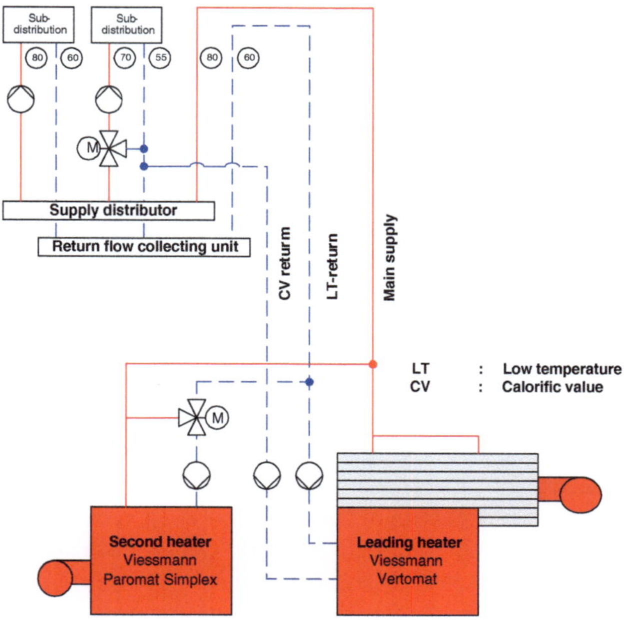

Fig. 6 Cascade heating system

3 Fuzzy Logic Control and LONWORKS®-Network-System for Optimization of the Heating System

3.1 Introduction

On the basis of implementation of the LONWORKS technology, integration of the Fuzzy Logic Control system should, in principle, is possible in every LON-Network in which the control and energy management of a cascade heating system is realized. In order to expand use of this **F**uzzy **L**ogic **C**ontrol (FLC) system to different types of plant, heater and burner types, parameterability of the system is necessary using commercially available network management tools.

The advantages of control and energy management of a cascade heating system using FLC compared to a conventional digital controller were demonstrated in the already completed project "The Switch strategy for cascade heating system and supply temperature control using a FLC system for multiple heater system, as well as development and commissioning of an "Engineering Tool" for system configuration" [6].

This FLC system is realized on a proprietary industrial system controller. Based on this project and the positive results which were achieved using "Intelligent control and energy management of an Air conditioning system with LON and FLC"[XX], the FLC for control and energy management of a cascade heating system on the basis of LONWORKS technology is implemented in this project. The controller nodes (communication participant in a LONWORKS) produced in this way is hereinafter referred to as a fuzzy (BA- system).

There is a cascade heating system (see in figure 6) present in the BA system on which trials were run with the FLC-System. This fuzzy controller was integrated into an existing BA. - System

3.2 The Fuzzy Logic Control System

Figure 7 shows Fuzzy Logic system for control and energy management of cascade heating system. The FLC first takes over autonomous supply temperature control of the heater. This is realized using fuzzy block 1. The second function to be taken over is switching control of the cascade heating system which is realized using fuzzy blocks 2 and 3. The FLC system for control and energy management of the cascade heating system is shown in Figure 3.

Fuzzy block 1 is responsible for the supply temperature control of the system. It calculates the output variable for "modulation" of the burner based on the input variables "set point value" and "process value". Every heater has its own temperature sensor. A fuzzy block 1 is assigned to supply temperature control for each heater which means that each heater autonomously controls the supply temperature.

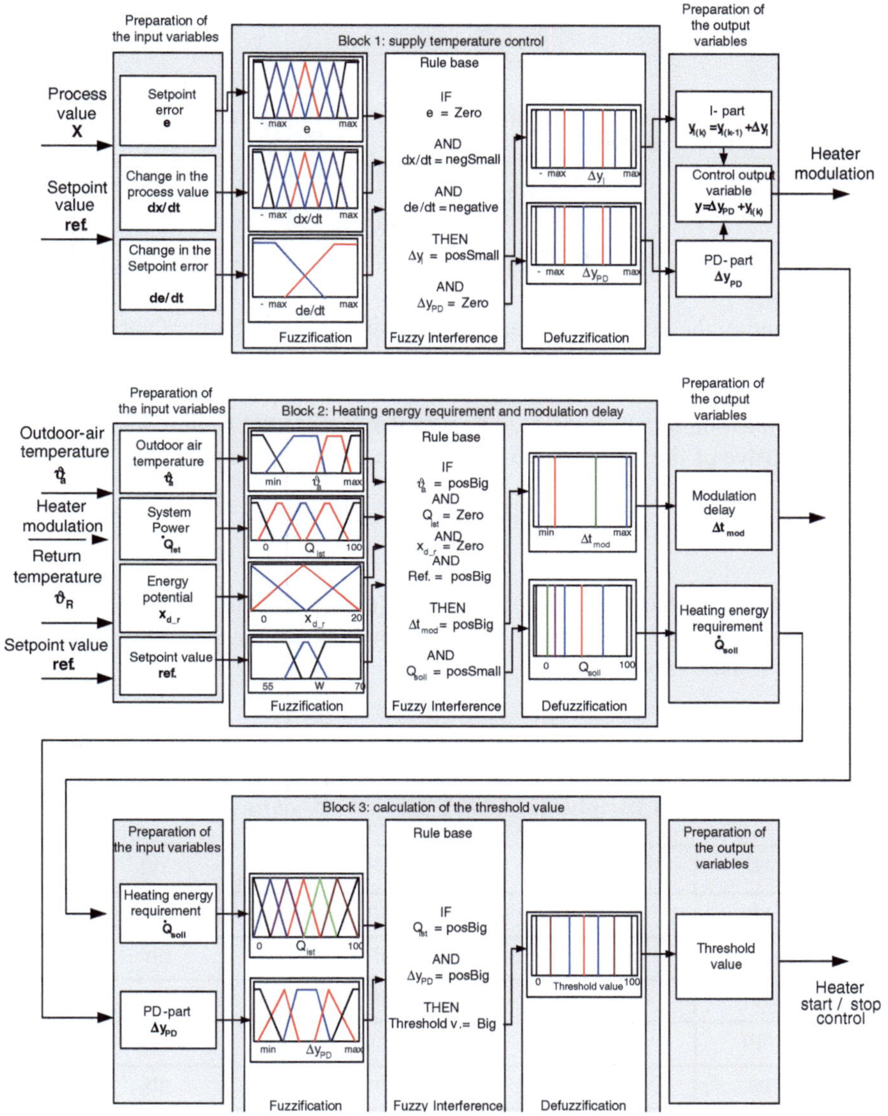

Fig. 7 Fuzzy Logic System for control and energy management of cascade heating system

The set point value for both heaters is determined from a MAX selection of all heating circuits. The input variables for the fuzzy block for supply temperature control are:

- **Set point error xd**
- **Change in the process variable dx/dt**
- **Change in the set point error de /dt**

The **set** point error is calculated using Equation 1.

$$e = ref - x \tag{1}$$

The derivative of the process variable is calculated using Equation 2:

$$\frac{dx}{dt} = \frac{x_{(k)} - x_{(k-1)}}{Tc} \tag{2}$$

With: x (k) ≡ process variable in cycle, x (k-1) ≡ process variable in cycle k-1, Tc ≡ scan time.

By adapting the maximum speed of change of the control output variable to the heater in question, fuzzy block 1 can be adjusted for various different dynamics. The definition range for these input variables is freely scalable for a user in order to be in a position to adjust the fuzzy controller to different heater characteristics. The **derivative of the set point error** is calculated using Equation 3:

$$\frac{de}{dt} = \frac{|e_{(k)}| - |e_{(k-1)}|}{Tc} \tag{3}$$

This input variable has the task of countering small variations in the control value within the range e= 0 [5]. Table 1 shows as an example for the Fuzzy Rules, the rule base for the Fuzzy Control 1.

Table 1 Rules for the PID- Fuzzy Control 1

Fuzzy Input variables				Fuzzy Output variables	
Nr.	**xd**	**dx/dt**	**Δxd**	**ΔyPD**	**Δyi**
1	nb	-	-	nb	nb
2	nb	pb	-	nb	nb
3	nm		-	nb	nb
4	nm	pb	-	nm	nm
5	ns	nb	-	nb	nb
6	ns	nm	-	pm	pb
7	ns	ns	-	pm	pm
8	ns	zo	-	ps	nm
9	ns	ps	-	ns	ns
10	ns	pm	-	ns	ns
11	ns	pb	-	nm	nm
12	zo	nb	-	nm	nm
13	zo	nm	-	pm	pm

Table 1 *(continued)*

14	zo	ns	-	pm	pm
15	zo	ns	n	pm	ps
16	zo	ns	p	zo	ps
17	zo	zo	-	zo	zo
18	zo	ps	n	zo	zo
19	zo	ps	p	zo	zo
Fuzzy Input variables				**Fuzzy Output variables**	
Nr.	**xd**	**dx/dt**	**Δxd**	**ΔyPD**	**Δyi**
20	zo	pm	-	zo	zo
21	zo	pb	-	nm	ns
22	ps	nb	-	nm	nm
23	ps	nm	-	pm	nm
24	ps	ns	-	pm	ps
25	ps	zo	-	pm	pm
26	ps	ps	-	pm	ps
27	ps	pm	-	zo	pm
28	ps	pm	-	nm	nm
29	pm	pb	-	nm	nm
30	pm		-	pm	pm
31	pm	pb	-	pm	nm
32	pb	pm	-	nb	pb

Fuzzy block 1 for supply temperature control has two output variables, these are:

- **Change in the PD part Δy_{PD}**
- **Change in the I part Δy_I**

The control output variable, that is the burner modulation, is calculated from the output variable according to equations 4 and 5:

$$y_{I(k)} = y_{I(k-1)} + \Delta y_I \tag{4}$$

$$y_{PID} = \Delta y_{PD} + y_{I(k)} \tag{5}$$

The **Change in the PD part realizes** the proportional and differential behavior of fuzzy block 1. The set point error influences the P behavior, the change in the control variable and the Derivative part considers the dynamic behavior of the Process. In the case of a change in the control variable of zero, the output factor "change in the PD part" is assigned a proportional control variable change.

The linguistic variables for the output are represented by singletons to reduce the computational effort. Free scalability of the output variables is again realized for parameterability of the fuzzy controller. The change in the Integral part mirrors the integral behavior of the controller in that it is added in every clock step. The integral behavior of the fuzzy PID controller is not realized using an integral term for the input variables, but instead with the output variable "change in the Integral part". **Fuzzy blocks 2** and **3** are responsible for switching control of the heaters and are used just once because they refer to the whole plant.

Fuzzy block 2 calculates the heating energy output requirement and modulation delay based on the input variables "outdoor air temperature", "modulation", "set point value" and "main return flow temperature".

The first input value i.e. Current Heating power demand value Q_t, is the most important factor for evaluation of the start / Stop - point of the heaters. The second input value has been calculated by the first Control block and determines the position of the Control range of the heaters. As soon as this value is higher than 40 %, the Start phase of the second heater will be released. The third input presents the set point error of the system as well as the dynamic behavior of the heating system because of the PID- characteristic of the controller. The fourth indicates the gradient of the thermal energy, which is necessary in order to keep the set point temperature of the system constant. The crisp value of this input is calculated by equation (9) as follow:

$$xd\_r = w - x\_r \tag{6}$$

With: w ≡ reference set point temperature, x_r ≡ system return temperature

Fuzzy Block 3 controls switching of the heaters on and off over the calculated threshold value. The threshold value is calculated based on the input variables "heating output requirement" (from fuzzy block 2) and "change in the PD part" (from fuzzy blocks 1 for the heater).

4 Implementation of the FLC for Control and Energy Management into a LONWORKS Node

The fundamentals for realization of the FLC for control and energy management of a cascade heating system on the basis of LONWORKS technology is use of suitable hardware on which the FLC system is implemented. The LONWORKS hardware, which appears very usable, is the Easylon® EMC^4-4-Channel Multipurpose Controller (EMC^4). The Neuron® 3150, which is implemented in the EMC^4, offers all of the resources needed for implementation based on its 8kByte SRAM and 48kByte FLASH as well as its 10 MHz clock rate. Figure 8 shows Schema of a Neuron Chip host device as the hardware of LON-Technology.

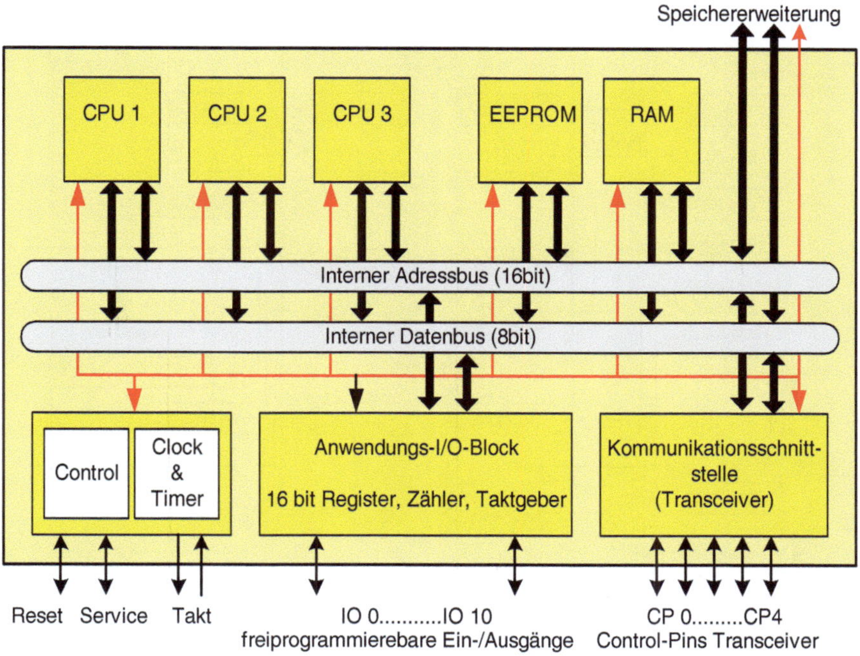

Fig. 8 Schema of a Neuron Chip host device as the hardware of LON-Technology

The software development environment primarily consisted of two programs. Echelon NODEBUILDER® 1.5 is used for the programming and loading of the source codes onto the Neuron chip. The management tool "Alex" from MK Control Systems (known today as SPEGA) is used to test the nodes (e.g. for bindings) and for later integration. The fundamentals for development of a concept for programming is that all input and output variables of the FLC should exclusively be made available on the network side in order to make integration of the building automation system as simple as possible, that is only through execution through bindings in the LONWORKS network. The existing automation stations (DX9200 industrial controller from Johnson Controls) are programmed in such a way, for this purpose, that they provide the required network variables for all input and output variables of the FLC. To commission the FLC the network integrator must primarily perform the bindings needed for each control circuit (process value, set point value and control variable) and activate the FLC. The fuzzy system is not mirrored on the Neuron for the integrator on the network side, but rather objects are realized instead which bring together the relevant functions (supply temperature control and switching of the heater) with network variables. From this it follows that three objects are realized on the fuzzy controller which are decisive for the function and integration of the fuzzy controller. Figure 9 Shows Binding Scheme fort Fuzzy Controllers with the existing automation stations.

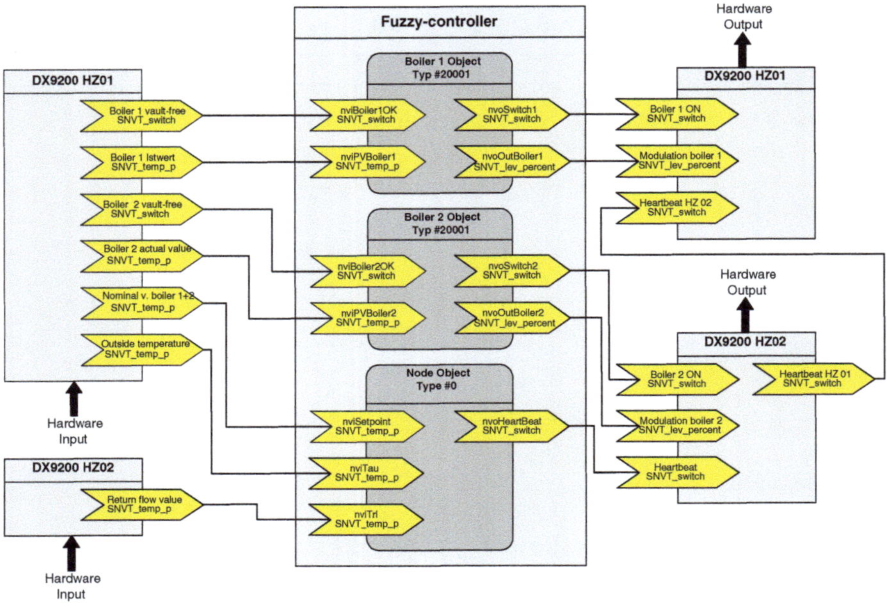

Fig. 9 Implementation of the Fuzzy Control system into a LON-Object

4.1 Description of the Embedded FLC-LON-System

The objects of the fuzzy controller: The objects are there to summarize together network variables and parameters into logical function units. When establishing the network variables one must take account of the fact that the standard network variables (SNVT) used in the fuzzy controller is also free for assignment in DX9200 or is there at all.

As a result of these circumstances it is not always possible to program to be LONMARK® compliant (that is following the FP "Burner and heater Controller"). It can only be verifiably designated as being interoperable in the current situation with the DX9200 used and the application used on it. However, use of SNVT_temp_p, SNVT_lev_percent and SNVT_switch for the most important Network Variables (NVs) allows one to assume interoperability with other I/O modules.

The Node Object – node management

Properties of the FLC systems are determined with the aid of variables and parameters on the node object. Parameters of the node object allow the fuzzy controller to be set up for any cascade heaters with two boilers. All plant-relevant data can be parameterized on this object.

The heater objects – controlling and regulation

All variables which are necessary for control and energy management of the cascade heating system can be found on the boiler objects. Using the parameters in the boiler objects, every boiler object can be set up for the heater in the heating cascade system. The leading heater is basically the first heater while the following heater is the second heater. The parameters for the heater objects take on a very important significance.

The fact that the fuzzy controller should also be integratable into other systems means that the input and output variables for the fuzzy blocks are implemented as a freely scalable parameter for heater supply temperature control.

It is therefore possible to react to a different dynamic for the heater being used. The heater objects can be adapted to all installed heater types and heater outputs using these parameters. The fuzzy controller with options for parameterization can be used universally on the node object through entry of the nominal output of the heater and the form of the characteristic curve of the burner.

5 Results of the System Behavior

The fuzzy PID Controller (fuzzy block 1) reduces the response time and the overshoot range of the process value compared to digital PID- controller. The unsharp points in time for switching on and off of the fuzzy system also reduces the frequency of switching of the following heater and the heater operates within the optimal efficiency range.

The fuzzy controller realized here can be integrated into all LONWORKS networks and therefore into different plants.

The **"Viessmann Vertomat-heater (lead heater)** " presents a higher degree of difficulty than the "Viessmann Paromat". To prevent strong overshooting of the supply temperature and ensure a smooth control behavior, the definition range for the maximum set point error e is scaled to e max= ±22K. Furthermore, the output variable "change in the PD part„ is scaled to a maximum of $\Delta\ y_{PDmax}= \pm 55\%$ and the "change in the integral part „to a maximum of $\Delta\ y_{Imax}\ \pm 2\%$, in order to prevent excessively strong "integration" of the integral part. Figure 9 shows set point step response for the "Viessmann Vertomat-heater by fuzzy control -block 1 for heater 1. The diagram shows the maximum overshoot range of 2K and control time of 360 seconds with the parameters mentioned for fuzzy block 1 for heater 1.

From Figure 10 one can clearly see the response after set point change for the supply temperature and the increase in modulation of the burner. The **"Viessmann Paromat"** presents a lesser degree of difficulty. Figure 10 shows a set point step response of the "Viessmann Paromat" by fuzzy control block 1 for heater 2.

From Figure 11 it is clear to see a strong increase in modulation after the set point change for the supply temperature.

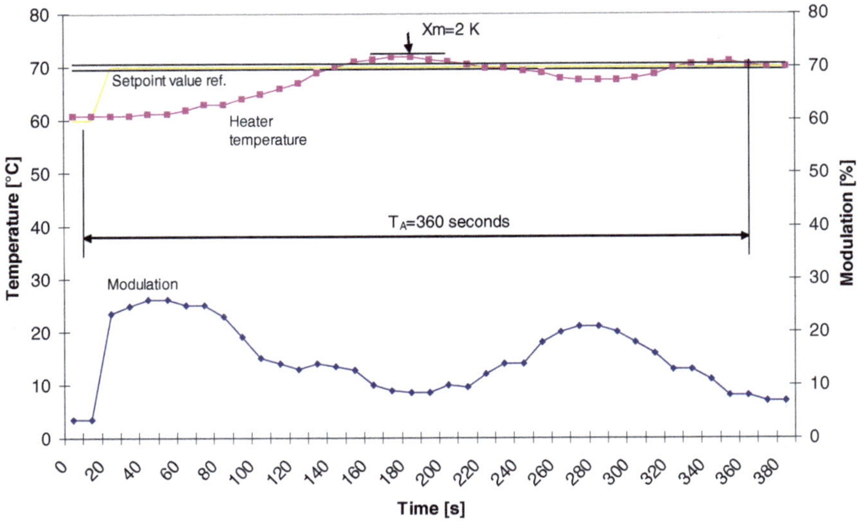

Fig. 10 Set point step response of the Fuzzy control system for the "Viessmann Vertomat"

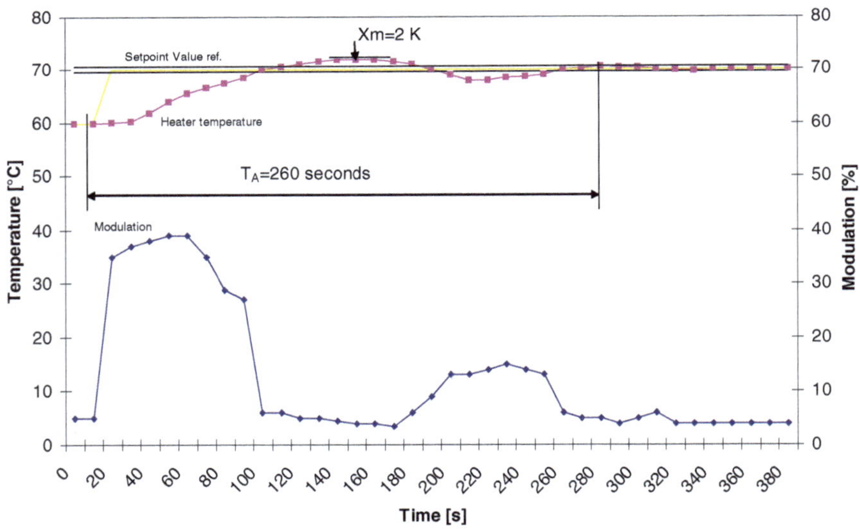

Fig. 11 Set point step response of the Fuzzy control system for the "Viessmann Paromat-heater"

5.1 Comparison of the Results of Boiler Temperature Controller Loops

Despite a different heater dynamic of the Vertomat-heater compared to the Paromat-heater, optimal control behaviour for both heaters is achieved through the design of the input and output variables of fuzzy block 1. The overshoot range is the same for both heaters, while the control time makes clear the difference in the difficulty of the controlled heaters.

6 Summary

The objective of this project was realization of control and energy management of a cascade heating system by Fuzzy Logic Control based on the open network standard LONWORKS. Integration of FLC should be achieved in an existing buildings automation system. To do this a FLC system was adapted for control and energy management of a cascade heating system, under the framework conditions which are dictated by the use of LONWORKS technology. The FLC system was implemented on a LONWORKS node. The solution is use of the Easylon® EMC^4 4-Channel Multipurpose Controller from Gesytec in Aachen. The Neuron 3150 chip in this hardware fulfills all requirements concerning computing power, storage capacity and programmability with the development environment available. The nodes were developed from a software point of view on the Node Builder. Combined use of Node Builder 1.5 with the binding tool "Alex" dispenses with the necessity to use a Lon Builder to test the network behavior of the node. The result is a fuzzy controller for control and energy management system of a cascade heating system with two heaters and modulated burners, which is available as a Neuron-Chip Hosted Device on the open network standard LONWORKS. Use of the LONWORKS technology allows exclusive parameterability of the FLC system over network management. The parameterability of the input and output variables of fuzzy block 1 for supply temperature control allows optimization of the system behavior of individual heaters with differing boiler dynamics. Integration and commissioning of the fuzzy controllers are exclusively possible on the network side through use of all LONWORKS network management tools available on the market. In this project the FLC system for control and energy management of a cascade heating system is implemented on **one** LON node. Extension of this development to a distributed system with a number of nodes is very possible. The Fuzzy Control system introduced here serves as a Control - and operation management system for a Cascade Heating System. Three different Fuzzy controllers have been realized, in order to optimize the system's thermal features from an economical and ecological point of view. To fulfill these requirements, analysis of the thermal behavior of the Building and the heating system was necessary in order to formulate proper input and output variables for the Fuzzy Controller. This operation strategy of the Cascade Heating System avoids the thermal loses of the system and reduces the start / stop frequency of the burner's to a minimum. The Supply temperature Control loop of the system is designed and commissioned as a non-linear Fuzzy PID - Controller for a non linear thermal process. This kind of controller can be described as a robust Control System. This control and operation management system provides a real demand oriented heating energy with a minimum of fuel consumption and therefore with a minimum of contaminated exhaust gas emissions. The most important features of this new system is to expand this system to different type of heating systems on Site, with just little changes of the Fuzzy Input Variables -scaling and reconfiguration of the System characteristics. There is no need to design new rule base for the Fuzzy system, and so the requirements of know how about Fuzzy Control for System engineers are very little. Therefore it is a success promising solution for the wide use of Fuzzy Logic Control and Open Network System within the Building automation and Building Energy Management systems.

References

1. Talebi-Daryani, R.: Introduction of LON-Technology for Building automation, Textbook in German, Cologne University of Applied Sciences (2005)
2. Germany's share in final energy consumption: Federal Ministry for Economics and Technology (October 27, 2011), `http://www.bmwi.de/BMWi/Redaktion/Binaer/Energiedaten/energiegewinnung-und-energieverbrauch5-eev-nach-anwendungsbereichen,property=blob,bereich=bmwi,sprache=de,rwb=true.xls`
3. Talebi-Daryani, R.: Text book for Building automation Part II, Cologne University of Applied Sciences (1995)
4. Talebi - Daryani, R., Olbring, M.: Application of fuzzy control for energy management of a cascade heating system, soft computing, multimedia, and image processing, vol. 11, pp. 618–625. TSI Press Series, Albuquerque (2000) ISBN 1-889335-13-4
5. Talebi - Daryani, R., Plass, H.: Application of fuzzy control for intelligent building part I: fuzzy control for an AC system, intelligent automation and control. In: Proceedings of the WAC 1998, pp. 745–750. TSI Press Series, Albuquerque (1998) ISBN 0-9627451-7-0
6. Talebi-Daryani, R., Pfaff, J.: Intelligent control and power management of air conditioning systems using Fuzzy logic and LOCAL OPERATING SYSYTEMS. In: World Automation Congress. Congress Proceedings-CD, Orlando Florida. TSI-Press (2002) ISBN: 1-889335-17-17
7. Rebel, A.: Programming and implementing of control and energy management of a cascade heating system by Fuzzy Logic Control and LonWorks. Unpublished Master Thesis, Cologne University of Applied Sciences, Laboratory for control engineering and energy management systems (2004)

Chapter 3

Knowledge Driven Approaches for Product Engineering

Diagnostics in Lithium-Ion Batteries: Challenging Issues and Recent Achievements

S.M. Mahdi Alavi[1], M. Foad Samadi[2], and Mehrdad Saif[1]

[1] Department of Electrical and Computer Engineering,
University of Windsor, Windsor, Ontario, Canada, N9B 3P4
[2] School of Engineering, Simon Fraser University,
Burnaby, British Columbia, Canada, V5A 1S6
msamadi@sfu.ca,{malavi,msaif}@uwindsor.ca

Abstract. This paper highlights some of the issues which have made diagnostics of Lithiumion (Li-ion) battery energy storage systems very challenging from the both chemical and control engineering perspectives. The application of standard observers, Kalman, and particle filters in state estimation of Li-ion batteries are fully reviewed. Recent achievements in this field, pros and cons of various approaches, and future direction for research are outlined.

Keywords: Diagnostics, Battery Energy Storage.

1 Introduction

The use of Lithium-ion (Li-ion) batteries has increased at an unprecedented rate in many devices such as cell phones, laptop computers, hybrid and full electric vehicles, etc. This is because of their significant advantages in terms of energy density, life time, no memory effects, and a slow self-discharging rate when not in use, [22]. However, many micro-scale failure mechanisms have been identified which degrade the Li-ion batteries performance. References [3], [4] and [17] provide comprehensive surveys of the identified failure mechanisms in Li-ion batteries. These failures may lead to abrupt or gradual battery degradation and in some cases to irreparable damage or dangerous failure conditions. Thus, it is very important to detect these failures as soon as they occur, and to take necessary actions and have appropriate protection mechanisms in order to maintain the battery performance at a satisfactory level and within the desired specifications and operating range.

The goal of this paper is to introduce *model based fault diagnosis* problem in Li-ion batteries. Fundamentals of the model-based fault diagnosis theory are firstly outlined in Section 2. Section 3 deals with modeling of Li-ion batteries. Both equivalent circuit and electrochemical model development approaches are reviewed. Section 4 highlights some of the issues which make diagnostics in Li-ion batteries very challenging from both chemical and control engineering perspectives. The application of standard observers, Kalman, and particle filters in state estimation of Li-ion batteries are fully reviewed in section 4.2. Along this line, recent achievements, pros and cons of various approaches, and future direction for research are outlined.

2 Fundamentals of Model-Based Fault Diagnosis

The objective of a model based fault diagnosis system is to monitor a dynamical system against incipient or complete failures and ensure that that system continues its operation safely, efficiently, with more autonomy, and less interruption in service for extended periods of time.

A complete fault diagnosis system has fault detection, isolation, estimation and accommodation capabilities and each of these tasks may be accomplished within a module. The Fault Detection (FD) module determines whether a fault has occurred in the plant. The fault isolation module determines the fault's type and its location. The fault estimation module estimates the extent of failure, and finally the fault accommodation is responsible for reconfiguration of the control system so that the system can continue to operate until such time that timely repair can be performed, [5].

Common approaches to fault diagnosis can be grouped into four broad areas:

- *Temporal redundancy* which uses limit and trend checking on inputs and outputs based on some priori information about the system to only detect failures. This approach is perhaps the most commonly used one in industry today. Expert and knowledge based schemes may also be grouped under this classification.
- *Hardware redundancy* which uses majority vote ruling logic for FD. Although popular in safety and mission critical systems, the cost alone is a major reason for not employing this scheme in variety of applications.
- *Analytical redundancy* or *model based* technique which uses the system's model and measurements to generate software quantities, called *residuals*. The residuals are sensitive to the faults, and processed for detection, isolation, estimation and accommodation.
- *Algorithmic redundancy* which is obtained by combining different algorithms from the same or different classes.

With an ability to detect variety of failures at a lower cost with a lot less calibration and testing, model-based fault diagnosis has been the focus of many research since 1970, [9]. Another important feature of the model-based fault diagnosis is that the concept is easily transferable from one application to another.

In a model-based fault diagnosis system, FD module plays an important role. The typical FD module consists of two parts, i.e. residual generation and evaluation, as shown in Figure 1. The box labeled "Plant" represents the dynamical system that is being monitored. The general nonlinear model of the plant is given by:

$$\dot{z}(t)=g(z(t),u(k),d(t),f(k)) \tag{1}$$

$$y(t)=h(z(t),u(t),d(t)) \tag{2}$$

where $z\in\Re^n, u\in\Re^m$ and $y\in\Re^p$ denote system states, input and output vectors, respectively. $d\in\Re^i$ and $f\in\Re^q$ are disturbance and fault vectors, respectively. g and h are nonlinear vector functions describing the plant dynamics, respectively.

The residual generation sub-module receives both output and input signals $y(t)$ and $u(t)$, and generates a residual signal which is sensitive to faults. The feature of the generated residual signal is that it is almost zero when there is no fault in the plant[1]; when a fault occurs, the residual signal becomes large. The task of the residual evaluation sub-module is to produce appropriate fault alarms. Typically, the fault alarm is activated when the energy of the residual signal becomes greater than a certain threshold value J_{th}. In most of FD techniques, J_{th} is set to the supremum of the energy of the residual signal over the run-time when there is no fault in the system [13]. For the purpose of fault isolation in the presence of multiple failures, a bank of residuals is generated, each of which dealing with one fault.

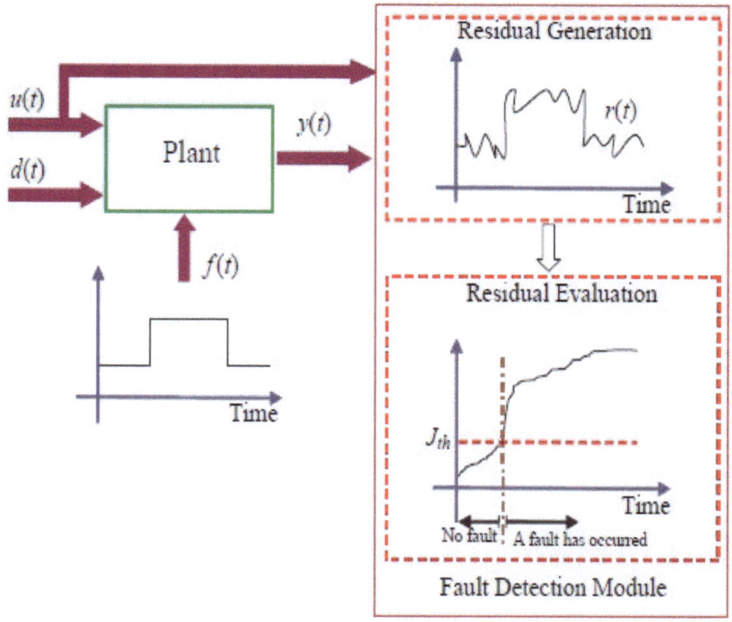

Fig. 1 The structure of Fault Detection (FD) module

[1] It should be noted that the residual signal is almost zero due to unavoidable noise and uncertainties in the plant, otherwise, under ideal conditions, it could be designed to be zero.

2.1 Residual Generation

Residual generation process is an important part in the design of a model-based FD module. The majority of the proposed techniques are either based on parity-relation or estimation concepts. In the parity-relation based technique, the fault detectability issue is converted into a matrix rank condition, [8]. Although, the parity-relation residual generation approaches are simple, they are suitable for application to linear systems with certain dynamics. The parity based residual generation becomes difficult when nonlinearity, model uncertainty, time-delay, etc., are added into the system structure, [6].

In the estimation based residual generation, an indirect estimate of the system output or fault is obtained by using observers or filtering techniques. The general model of the proposed observers estimating the system output with linear injection of the error, e, between the estimated and actual measurement is given by:

$$\begin{aligned} \dot{\hat{z}}(t) &= g(\hat{z}(t), u(t)) + \gamma\, e(t) \\ \hat{y}(t) &= h(\hat{z}(t), u(t)) \\ e(t) &= y(t) - \hat{y}(t) \end{aligned} \tag{3}$$

where $\hat{z} \in \Re^n$ and $\hat{y} \in \Re^p$ denote estimated states and output vectors. The observer gain γ is designed such that $e(t)$ becomes as small as possible when time goes to infinity, i.e., minimize e as $t \to \infty$. This residual generation/FD methodology is well-known as *output observer residual generation.*

Since the mid 1990s, more attention has been paid to direct estimate of the fault. Typically, a linear FD filter in the form of

$$\begin{aligned} \dot{w}(t) &= A_f \dot{w}(t) + B_f [u(t)\ y(t)] \\ r(t) &= C_f w(t) + D_f [u(t)\ y(t)] \end{aligned} \tag{4}$$

is designed which takes information from the system output and input y and u respectively, and generates a residual signal $r(t)$ which is sensitive to the fault. In (1.4), w denotes the FD filter state vector; A_f, B_f, C_f and D_f are FD filter parameters which are designed such that the error between the fault and the residual is minimized, i.e.,

$$\text{minimize } e(t) = r(t) \quad f(t) \tag{5}$$

where, f denotes the system fault.

There is an extensive body of literature addressing the robustness issue in relation to model uncertainty and nonlinearity, exogenous disturbances and noise, time-delay, packet drop in networked system, etc., which is beyond the scope of this paper. Interested readers are directed to consult [9] and [1] for more information.

In the following, application of the outlined FD techniques to Li-ion battery energy storage systems is investigated. First, mathematical models of the battery system are reviewed.

3 Mathematical Models of the Lithium-Ion Batteries

Two mathematical models for Li-ion batteries have widely been employed in the literature. These are: Equivalent Circuit Model (ECM) and Electrochemical Model (EM).

3.1 Equivalent Circuit Model (ECM)

In the ECM, the battery is modeled as an electric circuit consisting of a number of Direct Current (DC) voltage sources, resistors, capacitors, and nonlinear functions accounting for the un-modeled dynamics, as shown in Figure 2. The values of resistances and capacitances are obtained through electrochemical impedance spectroscopy, [22]. The dynamic of nonlinear function, denoted by h in Figure 2, also depends on many factors, the battery State of Charge (SOC), State of Health (SOH), temperature, etc.

The Li-ion battery ECM is very popular from a control engineering perspective for their simplicity, and Ordinary Differential Equations (ODEs) framework is used for modeling. They can easily be converted into state-space models where various controls and diagnostics tools are applicable. However, the validity and accuracy of the ECMs remain questionable for high power applications such as hybrid electric vehicles. For these kind of applications, at least a more complicated ECM is desired which leads to the computational burden and design complexities. More importantly, the ECM provides no physical interpretation about the chemical interactions taking place inside the battery, [10].

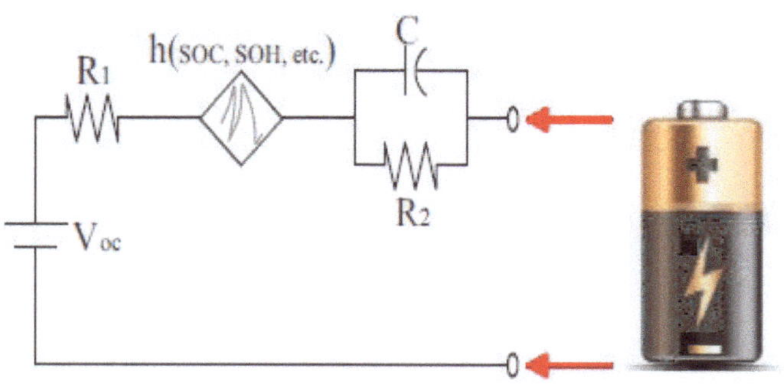

Fig. 2 A sample equivalent circuit model for the battery energy storage

3.2 Electrochemical Model (EM)

In the EM, the battery dynamic is formulated by using electrochemical principles. The Li-ion battery has four main components as shown in Figure 3, porous negative electrode, porous positive electrode, electrolyte and separator. The lithium ions, Li^+, leave the positive electrode and enter the negative electrode in the charge process, and vice versa during the discharge. The electrolyte enables these movements, and the separator is an electrical insulator that does not allow electrons to flow between the positive and negative electrodes.

Several electrochemical models with different degrees of complexity have been developed for Li-ion batteries, all of which are based on the model proposed by Newman and Tiedemann in [21]. By defining the battery parameters as in Table 1, the one-dimensional (1D-spatial) model of the Li-ion battery which only considers dynamics along the horizontal axis x and ignores the dynamics along y and z axes is given by [7]:

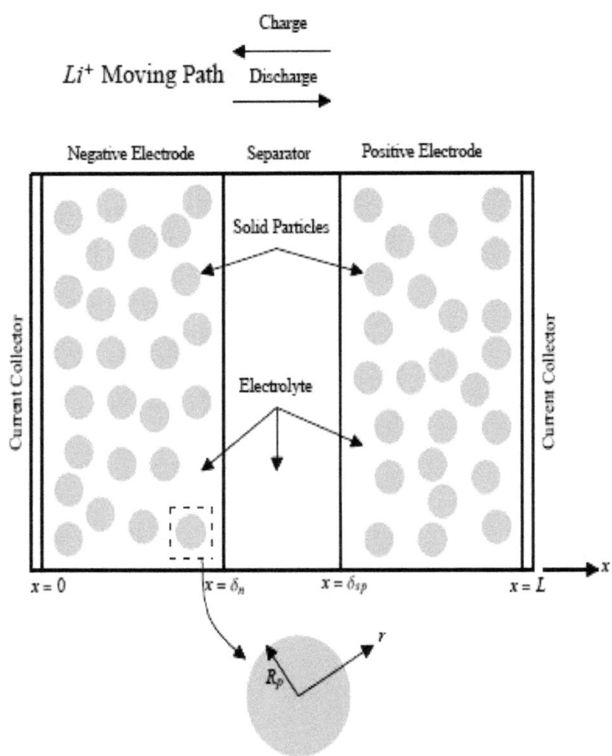

Fig. 3 A schematic of the Lithium-ion (Li-ion) batteries, [12]

$$
\begin{aligned}
\frac{\partial C_e(x,t)}{\partial t} &= \frac{\partial}{\partial x}\left(D_e \frac{\partial C_e(x,t)}{\partial x}\right) + \frac{1}{F\varepsilon_e}\frac{\partial\left(t_a^0 i_e(x,t)\right)}{\partial x} \\
\frac{\partial C_s(x,r,t)}{\partial t} &= \frac{1}{r^2}\frac{\partial}{\partial r}\left(D_s r^2 \frac{\partial C_s(x,r,t)}{\partial r}\right) \\
\frac{\partial \Phi_e(x,t)}{\partial x} &= -\frac{i_e(x,t)}{k} + \\
&\frac{2RT}{F}\left(1-t_c^0\right)\left(1+\frac{\partial \ln f_{c/a}}{\partial \ln C_e}(x,t)\right)\frac{\partial \ln C_e(x,t)}{\partial x} \\
\frac{\partial \Phi_s(x,t)}{\partial x} &= \frac{i_e(x,t)-I(t)}{\sigma} \\
\frac{\partial i_e(x,t)}{\partial x} &= \frac{3}{R_p}\varepsilon_s F j_n(x,t)
\end{aligned} \tag{6}
$$

where, t and r denote the time index and position in the radial coordinates, respectively. The molar flux $j_n(x,t)$ is computed according to the Butler-Volmer equation.

Table 1 ISDT Conferences (ISDT "Table description")

Symbol	**Definition**	**Unit**
i_e	Electrolyte current density	Acm^{-2}
i_s	Solid current density	Acm^{-2}
Φ_e	Electrolyte potential	V
Φ_s	solid potential	V
C_e	Electrolyte concentration	$molcm^{-3}$
C_s	Solid concentration	$molcm^{-3}$
C_{se}	Solid concentration at surface	$molcm^{-3}$
$C_{s,max}$	Maximum possible solid concentration at each electrode	$molcm^{-3}$
j_n	Butler-Volmer current density	Acm^{3}
j_0	Exchange current density	Acm^{-2}
U	Open circuit voltage	V
η	overpotential	V
F	Faraday's constant	$Cmol^{-1}$

Table 1 (*continued*)

R	Gas constant	$JK^{-1}mol^{-1}$
T_b	Temperature	K
I	Battery current	A
V	Battery voltage	V
$f_{c/a}$	Mean molar activity coefficient in the electrolyte	-
κ	Ionic conductivity of the electrolyte	$\Omega^{-1}cm^{-1}$
t_c^0	Transference number of the cations w.r.t. the solvent velocity	-
D_e	Electrolyte diffusion coefficient	cm^2s^{-1}
D_s	Solid state diffusion coefficient	cm^2s^{-1}
ε_e	Volume fraction of the electrolyte	-
ε_s	Active material volume fraction	-
σ	Conductivity of solid active material	$\Omega^{-1}cm^{-1}$
t_a^0	Transference number of the anion	-
R_f	Film resistance of the solid electrolyte interphase	Ω
α_a, α_c	Transport coefficients	-
R_p	Particle radius	cm
A	Electrode plate area	cm^2
δ	Electrode thickness	cm

in the both electrodes as follows:

$$j_n(x,t) = \frac{3}{R_p}\varepsilon_s j_0(x,t)\left[\exp\left(\frac{\alpha_a F}{RT_b}\eta_s(x,t)\right) - \exp\left(\frac{-\alpha_c F}{RT_b}\eta_s(x,t)\right)\right] \tag{7}$$

where,

$$j_0 = C_e(x,t)^{\alpha_a}\left(C_{s,max} - C_{se}(x,t)\right)^{\alpha_a} C_{se}(x,t)^{\alpha_c} \tag{8}$$

and,

$$\eta_s = \Phi_s(x,t) - \Phi_e(x,t) - U(C_{se}(x,t)) - FR_f\, j_n(x,t) \tag{9}$$

In the above equations, $C_{se}(x,t)$ is defined as the concentration at the surface or interface of the solid state, i.e.,

$$C_{se} = C_s(x, R_p, t) \tag{10}$$

Finally, the voltage of the cell is given by:

$$V(t) = \Phi_s(x = L) - \Phi_s(x = 0) - R_f I \tag{11}$$

In order to solve the above Partial Differential Equations (PDEs), the following set of boundary conditions is necessary.

$$\begin{gathered}
i_e(x=0,t)=0 \\
i_e(x=L,t)=0 \\
i_s(x=0,t)=\frac{I(t)}{A} \\
i_s(x=L,t)=\frac{I(t)}{A} \\
i_s(x=\delta_n,t)=0 \\
i_s(x=\delta_{sp},t)=0 \\
\frac{\partial C_s(x,r,t)}{\partial r}\Big|_{r=0}=0, \forall x \\
D_s\frac{\partial C_s(x,r,t)}{\partial r}\Big|_{r=R_p}=-\frac{j_n}{3\varepsilon_s F}R_p \\
\frac{\partial C_e(x,t)}{\partial x}\Big|_{x=0}=\frac{\partial C_e(x,t)}{\partial x}\Big|_{x=L}=0
\end{gathered} \tag{12}$$

Remark 1: *To simulate the battery dynamics, the above PDEs have to be solved simultaneously for the both positive and negative electrodes.*

4 Challenges and Achievements in Li-ion Battery Diagnostics

Diagnostics challenges in Li-ion batteries can be divided into two categories: those related to the battery chemistry, and systems/control challenges.

4.1 Challenges of Battery Chemistry

There are some electrochemistry challenges making diagnostics in Li-ion battery very difficult:

1. Many degradation (failure) mechanisms have been identified or postulated in Li-ion batteries. Table 2 lists a number of those. This issue makes finding a unique experiment or modeling strategy that elucidates degradation as a general phenomenon very challenging. This remains an open problem in the battery research area.
2. There is no mathematical model for many of the identified faults. This makes the task of residual generation very difficult, which adversely affects fault isolation and also accommodation.
3. Many different conditions may lead to the same fault. For instance, the loss of connectivity has been attributed to the movement of conductive carbon reducing electron transport within the electrode in [19], to particle fracture in [14], to precipitation of thick surface films in [23], to gas generation in [28], to loss of contact between active material and the current collector in [27] or between the current collector and the cell housing in [11], and to degradation of the binder in [15]. This is another issue making fault isolation really difficult.
4. Another major issue is related to the threshold value selection in the design of the residual evaluation submodule as mentioned in section 2. Clearly, it is highly dependent on the battery technology, i.e. what kinds of materials have been used inside the battery. These values should be modified by taking the battery's charge-discharge cycles and age into account. To date, there is no literature describing how to set up the fault detection threshold values in battery energy systems.

Table 2 A number of identified failures in Li-ion batteries

Loss of electrical contact between metallic grids and active materials
Current collector corrosion
Solid-Electrolyte Interface (SEI) layer
Electrolyte decomposition
Positive electrode dissolution
Electrode distortion, Disorder or fracture in lattice structure of electrodes
Loss of plate active surface area
Growth of large inactive materials
Plating
Loss of active material
Decrease in the diffusion coefficient in the negative electrode
Porosity change of the electrode
Change of particle size, Electrochemical grinding
Battery swelling
Increase of electrode's impedance
Binder decomposition

Addressing these issues has been the focus of many research themes during the past few years. Recently, it was shown in [16] that a general study of the Li-ion batteries degradation can begin with the measurement of Lithium ions transport and insertion into porous electrodes. From the above statements and by looking at the failure listed in Table 2, it is clear that a micro-scale measurement/observation of the electrochemical reactions is needed for the design of an efficient fault diagnosis in Li-ion batteries. In battery energy storage systems, the commonly available data are the battery current, voltage and temperature. These measurements themselves do not contain any microstructural information. Thus, an important question remains: *"is it possible to estimate internal states of the battery just by getting feedback from these measurements"*.

4.2 System and Control Related Challenges

A possible solution to the above question comes from the heart of estimation theory. By applying estimation theory, internal states of a dynamical system can be obtained by using limited measurement. Standard observers, Kalman and particle filters are most widely used estimators in the literature. They have successfully been developed and applied to many applications; chemical processes, robotics, automotive, power grids and flight systems to name but a few.

More recently, significant attempts have been made to develop and apply these techniques to battery energy storage systems, [2], [12], [18], [20], [24], and [25]. However, there are some challenging issues which make the problem very difficult to solve from the systems and control perspective.

The first fundamental issue arises from the fact that the battery dynamic consists of highly interconnected nonlinear PDEs. How this issue adversely affects the design of standard observers, Kalman, and Particle filters is addressed next.

The Battery Estimation Using Standard Observers

Observability conditions and stability/convergence of the estimation error are the most significant concerns in development of standard observers to the battery internal states estimation, [18], [20]. In the standard observers, the error between the actual and estimated measurement is injected into the estimator as in (1.3). The observer gain is then designed such that the estimation error converges to zero as time goes on. There is few literature dealing with the observer design in dynamical systems described with PDEs and only certain class of systems have been considered, [26]. In battery systems, estimation of the internal states by using the original PDEs and standard observer is still an unsolved problem.

In [18] and [20], the model of battery is firstly simplified, and then an observer with the linear injection error is designed. In [18], it is assumed that the Li concentration in the electrolyte is constant, i.e., C_e = constant, therefore the order of the PDE is reduced. In [20], the Single Particle Model (SPM) of the battery is employed, for which the whole electrode is considered as a single particle, i.e., the x variable is eliminated from the equations. It was shown in [12] that the battery model is weakly observable from the voltage measurement. It is almost impossible to estimate internal states of the battery in the both positive and negative electrodes by using just one observer. In other words, two separate observers should be designed, one for the positive electrode and one for the negative electrode. In order to improve the observability conditions, the SPM in [20] is reduced by approximating the cathode diffusion dynamics using its equilibrium.

In [18], the observer gain is designed such that the conservation of mass does not change during the system run-time. Therefore, the observer gains in both electrodes are exactly the same with an opposite sign. However, estimation error convergence has not been addressed yet. In [20], the observer is designed by using the backstepping design technique. The stability criterion is converted into some conditions in the form of the Klein-Gordon equation, for which its analytical solution does exist in the PDEs literature.

The Battery Estimation Using Kalman Filters

To use Kalman filters, model simplification and local linearizing around the equilibrium are necessary. In [12] and [25], the battery single particle and average models have been employed such that the use of extended and unscented Kalman filters become feasible. Therefore, there would always be an error between the estimated and actual states because of the simplifications that have been applied to the model. Another issue in the use of Kalman filters is related to their computational burden which is relatively high compared to the standard observers.

The Battery Estimation Using Particle Filters

Given a system with nonlinear and/or non-Gaussian noise, particle filters offer an appealing alternative approach compared to Kalman filtering-based methods for which no restrictive assumptions about the nature of the dynamics and form of conditional density has been made. Particle filters are based on Monte Carlo simulation and averaging technique, consisting of prediction and updating processes similar to Kalman filters.

A number of particles are selected at the beginning of estimation process. In the prediction stage, the posterior system states and output are calculated for all particles. The particles are weighted if needed. The error between the estimated and actual system outputs, and the Probability Distribution Function (pdf) of the error are then computed. In the updating stage, the likelihood of each a priori estimate is firstly normalized. Resampling process is then performed to select the particles that result in the best pdf error. The filtered particles are chosen as the priori states to be used in the prediction stage. The average of the filtered states forms a sub-optimal estimate of the system state.

Recently, the application of particle filters in states estimation of the Li-ion battery has been studied in [2] and [24]. Both full and reduced order models have been employed. The results are very satisfactory, however, estimation run-time of full order model is very high in comparison with the aforementioned Kalman filters and standard observers. Since particle filters are based on averaging, stability of the estimation error is also guaranteed provided that the battery is under control.

As a concluding remark, particle filtering technique is the only approach that is able to estimate the battery internal states by using its full-order model such that the stability of estimation error is also guaranteed. However, reducing its computational burden to be applicable to diagnostics in battery energy storage systems needs more research.

4.3 Further Issues

All the aforementioned estimation techniques have been concerned with the battery at the cell level. In many applications, a number of battery cells are appropriately combined together in order to supply more electric power. As it was seen in earlier sections, diagnostics in a battery cell itself leads to a number of issues which have not been fully addressed yet. In battery pack diagnostics, clearly, it is impossible to work with individual cells; apart from the design and computational complexities, huge numbers of sensors are also required which will increase the size, wiring, cost, etc. Thus, transferring of the condition monitoring and control knowledge from a battery cell to a battery pack based on the electrochemical model is still an open problem.

5 Conclusions

As discussed in this paper, state estimation is necessary for efficient diagnostics of Li-ion batteries, in particular for residual generation and evaluation. Some recent achievements in development of standard observers, Kalman and particle filters to battery state estimation have been reviewed. Challenging issues and open problems have been outlined. In summary, generalization of standard observes and Kalman filters to systems described by PDEs remain to be a problem. Although particle filter can address this issue, its very high computational burden is a big challenge in practice. Moreover, development of all of the proposed techniques to battery pack is another open problem which needs more research.

Acknowledgments. The authors would like to thank Professor G.-A. Nazri, Professor and Director of Energy Storage and Generation of Wayne State University, for his very fruitful comments during this work.

References

1. Alavi, S.M.M., Saif, M.: Fault Detection of Nonlinear Systems Over Lossy Network. In: Proc. IEEE Int. Conf. on Control Applications, pp. 964–969. Denver, USA (2011)
2. Alavi, S.M.M., Samadi, M.F., Saif, M.: Estimation of Lithium Transport Rate in Lithium-ion Batteries - A Particle Filtering Approach. In: Proc. of the 2012 IFAC Workshop on Engine and Powertrain Control, Simulation and Modeling (E-COSM 2012), France (2012)

3. Arora, P., White, R.E., Doyle, M.: Capacity fade mechanisms and side reactions in lithiumion batteries, J. Electrochem. Soc. 145(10), 3647–3667 (1998)
4. Aurbach, D., Zinigrad, E., Cohen, Y., Teller, H.: A short review of failure mechanisms of lithium metal and lithiated graphite anodes in liquid electrolyte solutions. Solid State Ionics 148, 405–416 (2002)
5. Blanke, M., Kinnaert, M., Lunze, J., Staroswiecki, M.: Diagnosis and Fault-tolerant Control, 2nd edn. Springer (2006)
6. Bokor, J., Szabo, Z.: Fault detection and isolation in nonlinear systems. Annual Reviews in Control 33(2), 113–123 (2009)
7. Chaturvedi, N.A., Klein, R., Christensen, J., Ahmed, J., Kojic, A.: Algorithms for Advanced Battery Management Systems. IEEE Control Systems Magazine 30(3), 49–68 (2010)
8. Chow, E.Y., Willsky, A.S.: Analytical redundancy and the design of robust failure detection systems. IEEE Trans. Automatic Control 29, 603–614 (1984)
9. Ding, S.X.: Model-based Fault Diagnosis Techniques- Design Schemes, Algorithms, and tools. Springer (2008)
10. Di Domenico, D., Creff, Y., Prada, E., Duchene, P., Bernard, J., Sauvant-Moynot, V.: A review of approaches for the design of Li-ion BMS estimation functions. In: Int. Scient. Conf. on Hybrid and Electric Vehicles RHEVE 2011, Rueil-Malmaison, France, December 6-7 (2011)
11. Dees, D., Gunen, E., Abraham, D., Jansen, A., Prakash, J.: Electrochemical Modeling of Lithium-Ion Positive Electrodes during Hybrid Pulse Power Characterization Tests. J. Electrochem. Soc. 155, A603–A613 (2008)
12. Di Domenico, D., Stefanopoulou, A., Fiengo, G.: Lithium-ion battery state of charge and critical surface charge estimation using an electrochemical model-based extended kalman filter. Journal of Dynamic Systems, Measurement, and Control 132(6), 061302 (2010)
13. Emami-Naeini, A., Akhter, M.M., Rock, S.M.: Effect of model uncertainty on failure detection: The threshold selector. IEEE Trans. Automatic Control 33(12), 1106–1115 (1998)
14. Gabrisch, H., Wilcox, J., Doeff, M.: TEM Study of Fracturing in Spherical and Plate-like LiFePO4 Particles. Electrochemical and Solid-State Letters 11(3), A25–A29 (2008)
15. Guerfi, A., Kaneko, M., Petitclerc, M., Mori, M., Zaghib, K.: $LiFePO_4$ water-soluble binder electrode for Li-ion batteries. J. of Power Sources 163(2), 1047–1052 (2007)
16. Harris, S.J., Timmons, A., Baker, D.R., Monroe, C.: Direct in situ measurements of Li transport in Li-ion battery negative electrodes. Chemical Physics Letters 485, 265–274 (2010)
17. Kanevskii, L., Dubasova, V.: Degradation of lithium-ion batteries and how to fight it: A review. Russian Journal of Electrochemistry 41(1), 3–19 (2005)
18. Klein, R., Chaturvedi, N.A., Christensen, J., Ahmed, J., Findeisen, R., Kojic, A.: Electrochemical Model Based Observer Design for a Lithium-Ion Battery. IEEE Trans. Control Systems Technology (to appear)
19. Kostecki, R., McLarnon, F.: Local-Probe Studies of Degradation of Composite $LiNi_{0.8}Co_{0.15}Al_{0.05}O_2$ Cathodes in High-Power Lithium-Ion Cells. Electrochemical and Solid State Letters 7, A380 (2004) (LBNL-55323)
20. Moura, S.J., Chaturvedi, N.A., Krstic, M.: PDE Estimation Techniques for Advanced Battery Management Systems - Part I: SOC Estimation. In: Proc. ACC 2012, Montreal, Canada (2012)

21. Newman, J., Tiedemann, W.: Porous-electrode theory with battery applications. AIChE Journal 21, 25–41 (1975)
22. Nazri, G.-A., Pistoia, G.: Lithium Batteries: Science and Technology. Springer (2009)
23. Safari, M., Morcrette, M., Teyssot, A., Delacourt, C.: A Multimodal physics-based aging model for life prediction of Li-ion batteries. J. Electrochem. Soc. 156(3), A145–A153 (2009)
24. Samadi, M.F., Alavi, S.M.M., Saif, M.: An electrochemical model-based particle filter approach for Lithium-ion battery estimation. In: Submitted to the 51 IEEE Conference on Decision and Control, USA (2012)
25. Santhanagopalan, S., White, R.E.: Online estimation of the state of charge of a lithium ion cell. Journal of Power Sources 161(2), 1346–1355 (2006)
26. Smyshlyaev, A., Krstic, M.: Adaptive Control of Parabolic PDEs. Princeton University Press (2010)
27. Stux, A.M., Swider-Lyons, K.E.: Li-Ion Capacity Enhancement in Composite Blends of $LiCoO_2$ and Li_2RuO_3. J. Electrochem. Soc. 152(10), A2009–A2016 (2005)
28. Wang, X., Sone, Y., Segami, G., Naito, H., Yamada, C., Kibe, K.: Understanding Volume Change in Lithium-Ion Cells during Charging and Discharging Using In Situ Measurements. J. Electrochem. Soc. 154(1), A14–A21 (2007)

Design of a Nanobiomaterial from Renewable Resources

Parisa Pooyan[1,3], Rina Tannenbaum[2], and Hamid Garmestani[3]

[1] The Woodruff School of Mechanical Engineering,
Georgia Institute of Technology,
Atlanta, GA, 30332, U.S.A
[2] Department of Mechanical Engineering and Biomedical Engineering,
Boston University,
Boston, MA, 02215, U.S.A
[3] School of Materials Science and Engineering,
Georgia Institute of Technology,
Atlanta, GA, 30332, U.S.A
parisa.pooyan@gatech.edu,
rinatan@bu.edu,
hamid.garmestani@mse.gatech.edu

Abstract. In recent years, a large emphasis has been placed on the use of renewable resources to less heavily rely on petroleum and to better utilize global energy needs. However, the lack of rigidity of nature's materials typically limits their mass production for high-tech applications. One promising approach to address this shortcoming is to introduce a composite material reinforced by high purity nanofibers found in nature. Cellulose nanowhiskers (CNWs), the most abundant biopolymer on earth, could integrate a viable nanofibrous porous candidate resulting in superior structural diversity and functional versatility for diverse applications from automotive industry to bioengineering design. Inspired by these fascinating properties, a fully cellulose-based composite was designed using the CNWs reinforcement and their oriented morphology. Comparable to carbon nanotubes or kevlar, CNWs introduced significant strength and directional rigidity to the composite even at 0.2 wt% yet doubled that under magnetic field of only 0.3T. The tendency of CNWs to interconnect with one another confirmed the formation of a three-dimensional rigid percolating network, fact which imparted an excellent mechanical rigidity to the entire structure at such low filler content. Hence, the green nanobiomaterial with an enhanced microstructure performance in this study could potentially increase the biomedical applications of cellulose-based materials.

Keywords: Nanocomposites, Mechanical Properties, Natural Polymers, Cellulose, Biomimetic Materials.

1 Introduction

The constant need to better allocate natural resources and to efficiently utilize energy supply has received much attention in today's research. As opposed to conventional materials, some renewable plant-derived sources can offer unique properties such as hierarchical structure, environmental compatibility, low thermal expansion, and flexibility for different custom-made applications through chemical modifications [1]. The lack of rigidity in traditional nature's materials however, limits their applications in industrial practice. To overcome the issue, a high purity bio-nanofiller can be used to reinforce the composite material and to introduce an enhanced microstructure performance. Among naturally derived polymers, cellulose has introduced major advantages given its renewable and environmentally benign nature, and its abundance and excellent biocompatibility. As a naturally occurring organic compound, cellulose is derived from D-glucose units condensing through β(1-4)-glycosidic oxygen bridges into a matrix of macro-cellulose fibers (Fig. 1).

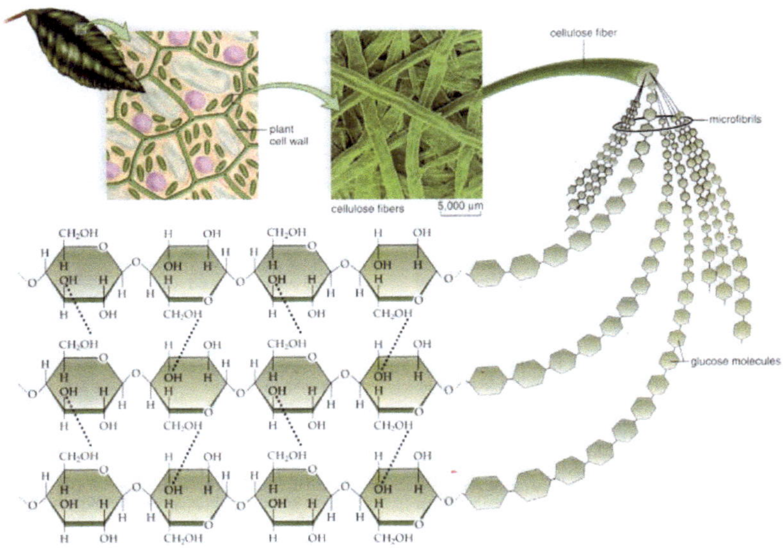

Fig. 1 The structural hierarchy of plant fibers and the spatial configuration of glucose monomers integrating a nanocrystalline biopolymer network [2]

Depending on the source of cellulose, the extracted cellulose nanowhiskers (CNWs) fabricated through a vigorous multi-stage chemical/mechanical processes, are generally 1-100 nm in diameter and 0.5 to 2 µm in length [3, 4]. Unlike other coiled and branched polysaccharides, the structural hierarchy of cellulose as shown in Figure 1 consists of linear polymer chains adopting a rather stiff rod-like conformation [3]. Laterally extended by hydrogen bonding, the associated cellulose chains develop a relatively stable polymer network, resisting a facile degradation in typical aqueous solvents. The hydrogen linkage holding the glucose residues intact from one chain to another also creates a rigid crystalline network imparting a significant strength and a directional rigidity to the polymeric structure. This extended rigid chain conformation along with the cooperative morphology of hydrogen-bonded layers results in CNWs significant load-carrying capacity when compared to other fiber reinforcing agents as summarized in Table 1. Not only CNWs offer an excellent mechanical properties but also they possess a non-toxic environmentally friendly nature [5] as opposed to other potentially toxic nanofibers such as carbon nanotubes [6, 7].

Table 1 The mechanical properties of cellulose nanowhiskers compared to other fiber: reinforcing agents [8]

Reinforcing Fibers	Tensile Strength (GPa)	Elastic Modulus (GPa)
Glass Fibers	4.8	86
Kevlar	3.0	130
Steel Wire	4.1	207
Graphite Whisker	21	410
Carbon Nanotubes	11-73	270-970
Cellulose Nanowhiskers	7.5	145

Given its tensile strength and elastic modulus (shown in Table 1), CNWs could integrate into a fibrous network and impart superior structural diversity and functional versatility. However, the difficulty associated with the surface interactions of CNWs and their acceptable level of dispersions within a polymeric matrix still remains as a major dilemma in the design of a cellulose-based composite [4]. As a result, the processing conditions, which control the interaction and dispersion of CNWs, can significantly affect the final performance of nanocomposite material. For instance, fabrication techniques such the pre-dispersion of nanofibers prior to mixing with the host matrix or the microstructure orientation by taking advantages of the susceptibility of CNWs to get aligned as exposed to a magnetic field can accordingly tune the mechanical and thermal performance of nanocomposite material [9-13]. In fact, the aligned microstructure not only could enhance the mechanical/ thermal properties of the system but also could extend the final applications of the material especially in biomedical field. For example, most naturally-occurring tissues exhibit a preferential alignment and a well-ordered structure, such as the parallel and aligned assembly in tendons, the concentric weaves in bone, the orthogonal lattices in cornea, the circumferential alignment of the smooth muscle cells of large arteries and the mesh-like architecture in skin [14-16]. A well-defined oriented microstructure could predominantly influence the cell adhesion while effectively could maintain the cellular phenotypic shape and its growth with respect to the fiber orientation for further tissue engineering applications [14, 17].

In order to take advantage of the excellent properties of cellulose and its derivatives, a nanocomposite of all-cellulose material was designed in the current study where cellulose acetate propionate matrix was embedded and entangled with the CNWs. This system was selected to introduce a fully functional green biomaterial where both matrix and reinforcing phase were based on natural resources and to further extend the applications of cellulose-based composites in different industries especially in bioengineering designs.

2 Experimental Section

2.1 *Materials*

Cellulose acetate propionate (CAP) with 2.5 wt.% acetyl and 46 wt.% propionyl content was purchased from Sigma-Aldrich, Milwaukee, WI. Microcrystalline cellulose (MCC) was acquired from Avicel-Aldrich. The spectroscopic grade acetone was purchased from VWR and preserved as directed. Cellulose nanowhiskers were isolated from MCC by acid hydrolysis with a 62 wt.% H_2SO_4 solution and a multistage procedure which was previously reported [11].

2.2 *Nanocomposite Design*

A clear solution of 5 wt.% CAP in acetone was prepared overnight. The CNWs were either directly added to the CAP solution, pre-dispersed in acetone prior to mixing with the host matrix, or pre-dispersed and aligned within a magnetic field. To better preserve the dispersion of the CNWs and to control their reaction with the matrix material, the CAP suspension was subjected to several minutes of sonication, followed by 2 hours of magnetic stirring. The aqueous suspension of the non-flocculated CNWs in CAP was then cast into a PTFE mold and was allowed to settle at room temperature to form a 200 μm film. The alignment of CNWs in the matrix was achieved by pouring the CNW-CAP suspension immediately after sonication into a mold, and applying a 0.3 T magnetic field for 1 hr at room temperature. For the experiments intended to probe the mechanical/thermal properties of CAP-CNW composites, the volume fractions of CNWs were varied to correspond to 0.2, 1, 3, 6, and 9 wt.%, this in order to better evaluate the effect of nanocrystal phase on the properties of cellulose-based nanocomposites, particularly at low filler contents. The uniformity and smoothness of the resulting membrane evidenced the homogeneous distribution of CNWs within the viscoelastic CAP medium.

2.3 *Characterization Methods*

The morphology of the aqueous suspension of CNWs was imaged using an AFM NanoScope (Multimode Scanning Probe Microscope (SPM), Veeco 3000).

A droplet of the suspension was initially dried on a glass slide prior to imaging and the scans were obtained in air with commercial Si Nanoprobe SPM tips of 1.6 μm in tapping mode. The cross-sections of nanocomposite films were imaged by a scanning electron microscope (LEO and ZEISS SEM) at an accelerating voltage of 5 kV. The specimens were initially frozen in liquid nitrogen for a few minutes before snapping-off the edge to remove the surface soft polymers and to preserve the nature of CNWs at the fractured section. Then, the snapped cross-sections were sputter-coated with gold for less than a minute prior to imaging. Also, the microstructure of nanocomposites were obtained by the same SEMs at 5 kV accelerating voltage. Prior to imaging, several drops of the CNW/CAP suspension were deposited on silicon wafers that were pre-cleaned with piranha solution (a typical mixture of 3 to 1- concentrated sulfuric acid, H_2SO_4 to hydrogen peroxide, H_2O_2) and ethanol, and allowed to quickly dry in an oven in order to remove the moisture from their surfaces. The silicon wafers was then sputter-coated as previously described.

Classical tensile tests were performed on the specimens of neat CAP and CNW-CAP composites fabricated at different CNW volume fractions. The specimens were cut into a standard dog-bone shape and tested using an MTS (Materials Testing Systems) Insight II at a nominal gage length of 20 mm and a crosshead speed of 1.2 mm/min. The data was collected at a rate of 20 Hz under a 100 N loading at body temperature (37 °C) to investigate the potential use of the fully cellulose-based material in bioengineering designs. Three specimens from each set of films at different filler concentrations were tested to validate the consistency of the reported data and the uniformity of the fabricated membranes.

Additionally, the weight loss and the thermal stability of the CNW-CAP composites were measured by thermogravimetric analysis (TGA) using a TGA Q50 from TA Instruments. The samples were heated from 40 °C up to 600 °C at a heating rate of 10 °C/min, under an argon atmosphere. The TGA measurements were tested on three specimens from different regions of each set of films to ensure the accuracy of the captured data and the homogeneity of the samples.

3 Results and Discussion

In composite science in general, design parameters such as the morphology/ geometry, the concentration/ volume fraction and the mechanical/thermal properties of each phase with their interface quality, can directly tune the performance of the entire composite system. In order to examine the effect of such parameters on the mechanical/ thermal performance of the designed nanocomposite in this study, three different processing conditions were applied as previously explained in details [11]. These methods include: (1) The direct mixing of freeze-dried CNWs with the CAP solution, (2) The pre-dispersion of CNWs in acetone suspension followed by mixing with the CAP solution, and (3) The orientation of CNW-CAP microstructure upon exposure to an externally-applied magnetic field of 0.3 T. The general observation was that the processing method had a direct impact on the morphological characteristics of CNWs and the manner by which they resided within the host matrix and ultimately, on the mechanical performance of nanocomposite as it was also reported in the literature [4]. For instance, the tight agglomeration of the synthesized CNWs in Fig. 2 evidenced the intermolecular hydrogen bonding which had to be accordingly interfered during the fabrication of the CNW-CAP composite to inhibit the subsequent whisker flocculation.

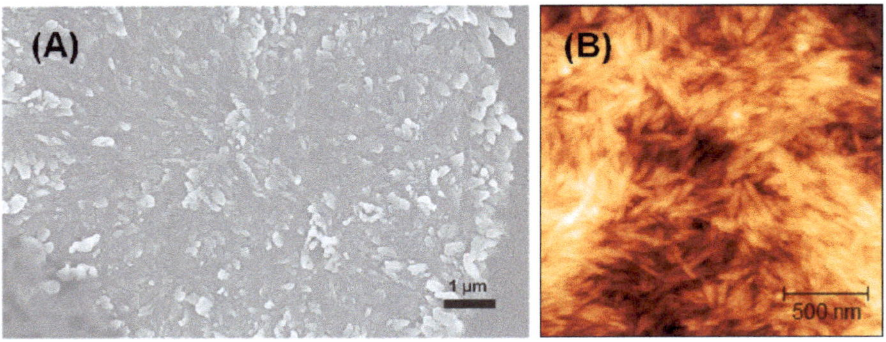

Fig. 2 The morphology of CNWs in an aqueous solution: (A) the SEM image representing the colloid stability of nanofibers (B) the AFM image illustrating the aggregation of nanofibers via hydrogen bonding

The resulting uniform microstructure of CNW-CAP composite with no indication of CNW aggregations confirmed the significant effect of pre-dispersion technique on the nanocomposite fabrication (Fig. 3). This can also be observed from the tensile stress-strain curve in Fig. 4: A where the various processing protocols changed the mechanical behavior of the designed cellulose-based composite.

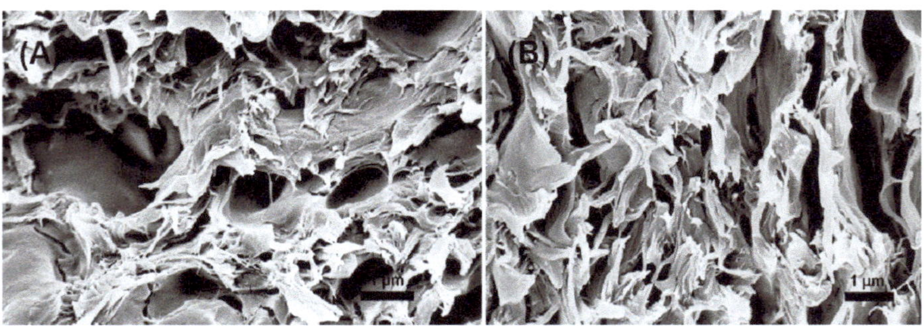

Fig. 3 SEM images of the CNW-CAP composite at 0.2 wt. % nanofibers using the pre-dispersion processing method: (A) a non-oriented microstructure (B) an aligned structure as exposed to a magnetic field of 0.3T

Besides fabrication techniques, other design parameters such as the volume fractions of nanofiller can also determine the best optimum nanoncomposite performance. For the CNW-CAP composite studied here, it turned out that a considerable enhancement was observed in the nanocomposite tensile behavior (Fig. 4b) where the filler concentration was up to about 3 wt% using samples that were fabricated under the pre-dispersion processing condition (method 2 described previously). Also, the pore distribution in different films was further investigated to ensure the accuracy of the subsequent mechanical comparisons [18].

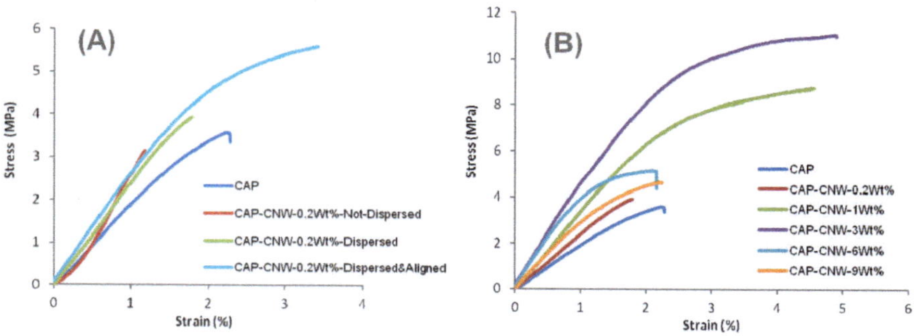

Fig. 4 The stress-strain curve of CNW-CAP composites obtained from classical tensile test at body temperature (37 ^{0}C) by considering (A) different fabrication techniques at 0.2 wt.% nanofiber. (B) different nanofiber volume fractions

Similar to the tensile measurements, the thermal behavior of CNW-CAP composite was further investigated for samples fabricated under different processing conditions and for samples with changes in the filler volume fractions as it was described earlier. From Fig. 5a, the pre-dispersion and alignment methods notably reduced the formation of inhomogeneous regions such as air bubbles, while also inhibited the CNW flocculation during the nanocomposite fabrication.

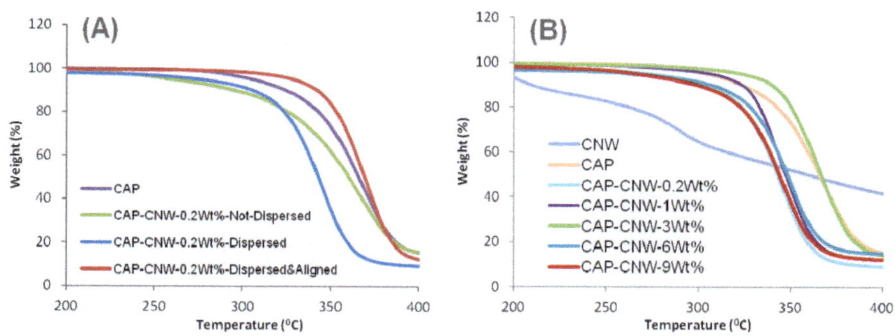

Fig. 5 The thermal degradation of CNW-CAP composites from TGA thermograms by taking (A) different processing methods at 0.2 wt. % (B) various range of nanofiber concentrations

Also, the smooth TGA profiles of the fabricated nanocomposites with no indication of a separate degradation stage as opposed to the pure CNWs shown in Fig. 5b suggested the successful uniform grafting of the CNWs within the host matrix.

In summary, the formation of a rigid percolating network of CNWs within the matrix introduced an unusual stiffening effect on the cellulose-based nanocomposite at the low filler content below 3 wt.% as it was studied in the previous work [18]. In addition, the likelihood of filler agglomeration due to the hydrogen bonding interactions among the nanofibers could further explain the degradation in mechanical/ thermal performance of the nanocomposites at higher concentration of CNWs (beyond 3 wt.%). It is also expected to inhibit the CNW aggregation and to enhance the nanocomposite mechanical/ thermal performance at higher filler contents by using a stronger magnetic field higher than 0.3 T which was investigated in this study [12, 13]. Moreover, the effect of fiber alignment at the lower concentration of 0.2 wt.% was closely comparable to the optimized amount of nanowhiskers at 3 wt.% without the magnetic alignment, as is evident from the tensile tests in Fig. 4: and the TGA thermograms in Fig. 5. Given this fact, the fiber rearrangements due to the magnetic alignment could potentially introduce a favorable filler/filler interaction, avoiding the possible adverse effect of the fiber aggregations, while at the same time readily provide a better interfacial adhesion between the filler and the host matrix. This could also effectively change the optimum volume fraction of fillers, which in this study was predicted to be 3 wt.%.

4 Conclusion

The science of imitating nature with a growing aspect in multidisciplinary fields has now a leading role in the fabrication of novel materials with remarkable performance. Biopolymer composites of high purity bio-nanofillers can arrange in intricate ways to offer a combination of light weight, strength, and biofunctionality in a diverse range of applications. Cellulose nanowhisker (CNW) as an attractive fiber-reinforcing agent can present a viable biomaterial due to its versatility in properties and relatively low cost in fabrication. With these inherent advantages of cellulose and its derivatives, a fully cellulose-based material was designed where the CNWs were embedded in a matrix of cellulose acetate propionate. The well-dispersed CNW phase within the host matrix imparted considerable mechanical/thermal stability to the entire composite system at only 0.2 wt.% and substantially enhanced these properties upon the orientation of nanofibers. The aligned features not only improved the directionality of nanoparticles within the medium, but also drastically lowered the optimum amount of CNWs required to obtain the best composite performance. The excellent performance at such low filler content is mainly due to the formation of a three-dimensional rigid percolating network of CNWs within the host matrix. Thereby, the all-cellulose nanocomposite designed in the current study with an oriented microstructure and tunable mechanical/ thermal properties could open new perspectives in the self-assembly of nanobiomaterial for biomedical applications while it could make the design of the next generation of fully green material a reality.

Acknowledgments. The Institute of Paper Science and Technology at Georgia Tech is gratefully acknowledged for the funding of this project. Also, many thanks are expressed to Il Tae Kim for his help and support throughout the experiments and the microscopy imaging.

References

1. Eichhorn, S.J., Gandini, A.: Materials from Renewable Resources. Mrs Bulletin 35, 187–190 (2010)
2. Bouwencyclopedie. Cellulose, `http://wwwjoostdevreenl/shtmls/celluloseshtml`
3. Hubbe, M.A., Rojas, O.J., Lucia, L.A., Sain, M.: Cellulosic nanocomposites: A review. BioResources 3, 929–980 (2008)
4. Samir, M., Alloin, F., Dufresne, A.: Review of recent research into cellulosic whiskers, their properties and their application in nanocomposite field. Biomacromolecules 6, 612–626 (2005)
5. Klemm, D., Schumann, D., Kramer, F., Heßler, N., Koth, D., Sultanova, B.: Nanocellulose Materials – Different Cellulose, Different Functionality. Macromolecular Symposia 280, 60–71 (2009)
6. Lam, C.W., James, J.T., McCluskey, R., Arepalli, S., Hunter, R.L.: A Review of Carbon Nanotube Toxicity and Assessment of Potential Occupational and Environmental Health Risks. Critical Reviews in Toxicology 36, 189–217 (2006)
7. Lam, C.W., James, J.T., McCluskey, R., Hunter, R.: Pulmonary toxicity of single-wall carbon nanotubes in mice 7 and 90 days after intratracheal instillation. Toxicological Sciences 77, 126–134 (2004)
8. Yano, H.: Cellulose nanocrystals, `http://forestproductsorstedu/faculty/simonsen/`
9. Kimura, F., Kimura, T., Tamura, M., Hirai, A., Ikuno, M., Horii, F.: Magnetic alignment of the chiral nematic phase of a cellulose microfibril suspension. Langmuir 21, 2034–2037 (2005)
10. Kvien, I., Oksman, K.: Orientation of cellulose nanowhiskers in polyvinyl alcohol. Applied Physics a-Materials Science & Processing 87, 641–643 (2007)
11. Pooyan, P., Kim, I.T., Tannenbaum, R., Garmestani, H.: Design of a cellulose-based nanocomposite as a potential bio-scaffold for tissue engineering (submitted, 2012)
12. Revol, J.F., Godbout, L., Dong, X.M., Gray, D.G., Chanzy, H., Maret, G.: Chiral nematic suspensions of cellulose crystallites; phase separation and magnetic field orientation. Liquid Crystals 16, 127–134 (1994)
13. Sugiyama, J., Chanzy, H., Maret, G.: Orientation of cellulose microcrystals by strong magnetic fields. Macromolecules 25, 4232–4234 (1992)
14. Murugan, R., Ramakrishna, S.: Design strategies of tissue engineering scaffolds with controlled fiber orientation. Tissue Engineering 13, 1845–1866 (2007)

15. Stevens, M.M., George, J.H.: Exploring and engineering the cell surface interface. Science 310, 1135–1138 (2005)
16. Xu, C.Y., Inai, R., Kotaki, M., Ramakrishna, S.: Aligned biodegradable nanotibrous structure: a potential scaffold for blood vessel engineering. Biomaterials 25, 877–886 (2004)
17. Dugan, J.M., Gough, J.E., Eichhorn, S.J.: Directing the Morphology and Differentiation of Skeletal Muscle Cells Using Oriented Cellulose Nanowhiskers. Biomacromolecules 11, 2498–2504 (2010)
18. Pooyan, P., Tannenbaum, R., Garmestani, H.: Mechanical behavior of a cellulose-reinforced scaffold in vascular tissue engineering. Journal of the Mechanical Behavior of Biomedical Materials 7, 50–59 (2012)

The Influence of Adding Porous Interlayer in the Brazing of Ceramic to Metal

Mohd Hamdi[1,2], Farazila Binti Yusof[1,2], Mohd Fadzil[1], Tuan Zaharinie[1], and Tadashi Ariga[3]

[1] Department of Engineering Design and Manufacture, Faculty of Engineering, University of Malaya, 50603 Kuala Lumpur, Malaysia
[2] Centre of Advanced Manufacturing and Materials Processing (AMMP), University of Malaya, 50603 Kuala Lumpur, Malaysia
[3] Department of Materials Science, School of Engineering, Tokai University, 1117 Kitakaname, Hiratsuka-shi, Kanagawa-ken, 259-1292 Japan
{hamdi,farazila,ibnjamaludin,tzaharinie}@um.edu.my, ttariga@keyaki.cc.u-tokai.ac.jp

Abstract. In the brazing of ceramic to metal, large coefficient of thermal expansion (CTE) mismatch between both materials will induce high residual stresses, reducing the strength of the joint. In addition, low wettability between ceramic and metal could increase weak phases at the joint interface. A porous interlayer is introduced in the brazing process to overcome the effect of residual stresses and formation of weak phases between ceramic and metal, in this study, the effect of porous interlayer in the brazing of sapphire to Inconel 625 and the brazing of diamond to stainless steel were investigated. Active filler material containing titanium (Ti) was utilized to enhance the wettability of the ceramic-metal joint. It was anticipated that utilizing the porous interlayer for direct brazing would reduce the thermal expansion between both materials. Preliminary experimental results have shown that good bonding between sapphire/filler alloy/porous layer/Inconel 625 was achieved and no voids were detected in the brazing layer. Similar observation was also detected in the brazing of stainless steel/filler alloy/porous interlayer/ diamond.

Keywords: brazing, porous interlayer, ceramic, metal.

1 Introduction

Nowadays, the demand on dissimilar material joints have increased from the viewpoints of energy consumption, environmental concern, high performance and cost savings. Unique dissimilar materials combinations between ceramics and metals would provide significant advantage in numerous applications. The inherent properties of ceramic such as the ability to withstand elevated temperatures and good resistance to corrosion make them highly attractive for demanding engineering applications. On the other hand, metals have common properties such as high strength, excellent heat and electrical conductivity [1], thus the combination of both materials will offer unique opportunity for new engineering applications.

As the applications for ceramic-to-metal joint increases, the need for improvement in the joining method to obtain highly reliable joining and strength becomes more crucial. Traditionally, brazing or soldering processes are used to join ceramics to metals. For brazing ceramic to metal, the ceramic surface must be coated with a metallic layer in order to achieve good bonding between both materials. However, this technique is somehow difficult to be implemented for some application such as for nuclear reactor components since it uses intermediate metals (ex: manganese) which exhibit poor corrosion and oxidation to the conventional reactor environment. Similarly, for silicon-based pressure sensors, there is a need to coat the ceramic sensing media with an anti-corrosion film before it can be assembled for brazing to metal part, a process which is difficult to be achieved [3].

Currently, in the brazing of ceramic to metal, a reactive filler material is utilized to achieve good bonding. Proper selection of brazing temperatures and duration are also important to avoid brazing failure during the high-temperature process. However, in high temperature brazing of ceramic to metal, there is a great possibility of post-process cracking (ex: macro and micro crack) and joining failures (adhesive and cohesive failure) occurred. These failures may be the results of significant coefficient of thermal expansion (CTE) mismatch between ceramics and metals. The differences in CTE could lead to high stresses and subsequently intensified thermal gradients that arise from differences in thermal diffusivity. In general, ceramic posses high elastic moduli and low-relaxation characteristic that prevent redistribution of the stresses. Obviously, the ceramic would not be able to resist fracture if high stresses are present at the brazing interface. To overcome this problem, a buffer layer, usually in the form of a porous media is sandwiched at the interfaces of ceramic and metal. According to A.A. Shizardi et al. [4], the utilization of a porous media between the ceramic and metal could give a smoother transition in thermal properties during heating and cooling process. In addition, brazed sample with interlayer addition was found to have a higher bonding strength than the sample without interlayer [5].

In this study, preliminary investigation were conducted on the addition of porous interlayer on direct brazing of sapphire to Inconel 625 and diamond to stainless steel 304 (SUS304). The filler material used was an Ag-base filler metal foil containing 2 mass% Ti. It was anticipated that utilizing a porous interlayer for direct brazing would reduce the thermal expansion between both materials.

The microstructure observations as well as micro hardness evaluations of joined interface and interfacial reaction between sapphire/Inconel 625 were discussed in detail. For the diamond/stainless steel 304 brazed joint, the microstructure and shear strength were evaluated accordingly.

2 Experimental Procedure

2.1 Brazing of Sapphire/Inconel 625

A sapphire specimen (single crystal α-Al_2O_3) with 99.999% purity is selected for brazing with Inconel 625. The sapphire is in a disk form with a 15 mm diameter and 1mm thickness. The dimension of Inconel 625 is 23 mm x 23 mm with a thickness of 0.5 mm. All the materials were received from Yamatake Corporation, Japan. The porous interlayers were formed from two sheets of porous pure copper (Cu) and pure nickel (Ni) with the thickness of 400 μm each. The porous sheets were sandwiched and rolled together, reducing the combined thickness to 400 μm forming a sheet of porous interlayer of Cu/Ni. The eutectic filler alloy 70Ag-28Cu-2Ti in sheet form having a thickness of 100 μm was used in the experiment. Four layers of filler alloy were used to match the thickness of the porous Cu/Ni interlayer.

The specimen was stacked in a sandwich configuration and clamped in a jig to hold the various layers in place, as shown in Fig. 1. The interlayer sheet was oriented so that the Cu side was placed facing the sapphire while the porous Ni faced the Inconel 625. The brazing process was carried out under vacuum conditions (10^{-4} Pa) using a Tokyo Vacuum furnace. The heating cycle involved a direct heating up to a brazing temperature of 865°C with a 5°C/min heating rate. Upon reaching the brazing temperature, it was held for 30 minutes before cooling down to room temperature with a cooling rate of 3°C/min.

The cross sections of brazed specimens and elemental reaction at the interface were analyzed using scanning electron microscopy (SEM) coupled with energy dispersive spectroscopy (EDS).

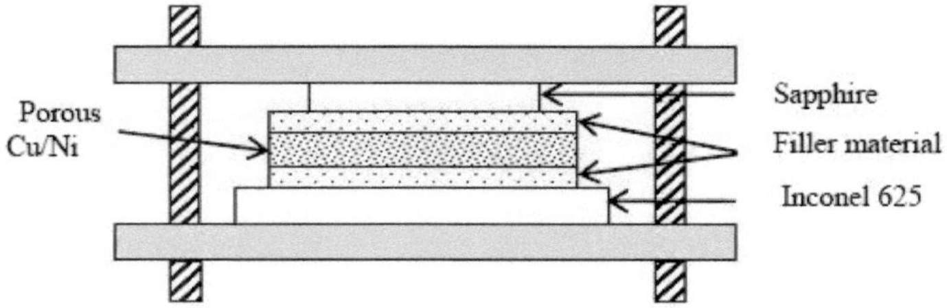

Fig. 1 Schematic diagram of Sapphire/Inconel 625 brazed experiment

2.2 *Brazing of Diamond Particles/SUS304*

Industrial grade diamonds with particles sizes of 40-50 mesh and stainless steel 304 with 2 mm thickness were used in this experiment. The400 μm porous interlayer used consisting of pure copper (Cu) and pure nickel (Ni), and the eutectic filler alloy, 70Ag-28Cu-2Ti in sheet form (thickness of 100 μm) is similarly prepared as those used in the sapphire/Inconel 625 experiment. The samples were layered on top of one an other as shown in Fig. 2. Stainless steel 304 (SUS304) was placed on the outer most layer while the while diamond particles, porous metal layer and filler materials were arranged in between.

The brazing process involved direct heating of the samples using a chamber furnace with a full flow of argon gas. The brazing temperatures selected were 880°C, 920°C and 960°C while the heating rate and holding time was kept constant at 5°C/min and 15 minutes respectively. The bonding shear strength of brazed samples was examined using a universal testing machine (UTM) and the microscopic analysis was accomplished using scanning electron microscopy (SEM) coupled with energy dispersive spectroscopy (EDS).

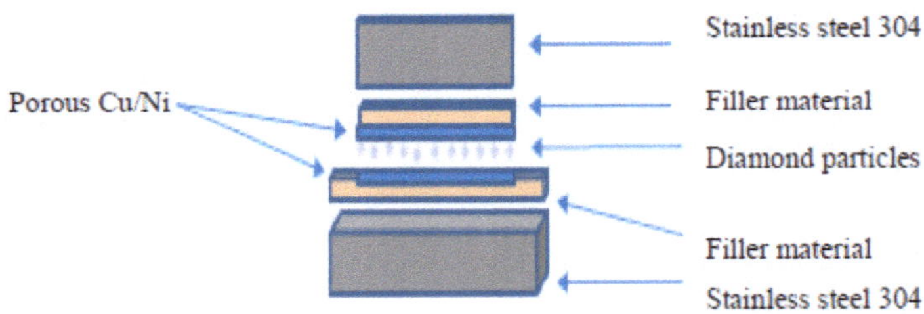

Fig. 2 The arrangement of the sample in the brazing diamond particles with SUS304

3 Results and Discussion

An optical micrograph of rolled porous (Cu/Ni) interlayer is shown in Fig. 3. It was observed that the porosity in the interlayer is still present even after the rolling process. The sizes of pores were in the range of 200 – 300 μm.

3.1 *Brazing of Sapphire/Inconel 625*

The brazing of sapphire/Inconel 625 was successfully achieved with firm adhesion and uniform brazing interlayer, which was free from cracks and voids. The microstructure examination of the cross-sectional brazed specimens is shown in Fig. 4(a). The results indicated that the porous Cu/Ni interlayer had diffused into the eutectic brazing filler and had left no voids at the interface. Thus, it can be assumed that the integrity of the joint is sufficiently adequate and porous Cu/Ni are able to accomodate the thermal expansion between sapphire and Inconel 625 and maintained the gap between sapphire and Inconel 625. According to A.A. Shirzadi et al. [4], a thicker foam layer would have abetter ability to accommodate thermal expansion differences between ceramic and metal. However it would reduce the shear strength of the joint and more filler alloy would be required to compensate the voids.

Fig. 4 (b) shows that a thin, continuous, diffusion layer was formed at interface of brazed layer and Inconel 625. However, no clear reaction layer was formed at the interface of the brazed layer and sapphire, as shown in Fig. 4(c), although a good wetting was successfully achieved. Three distinct zones can be identified from the cross section of the interface designated as I, II and III in Fig 4. The transitions between zones are mainly due to changes in microstructure and EDS line analysis as shown in Fig. 5. The following results were obtained from the analysis of the joint interface:

a) **Zone I**
Brazing interlayer adjacent to Inconel 625/brazing interlayer. This layer consisted of rich Ni and Ti elements and moderate content of Cu and Ag elements. Uniform diffusion layer had occurred between the metal alloy and the brazing filler material.

b) **Zone II**
Middle brazing interlayer consisted of rich Ag-Cu content. The porous interlayer of Cu-Ni had completely dissolved in the brazing filler. Significant amount of Cu and Ni element were detected as shown in the EDS line analysis.

c) **Zone III**
Brazing interlayer adjacent to the sapphire and eutectic filler materials. Rich Ni and Ti elements contents were detected.

Fig. 5 shows the result of the EDS line analysis of the Inconel 625/sapphire joint. In the figure, the dark phase indicates Cu rich areas, which may have been produced from brazing filler itself and porous Cu, while the light area indicates silver rich phase. The eutectic solidification of filler during the brazing process had formed a little island of copper (dark regions) distributed at the middle of the brazing interlayer. Insignificant amount Ni-Ti was also detected in the region. The analysis of the brazing interface on the sapphire side shows significant amount of Ti-Ni elements as compared to Ag-Cu. It is speculated that Ni and Ti elements have a tendency to migrate and react towards the adjacent ceramic. This result is consistent with the findings of Santella et al. [6], where the concentration of Ti is much higher than Cu in the layer adjacent to the ceramic as compared to the middle of brazing layer in which the Ti and Cu at% contents were almost of similar. It is also presumed that TiO_x and Ti_3Cu_3O might have been formed in the brazed region as seen in the EDS line analysis. S. Mandal et al. [7] reported that the formation for Ti_3Cu_3O occurred immediately after a thin layer of TiO phase in the brazing of Al_3O_2. Referring to the EDS analysis in Fig. 5, a significant presence of Cu, Ag, Ni and Ti elements were observed near the sapphire side. It is also speculated that the Ti and Cu elements had diffused towards the interfacial zone and the diffusion phenomenon is dependent on the brazing temperatures and composition. The addition of porous Cu/Ni interlayer may enhance the diffusion of the main element towards the ceramic side. Therefore the formation of TiO and Ti_3Cu_3O phases may have occurred in this brazing process. According to S. Mandal et al. [6], those phases are important to retain good adhesion between filler metal and Al_3O_2 since TiO solely cannot compensate the mismatched thermal expansion strains.

The micro hardness of the bonding interface was determined using a Vickers indenter with 25g load. Fig. 6 shows the results of the micro-indentations performed on Inconel 625, brazing interlayer and the sapphire sections. Small variations of micro hardness values were observed in the joined region, which can be attributed to the microstructure of the brazing interlayer. This interlayer consisted of eutectic type structure, which contains hard and brittle intermetallic phases. Lower microhardness value near the sapphire might be attributed to the absence of hard intermetallic phases between sapphire and brazed layer. Similarly, low microhardness value was observed in the reaction layer near Inconel 625 side. This is probably due to diffusion of filler material into the base metal, which causes decrease in hardness.

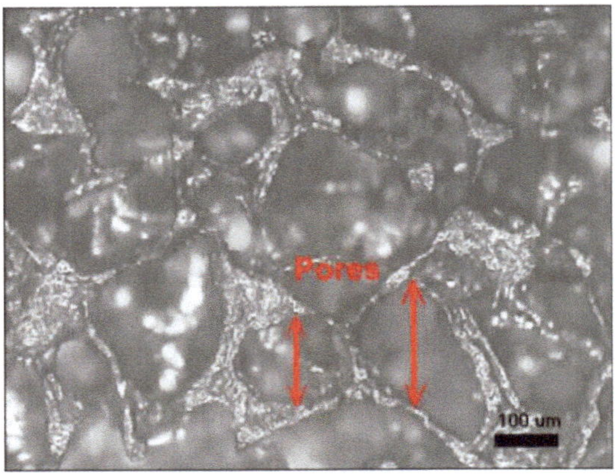

Fig. 3 Optical micrograph of porous (Cu/Ni) after rolling process

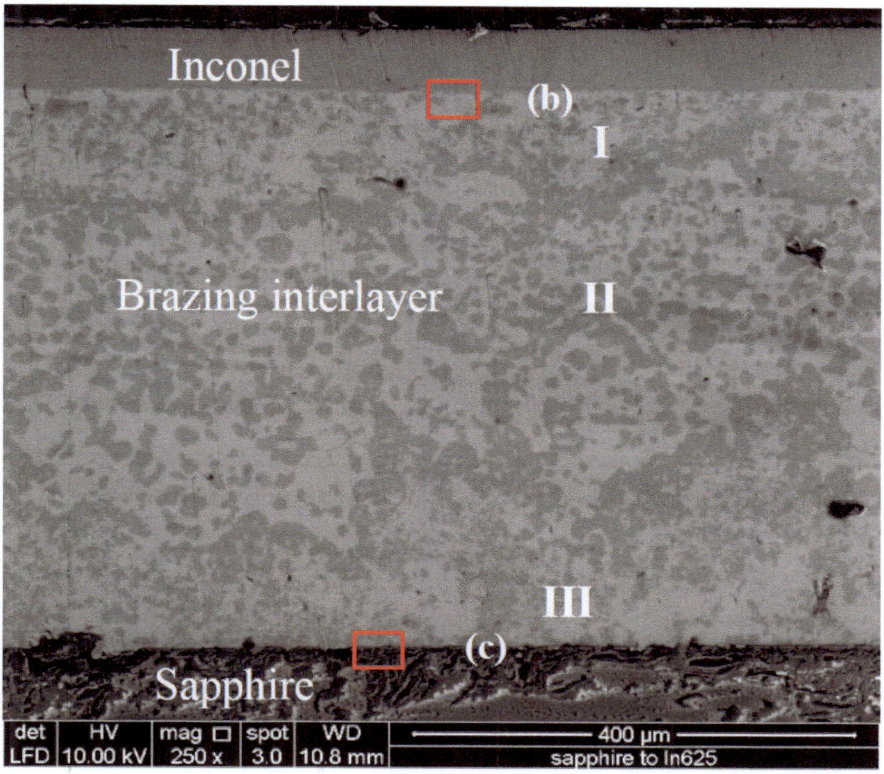

Fig. 4(a)

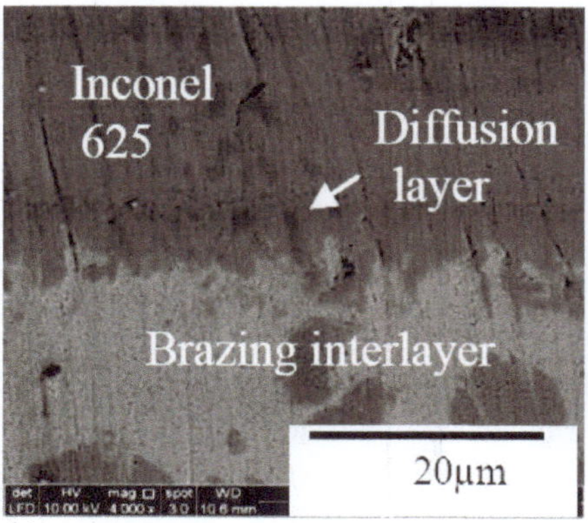

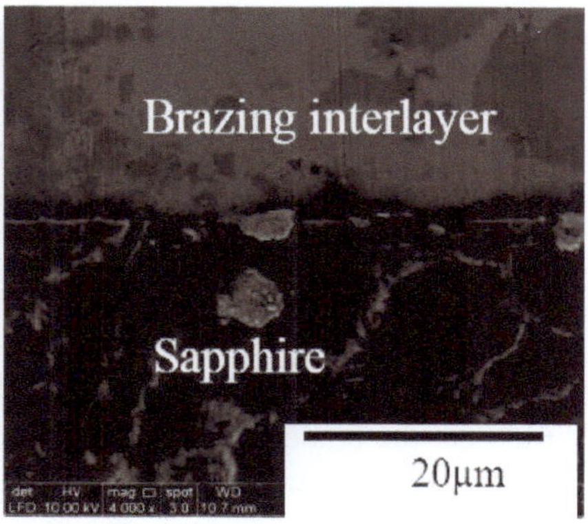

Fig. 4(b) Fig.4(c)

Fig. 4 (a): SEM micrograph of the sapphire/Inconel 625 interface, brazed at 865°C with soaking time of 30 minutes; (b): magnified brazing interlayer showing reaction layer occurred between filler alloy and Inconel 625; and (c): magnified brazing interlayer with sapphire

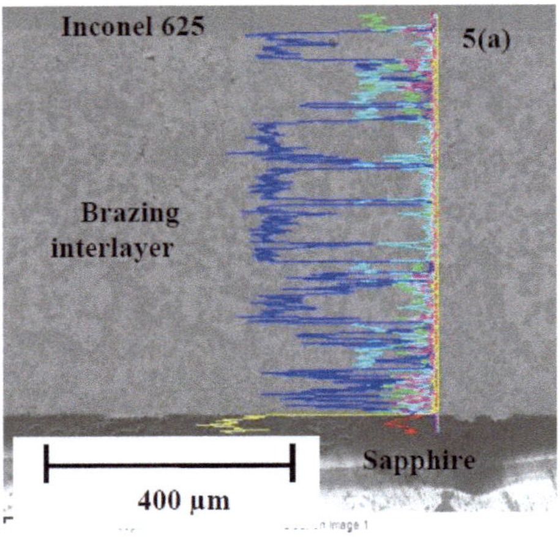

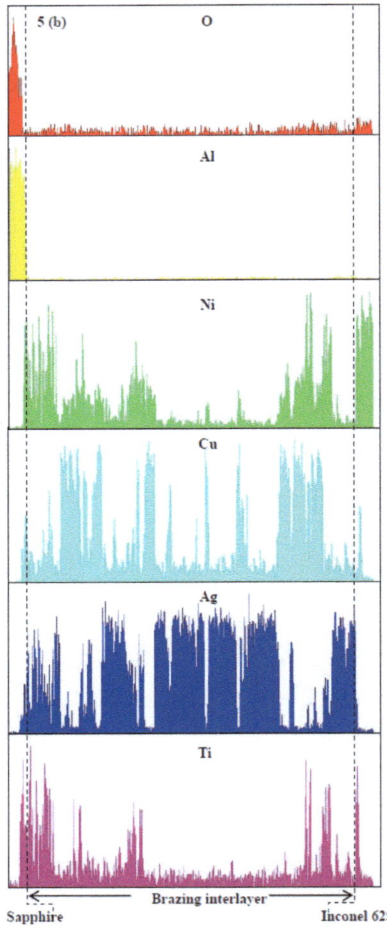

Fig. 5(a) and (b) Interface of Inconel625/ brazing layer/Sapphire and EDS line analysis profiles of the element

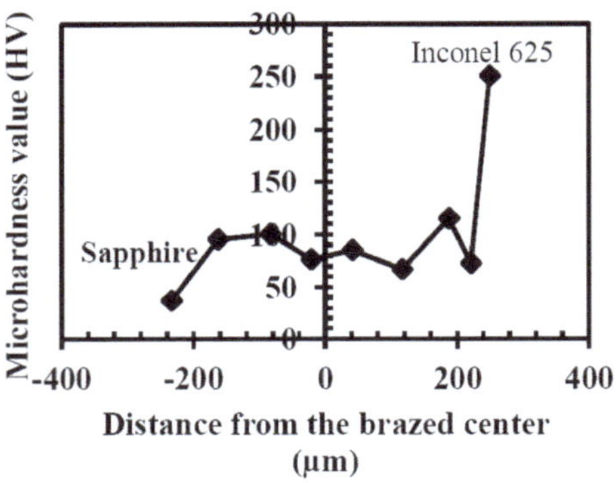

Fig. 6 Variations of micro hardness from the center of brazing interlayer

3.2 Brazing of Diamond/SUS304

The brazing of diamond particles to SUS304 was successfully obtained for all the samples at the brazing parameters chosen. Figure 7 shows that sound joining were observed for all samples, in which the diamond particles were tightly mounted inside the brazing filler and no significant appearance of voids can be detected. The elemental analysis of the joint brazed at 960°C was examined near the interface (indicated by the red box) using SEM-EDS and the spectrum is shown in Fig. 8. Significant presence of Ti, Ni, Ag and Cu elements were observed near the interface layer of diamond and the brazing filler material. According to J.C. Sung and M. Sung [8], the active Ti element will migrate and react with the diamond to form carbide (in this case, TiC). The reaction between Ti and C would enhance the wetting phenomenon of the brazing alloy on the diamond surface. In addition, strong carbide bonds will also hold the diamond firmly in place at the atomic level.

The presence of Cu element in the interface between diamond and brazing filler is believed to have enhanced the impact strength. J.C. Sung and M. Sung mentioned that the diamond's impact strength may be reduced due to the formation of carbide during brazing process. However, the presence of the Cu layer might control the formation of carbide without sacrificing the impact strength. They have proved that the adherence of the diamond plated with Cu is strengthened without compromising the impact strength.

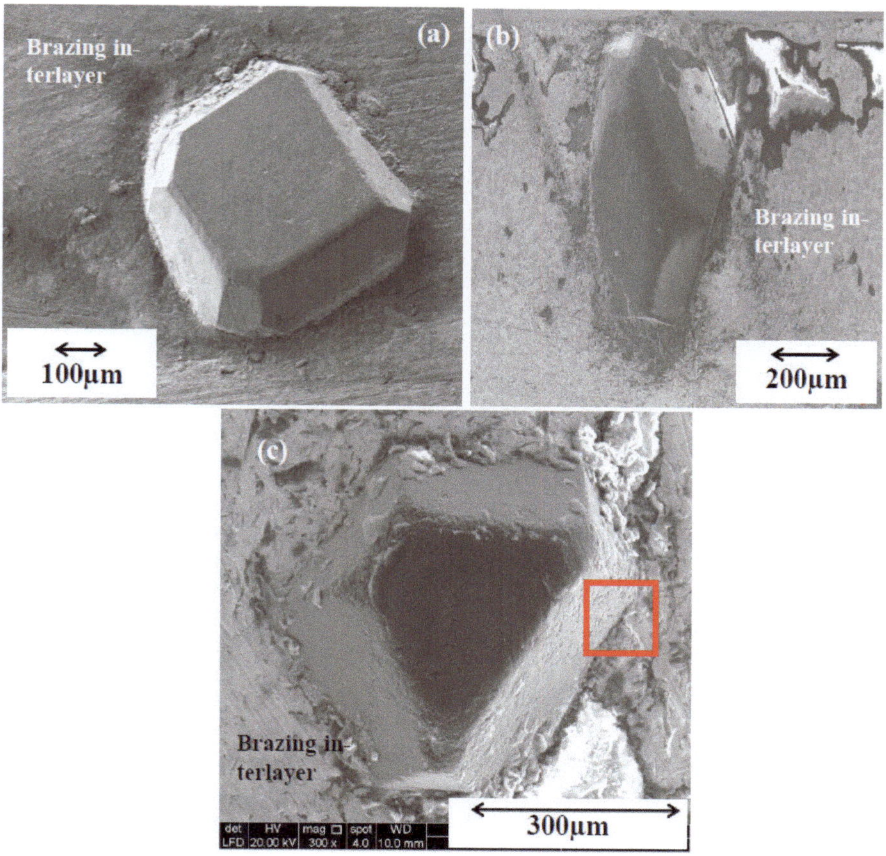

Fig. 7 SEM morphology of the diamond/SUS304 brazed joint with constant brazing time and brazing temperature of; (a) 880°C; (b) 920°C; and (c) 960°C

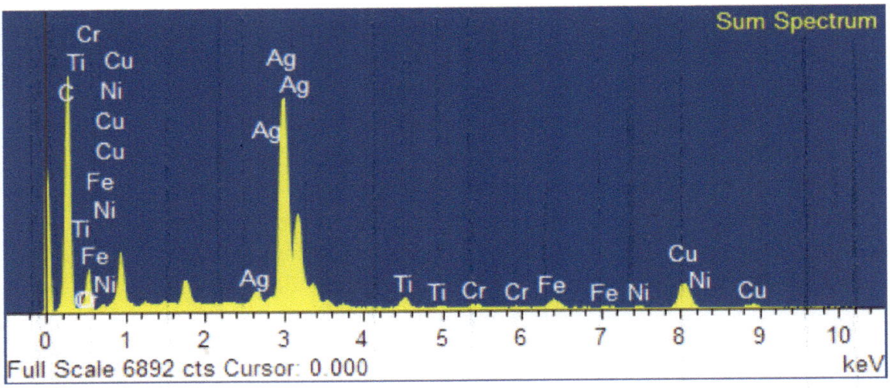

Fig. 8 The EDS spectrum of the diamond/SUS304 brazed sample with heating temperature of 960°C

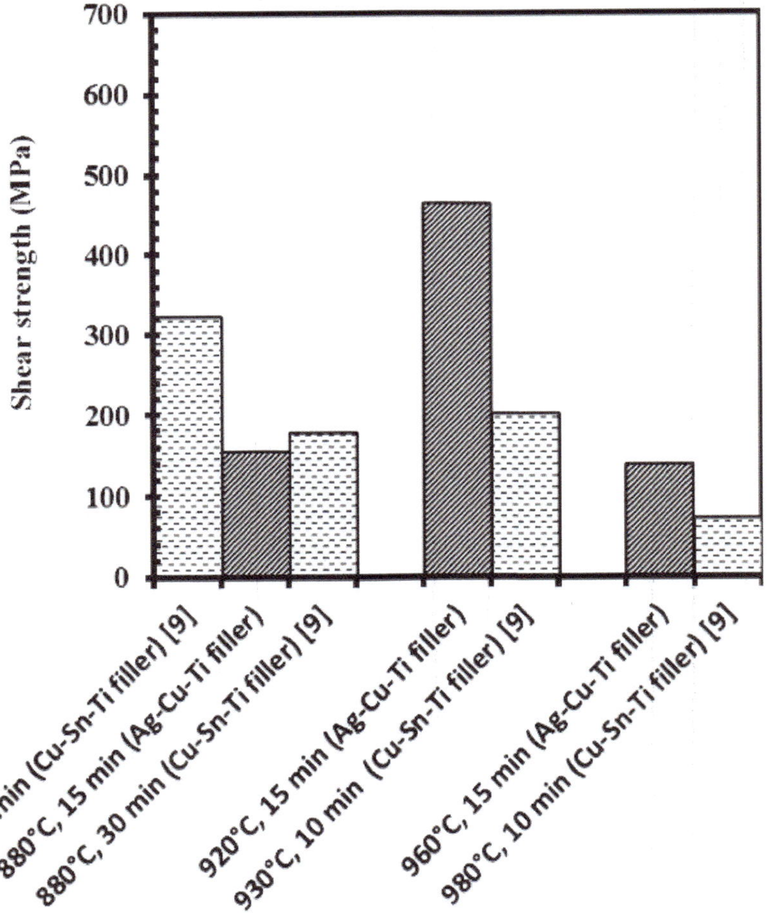

Fig. 9 The comparison of shear strength for different brazing parameters

The shear strengths of the brazed samples were determined using an Instron Universal tensile machine (UTM) and the values are listed in Fig. 9.The shear strength data was compared with the results obtained by S. Buhl et al. [9]. The highest shear strength was obtained for the brazing temperature of 920°C while slightly lower shear strengths had occurred at the brazing temperatures of 880°C and 960°C. This result slightly differs from the values obtained by S. Buhl et al. [9] where they have found that the shear strength value decreases with increasing brazing temperature. However in their study, the porous interlayer was not utilized. It is speculated that the bonding strength of the diamond and the SUS304 is strongly dependent on the interlayers at the diamond/filler alloy and filler alloy/SUS304 interfaces. The addition of porous Cu in the brazing filler it is believed to have reduced high stress induced from the interaction between the diamond particles and filler alloy.

4 Summary

The following conclusions can be drawn from the study on the influence of adding porous interlayer in the brazing of Sapphire/Inconel 625 and diamond/SUS304:

1. The addition of a porous interlayer isan effective method in preventing thermal expansion mismatch between sapphire/brazing filler/Inconel 625 and SUS304/filler/diamond/filler/SUS304. This prevented cracks during cooling and as a result, good adhesions and sound joints were obtained.
2. No significant presence of pores at the brazing interlayer which shows that all the filler materials were able to compensate and diffuse into in the porous layer.
3. It is believed that the shear strengths of the brazed samples were increased with addition of porous interlayer.

Acknowledgments. This research was supported by Ministry of Higher Education Malaysia underthe High Impact Grant (HIR-MOHE D000001-16001).

References

1. Hatch, J.E.: Aluminum: Properties and physical metallurgy. ASM International (1984)
2. Kapoor, R.R., Eager, T.W.: Ceramic Engineering Science Proceeding 10(11-12), 1613–1630 (1989)
3. Ishihara, T., Sekine, M., Ishikura, Y., Kimura, S., Harada, H., Nagata, M., Masuda, T.: Journal of Transducer, 503–506 (2005)
4. Shirzadi, A.A., Zhu, Y., Bhadeshia, H.K.D.H.: Materials Science and Engineering A 496, 501–506 (2008)
5. Xiong, H.P., Mao, W., Xie, Y.H., Guo, W.L., Li, X.H., Cheng, Y.Y.: Materials Letters 61, 4662–4665 (2007)

6. Santella, M.L., Horton, J.A., Pak, J.J.: Journal of the American Ceramic Society 73(6), 1785–1787 (1990)
7. Mandal, S., Kumar Ray, A., Kumar Ray, A.: Materials Science and Engineering A 383, 235–244 (2004)
8. Sung, J.C., Sung, M.: International Journal of Refractory Metals & Hard Materials 27, 382–393 (2009)
9. Buhl, S., Leinenbach, C., Spolenak, R., Wegener, K.: International Journal of Refractory Metals & Hard Materials 30, 16–24 (2012)

Influence of Milling Atmosphere on the High-Energy Ball-Milling Process of Producing Particle-Reinforced Aluminum Matrix Composites

Steve Siebeck, Daisy Nestler, Harry Podlesak, and Bernhard Wielage

Chemnitz University of Technology,
Faculty of Mechanical Engineering,
Institute of Materials Science and Engineering,
D-09107 Chemnitz, Germany
steve.siebeck@mb.tu-chemnitz.de

Abstract. High-energy ball milling (HEM) with subsequent consolidation is a suitable method to create particle-reinforced aluminum materials. In addition to other parameters, the used PCA (process control agent) as well as the atmosphere significantly influence the milling procedure. The present article deals with the influence of different milling atmospheres (air, argon, nitrogen) on the high-energy ball-milling process when milling an Al alloy with SiC particles. The investigations show that the reaction of the ground material with air, when rinsed with air, changes the milling behavior of the aluminum powder significantly. Unlike with inert atmospheres, the use of a process control agent (PCA) is therefore no longer necessary.

Keywords: aluminum, silicon carbide, microstructure, oxide formation, AMC, in-situ oxidation.

1 General Introduction

Aluminum reinforced with hard particles is expected to show improved mechanical properties in comparison to the unreinforced alloy. A fine dispersion of particles in the metal matrix and an appropriate interface state are required. In this context, the powder-metallurgical production of MMCs has advantages over casting methods. High temperatures, as in the processing in the molten state, can be avoided. So, the diffusion and chemical reaction between the matrix and hard particles is limited or prevented. A suitable method for producing particle-reinforced aluminum alloys is the high-energy ball milling (HEM) in combination with a subsequent consolidation. The milling process is influenced by a lot of parameters such as the milling atmosphere. In order to avoid undesirable reactions of the milling material, inert gases are applied frequently [1, 2]. Selective phase transformation through a reactive ambient medium such as air is another way. In the case of aluminum powder, oxides developing on the powder surface can turn into finely dispersed reinforcement particles during the milling.

In our previous publications [3-6], we discussed materials which were milled by means of a closed chamber with a constant small amount of process control agent (PCA). This work deals with the influence of rinsing with gas, and varying the amount of PCA on the HEM process and the resulting powder for the material pairing present.

2 Experiments

Spherical Al powder of a diameter $\leq$ 100 μm (EN AW 2017) was used as feedstock material for the matrix. SiC of submicron- and micron-grain size was chosen as reinforcement particles. In this work, the milling behavior with and without SiC particles was investigated to determine the influence of the milling atmosphere on the formation of the composite powder and the composition of the matrix material. The starting point was the milling atmosphere, i.e. the residual air still in the chamber after charging, which was described in previous publications [4-6]. A supply and discharge of gases during the process did not take place. In contrast, the other test setups worked with gas rinsing, where the milling chamber was initially flooded and continuously rinsed during milling. To control the gas flow, a bubble counter was used. Nitrogen and argon were used as inert gases in addition to air. The high-energy ball milling was carried out with a Simoloyer CM08 mill (Zoz) with steel equipment. The milling parameters used in all experiments are listed in Tab. 1. A PCA was added to limit the welding of the particles and to avoid unwanted adhesions on the rotor, the balls and the chamber wall. The use of stearic acid is very common [7–9]. In the framework of this contribution, 0 and 0.5 wt.-% stearic acid were applied as PCA in addition to 0.13 wt.-% as used in our previous publications. Thus, the influence of stearic acid on the ground material and possible interactions with the milling atmosphere were examinable. The prepared powder cross-sections were first characterized by means of light microscopy (LM, Olympus PMG 3). The main focus of the investigations was on the particle-reinforced powders because the degree of dispersion and the distribution of the SiC particles in the matrix material can be very well represented by light microscopy. However, the size and shape of the composite powders are also important characteristics from which conclusions can be drawn about the influence of the milling atmosphere and the PCA. The samples were further analyzed by scanning electron microscopy (SEM, Zeiss Leo1455VP) and energy dispersive X-ray microanalysis (EDXS, EDAX Genesis).

The composite powder formation is described in detail in [4, 5]. Initially, the spherical aluminum particles are deformed into flat particles. Simultaneously, the reinforcements attach to the surface of these flakes. In the next stage, the effect of

Table 1 Milling parameters

Parameter (Simoloyer ® CM08)	Value
Mass of steel balls	8 kg
Ball diameter	4.6 mm
Ground material / powder mass	0.8 kg
Rotor speed	400 - 700 1/min (cyclic)
Milling time	4 h

cold welding starts and leads to the formation of larger composite particles with lamellar structure. The resulting structure is a mixture of alternating reinforced and unreinforced lamellas. Free particles are no longer existent at that stage. A steady deformation of the composite particles causes an increase in mixing and thus an improvement of the dispersion degree. The chronological procedure of the composite powder formation depends on the milling parameters. Strong welding effects due to high rotational speeds lead to premature formation of large, poorly mixed particles. They also lead to an increase in the composite powder grain size fraction.

The described process can be significantly influenced by the amount of PCA. In the early milling state, a large proportion of PCA leads to an increased flattening of the metallic powder without attaching of the SiC particles. The effect of cold welding is hindered by the PCA until it is degraded. Thus, the composite powder formation is delayed. The effect is particularly evident when milling aluminum powder without SiC particles. It results in extremely flattened particles (Fig. 1).

Fig. 1 Optical micrograph of milled AA-2017 without reinforcements and 0.5 wt.-% stearic acid after 4 h of milling time

During milling at residual air atmosphere (closed vessel) without PCA, the cold welding causes a strong coarsening of the powder (Fig. 2). This limits the possible milling time and thus the desired fine distribution of hard particles inside the powder particles. Increasing the amount of PCA reduces the effect of cold welding and so limits the powder coarsening. When using an inert milling atmosphere, similar effects can be observed.

A completely different behavior of the process is detectable when using air rinsing (Fig. 3). The influence of air reduces the cold welding to a favorable level. Therefore, the PCA has no significant influence under these conditions. The composite powder formation works very well even without PCA (Fig. 3a). Adhesions to the milling tools do not take place. The formation of oxide on the surface of the aluminum particles is assumed to inhibit the tendency to cold welding similar to the PCA. In the literature, alumina is occasionally used as PCA [10].

It is likely that the atmosphere and the PCA cause changes in the composition or contaminations of the ground material. For large amounts of stearic acid (up to 5 wt.-%), an influence was already detected by means of thermogravimetric measurements [11]. Own TGA investigations on powders after milling with 0.5 wt.-% of stearic acid have not yielded any useful results.

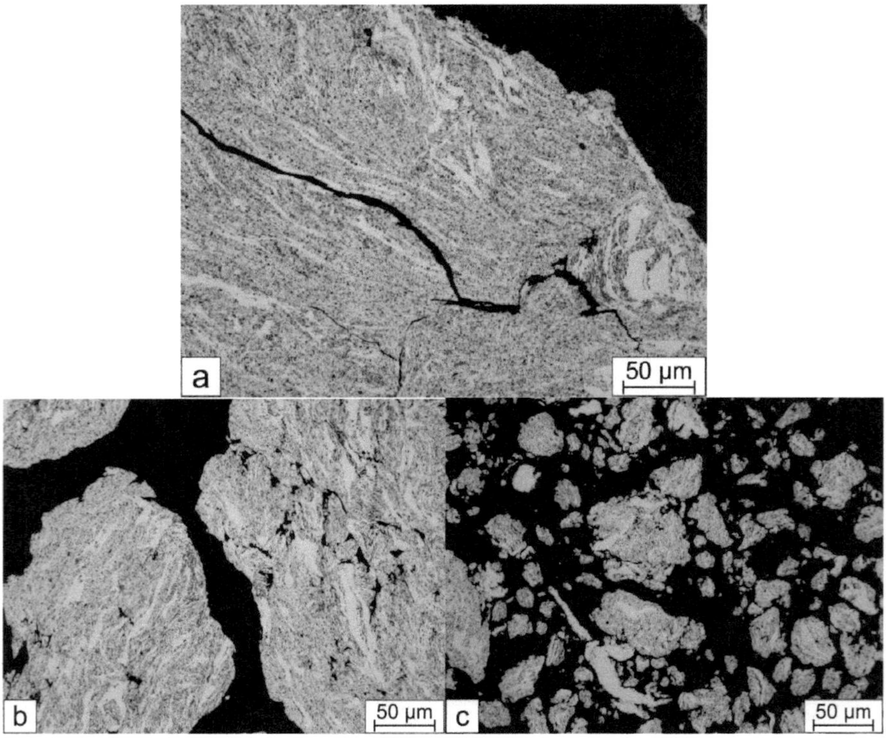

Fig. 2 Optical micrograph of milled AA-2017 with 10 vol.-% SiC under residual air: a) without PCA, after 3 h of milling time b) with 0.13 wt.-% stearic acid after 4 h of milling time c) with 0.5 wt.-% stearic acid after 4 h of milling time

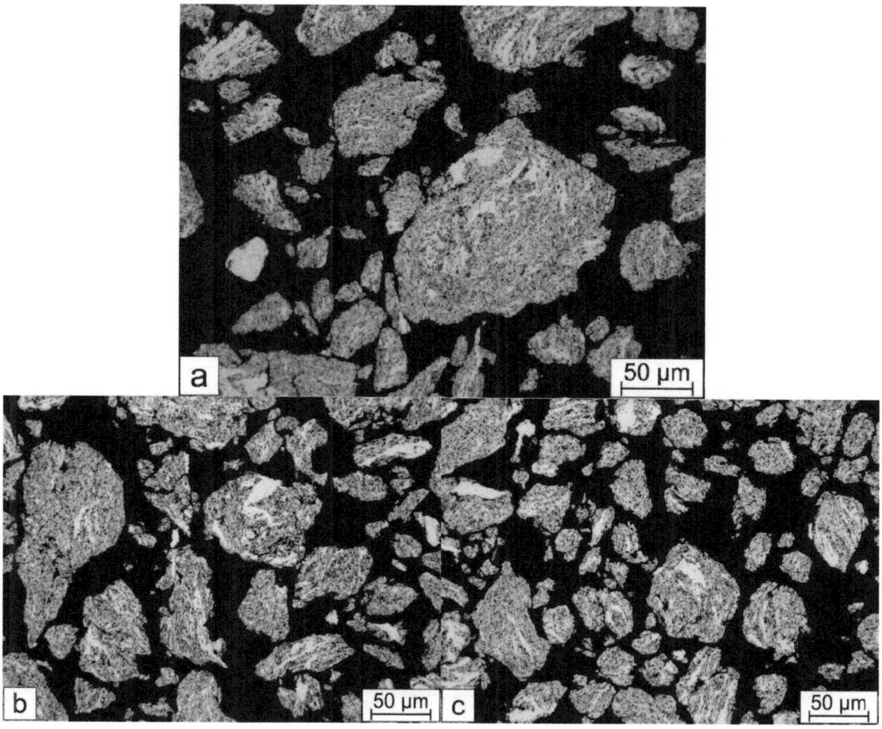

Fig. 3 Optical micrograph of milled AA-2017 with 10 vol.-% SiC under rinsing air: a) without PCA, after 4 h of milling time b) with 0.13 wt.-% stearic acid after 4 h of milling time c) with 0.5 wt.-% stearic acid after 4 h of milling time

On the basis of EDXS measurements, the oxygen concentration of the powder was tested. Analyses were performed on powder cross-sections in the core of the powder particles. Tab. 2 clearly shows the influence of the milling atmosphere on the oxygen content in the composite. As expected, the oxygen content is multiplied by using air rinsing compared to inert gas rinsing. It has to be noted that the PCA also inserts oxygen into the ground material. The extent of this effect depends on the milling conditions. This is particularly evident in milling tests with a closed milling chamber. Due to the high pressure in the chamber, the evaporation of the stearic acid only takes place at higher temperatures. Additionally, the gaseous stearic acid cannot leave the milling chamber, which would be possible with gas rinsing. In this respect, for the closed as well as the rinsed experimental setup, different amounts of PCA are involved in the milling process. It is assumed that the oxygen content in the case of milling with a closed chamber originates from both the residual air and the stearic acid. This has been confirmed by the comparison of the oxygen concentrations of 2.4 and 2.9 wt.-% for the PCA amounts 0.13 and 0.5 wt.-% respectively.

Table 2 EDXS measurements: oxygen concentration in milled AA-2017 powder, depending on the atmosphere at constant grinding parameters and 0.13 wt.-% PCA

Milling atmosphere	Oxygen content [wt.-%]
Closed chamber (residual air)	2.4
Air rinsing	4.8
Argon rinsing	0.9
Nitrogen rinsing	0.9

3 Summary

The influence of the milling atmosphere and the amount of PCA on the high-energy milling of an aluminum alloy with SiC particles (0.2 to <2 microns) was studied. As expected, the comparison of the inert gases nitrogen and argon shows no significant differences with respect to the grinding behavior. However, the use of rinsing air results in different behavior. Whereas the use of inert gases and a closed milling chamber principally only works with a PCA in order to limit excessive adhesions on the milling tools as well as powder coarsening, this can be omitted because of the separating effect of the in-situ-formed alumina. Accordingly, air rinsing shows the clearest changes in the increase of the oxygen content during milling without SiC.

Acknowledgments. The authors would like to thank the Deutsche Forschungsgemeinschaft (DFG) for supporting the research project SFB 692 A2-1. Further thanks go to G. Engelhardt and A. Graf.

References

1. Zhao, N., Nash, P., Yang, X.: The effect of mechanical alloying on SiC distribution and the properties of 6061 aluminum composite. Journal of Materials Processing Technology 170(3), 586–592 (2005)
2. Yang, Z.-G., Shaw, L.L.: Synthesis of nanocrystalline SiC at ambient temperature through high energy reaction milling. Nanostructured Materials 7(8), 873–886 (1996)
3. Wagner, S., Podlesak, H., Siebeck, S., Nestler, D., Wagner, M.F.X., Wielage, B., Hockauf, M.: Einfluss von ECAP und Wärmebehandlung auf Mikrostruktur und mechanische Eigenschaften einer SiC-verstärkten AlCu-Legierung. Effect of ECAP and heat treatment on microstructure and mechanical properties of a SiC reinforced Al-Cu alloy. Materialwiss. Werkstofftech. 41(9), 704–710 (2010)
4. Podlesak, H., Siebeck, S., Mücklich, S., Hockauf, M., Meyer, L., Wielage, B., Weber, D.: Pulvermetallurgische Erzeugung von SiC- und Al2O3-verstärkten Al-Cu-Legierungen. Materialwiss. Werkstofftech. 40(7), 500–505 (2009)
5. Nestler, D., Siebeck, S., Podlesak, H., Wagner, S., Hockauf, M., Wielage, B.: Powder Metallurgy of Particle-Reinforced Aluminium Matrix Composites (AMC) by Means of High-Energy Ball Milling. In: Fathi, M., Holland, A., Ansari, F., Weber, C. (eds.) Integrated Systems, Design and Technology 2010Integrated Systems, Design and Technology 2010, vol. 58, pp. 93–107. Springer, Heidelberg (2011)

6. Wielage, B., Nestler, D., Siebeck, S., Podlesak, H.: Untersuchungen zur Herstellung siliziumkarbid-partikelverstärkter Aluminiumpulver durch Hochenergiekugelmahlen. Materialwiss. Werkstofftech. 41(6), 476–481 (2010)
7. Benjamin, J., Bomford, M.: Dispersion strengthened aluminum made by mechanical alloying. Metallurgical and Materials Transactions A 8(8), 1301–1305 (1977)
8. Moon, K.I., Lee, K.S.: Development of nanocrystalline Al-Ti alloy powders by reactive ball milling. Journal of Alloys and Compounds 264(1-2), 258–266 (1998)
9. Zhou, F., Liao, X.Z., Zhu, Y.T., Dallek, S., Lavernia, E.J.: Microstructural evolution during recovery and recrystallization of a nanocrystalline Al-Mg alloy prepared by cryogenic ball milling. Acta Mater. 51(10), 2777–2791 (2003)
10. Suryanarayana, C.: Mechanical alloying and milling. Prog. Mater. Sci. 46(1-2), 1–184 (2001)
11. Kleiner, S., Bertocco, F., Khalid, F.A., Beffort, O.: Decomposition of process control agent during mechanical milling and its influence on displacement reactions in the Al-TiO2 system. Mater. Chem. Phys. 89(2-3), 362–366 (2005)

Numerical Simulation of Scratch Tests for the Verification of Material Models for Particle-Reinforced Coatings

Tobias Müller, Daisy Nestler, Thomas Lampke, and Bernhard Wielage

Chemnitz University of Technology,
Chemnitz, Germany
tobias.mueller@mb.tu-chemnitz.de

Abstract. Material models are the basis of most numerical simulations in mechanical engineering. In the field of elastic deformation, the material models are quite simple but plasticity and material destruction are very difficult to calculate. For developing and testing material models, an easily and quickly accomplishable test is necessary to verify the results. The scratch test is a good choice for the modeling of surfaces. For that test, a diamond tip (indenter) moves onto the material with either constant or progressive normal force. First, the material will be deformed elastically; with increasing force, plastic deformation occurs, leading to crack and chip formation, depending on the ductility of the material. The result can be directly compared to the simulation of this test.

The article describes problems and solutions during the simulation of the scratch test for the steel C45 (normalized) with and without a nickel coating. The comparison with experimental results shows that this approach is successful. In the future, a closed material model for this type of stress will be developed.

1 Introduction

The scratch test is normally used to determine the quality and performance of material surfaces. Therefore, the surface will be scratched by a diamond (mostly with a Rockwell geometry) with constant or progressive load. This test is an important tool for characterizing the adhesive strength of coatings especially for thin PVD/CVD coatings [1, 2] or electro-deposited layers [3], but is used also in thermally sprayed coatings [4]. Increasingly, the use of this method is to determine the scratch energy density and to quantitatively assess the abrasive wear behavior. The scratch tester detects plastic and elastic penetration depth, normal force and friction force, and acoustic signals resulting from crack formation. Along with the visual evaluation, it results in a comprehensive description of the surface properties at scratch load [5-8]. Figure 1 draws an overview of the scratch tester with directions of movements and forces.

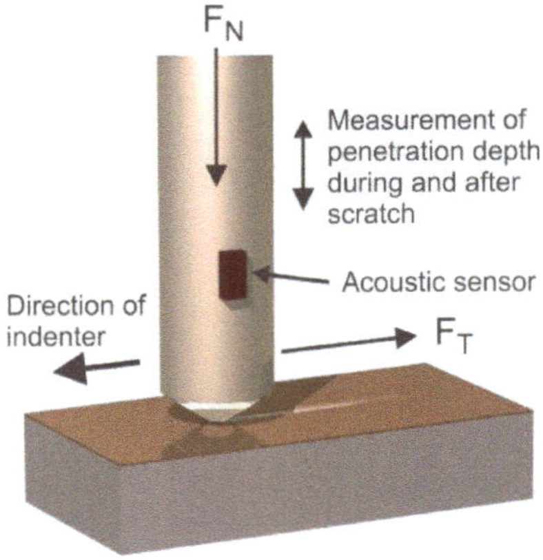

Fig. 1 Schematic of scratch tester

The scratch test is qualified for fast and simple testing of surfaces. But the material behavior is examined only from the outside. The stress state inside the specimen in particular in the interfaces is normally unknown. The simulation of the scratch process can provide a better understanding of the mechanisms inside the bulk material and in one or more coatings during a scratch. The stress state inside the system and the resulting material destructions or delaminations can be observed. This can provide new evidence for the development of coating systems. Similarly, the complex simulations of plasticity and destruction can be easily validated. Figure 2 shows the correlation of simulation, coating process and scratch test.

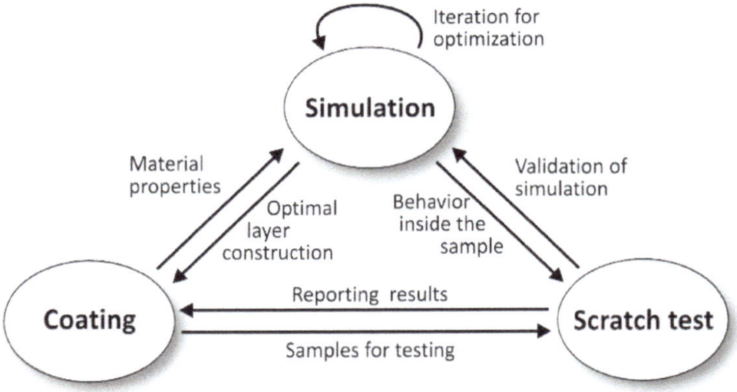

Fig. 2 Correlation between simulation coating, scratch test and simulation

The simulation of a scratch test is a great challenge even with modern FEM tools. The reason is the strong deformation of material during the scratching and the potential cracking and chipping. Thus, the scratch test can also be used to support the validation of new material models in the field of coating, large deformation and wear simulation.

First, uncoated homogeneous material is simulated and tested. At the moment, the simulation of different homogeneous layers on a ductile base material is performed. The aim of this research is the simulation of particle-reinforced coatings such as electroless nickel-based composite coatings.

2 Simulation in Detail

Initial simulations were performed using the simulation software Abaqus. Now the FEM system Marc-Mentat is used for a simulation with and without coatings. Marc-Mentat includes better remeshing techniques which are very important for the computation of high deformations.

2.1 Model Assembling

The scratching of the material takes place by a diamond (Rockwell C) with an opening angle of 120° and a tip radius of 0.2 mm (Fig. 3). The diamond is a stiff (rigid) model, which means it does not deform under the action of forces and temperatures.

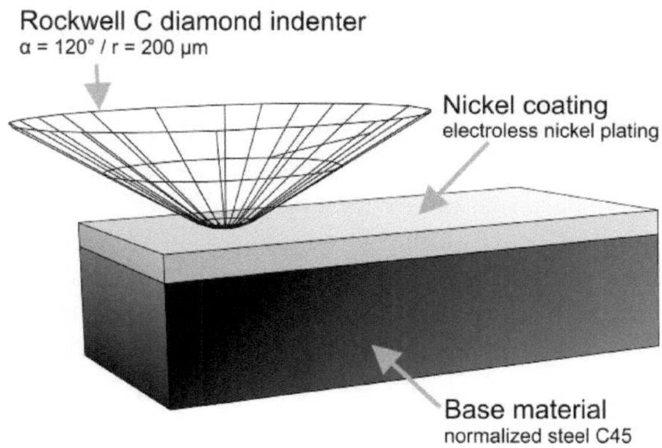

Fig. 3 Geometrical modeling of scratch test with one layer

Since the scratch operation proceeds approximately symmetrically (with ideal material exactly symmetrical), it could be assumed that only half of the sample was modeled. Due to numerical problems in calculations with models for material destruction and remeshing, it is reasonable to work with the full model.

In simulation and experiment, the scratch process was performed for a length of 10 mm with a progressively applied force of 1 - 100 N. This is the default procedure for a scratch experiment. To reduce the simulation complexity and time, the scratch length was shortened to 3 mm. Fig. 4 shows different measurements with 3 and 10 mm scratch length. The penetration depth is within the tolerance range and depends essentially on normal force.

The size of the simulated material cut-out is 2 x 1 x 5 mm (W x H x D) for 3 mm scratches. The coating is firmly connected with the base material till a defined

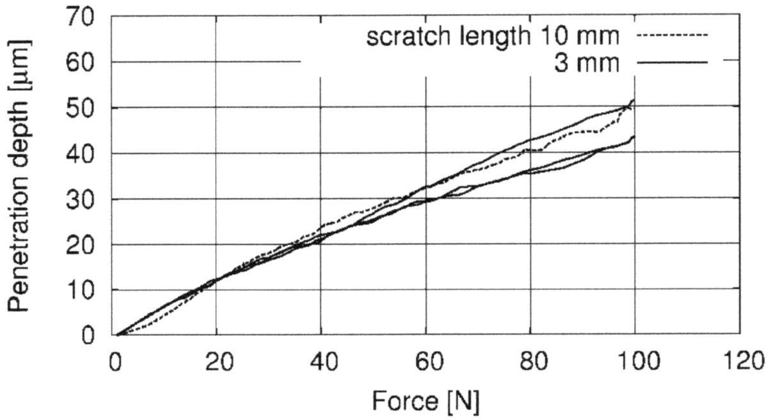

Fig. 4 Scratch tests with different scratch lengths

stress state is reached. The material models that characterize this behavior are described in the following section. The modeling of the interface with a functional interface layer does not work according to the remeshing algorithms.

2.2 *Material Models*

The behavior of ductile materials under stress can be divided into three sections. For small strains up to the yield point, the material deforms only elastically. For larger strains, plastic deformation occurs. For metals, this leads to strain hardening (dislocation multiplication), resulting in an increase in yield strength. In the third state, the strain begins to decrease. Eventually, gradual destruction of the material occurs, up to fracture. Each of these states is represented in the simulation system by different material models. An overview of the material behavior in these states is given in Fig. 5.

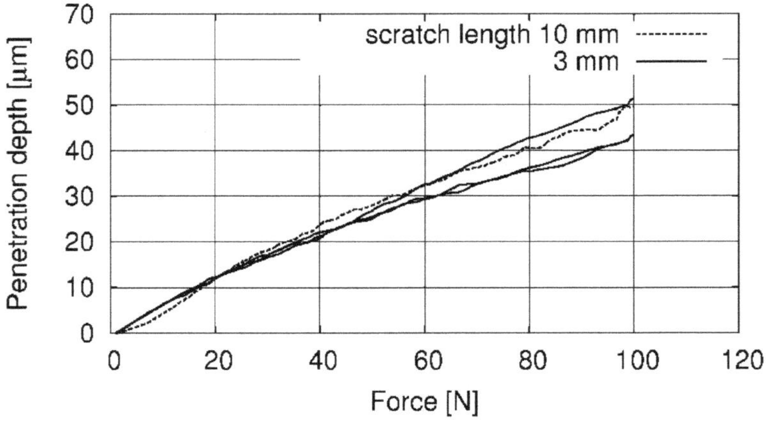

Fig. 5 Different material models for different material states

In the elastic range, Hooke's law is used, till the strain drops after a plastic deformation. The yield condition is determined due to the stress deviator, i.e. the deviation from the hydrostatic axis. Therefore, the yield condition describes a surface in the stress space. For the results presented here, the von-Mises yield criterion is used, which is represented by a cylindrical surface in the stress space. If the yield condition is satisfied, the material state cannot be clearly defined by the value of stress. In this case, the strain tensor and its scalar equivalent, the effective elastic-strain increment, define the states of the material sections.

The simulation of the tribological contact between indenter and specimen requires special models. The often used Coulomb model with static and dynamic friction is difficult to solve in an FEM system. Fig. 6 shows the behavior of the Coulomb model and a "softer" model which uses the arc tangent function. F_t is the tangential force, v_r the velocity between the friction partners. In the Coulomb model, a movement starts with the force $\mu * F_n$. This discontinuity is the reason for the problems in solving the equations.

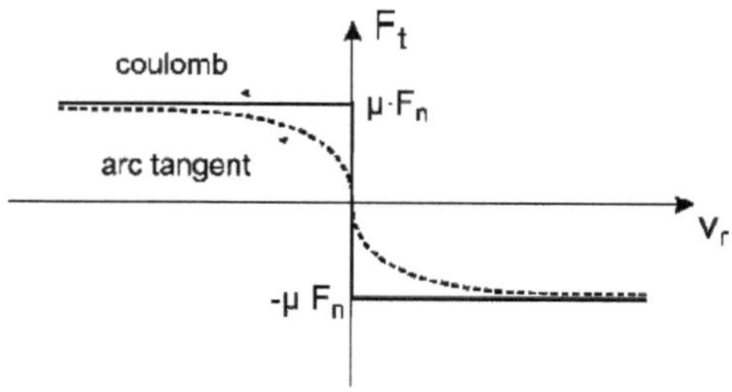

Fig. 6 Coulomb and arc tangent model in comparison

The arc tangent model

$$||\mathbf{F}_t|| = -\mu_{gleit} \cdot F_N \cdot \frac{2}{\pi} \arctan\left(\frac{||\mathbf{v}_r||}{S_c}\right) \cdot \frac{\mathbf{v}_r}{||\mathbf{v}_r||} \tag{1}$$

smooths the hard edges with the parameter S_c. Now there is a dependency between velocity and force, and stick-slip effects will be eliminated.

For simulations with coatings, a mechanism for layer debonding has to be implemented. In its initial state, the layer is attached to the base material. If the condition

$$\left(\frac{\sigma_n}{S_n}\right)^m + \left(\frac{\sigma_t}{S_t}\right)^n > 1 \tag{2}$$

is true in an FEM node, the node is counted as disconnected. S_n, S_t, m and n are the interface parameters, σ_n and σ_t are the normal and tangential stresses in the interface. If m and n are assumed as two, this construct can be imaged as an ellipse in the stress space.

2.3 Remeshing

The underlying algorithm for the FE computation is called Lagrange formalism. It requires each node to be linked to a unique position in the material. For large deformations, like the plastic deformation in a scratch test, the mesh begins to degrade and the elements become misshapen. In order to circumvent these problems, the mesh will be re-created between the time steps of the simulation. The node and element states have to be assigned with interpolation to the new mesh. The trigger for remeshing is the aspect ratio, stress or strain gradients or penetration depths in contact areas. The disadvantage of remeshing is the growing inaccuracy caused by the interpolation algorithm.

The FEM system Marc contains a tool that can control the density of the mesh in “mesh boxes” (Fig. 7). These mesh boxes can be moved and resized in every remeshing step. Thus, the mesh density can be controlled in the simulation time. This saves computing time and increases the potential for denser meshes in areas with high gradients.

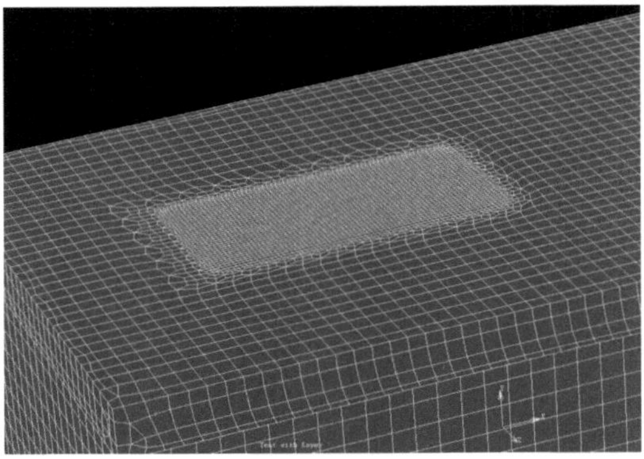

Fig. 7 Different mesh densities in different areas controlled by mesh boxes

3 Results and Comparison with Measurements

The following sections describe the results of simulation compared with different measurements. As a good parameter for comparison, the penetration depth is shown in relation to the normal force.

3.1 Simulation without Coating

The simulation of scratch tests in homogeneous materials without coating works well for different ductile materials. The simulation results are comparable to the measurements. The basis of a good simulation is the knowledge of the stress-strain behavior from elastic deformation to damage of the material. Fig. 8 shows the results of measurement and simulation for two different metals. The spikes in the simulation charts result from the remeshing.

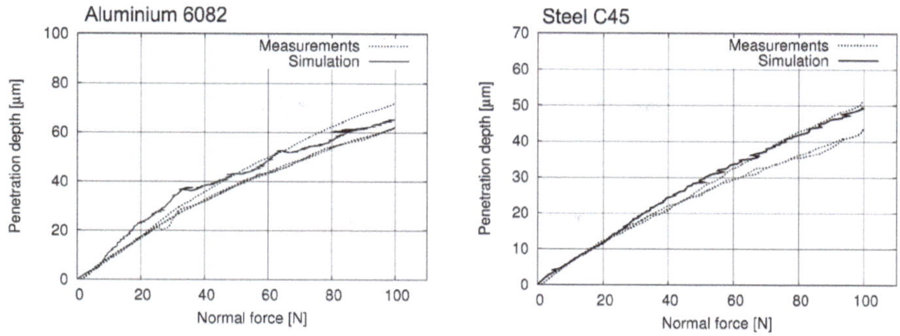

Fig. 8 Comparison between measurement and simulation of 3 mm scratches without coating

A cross-section at a specific position of a 10 mm scratch is pictured in Fig. 9 (left). There is also a good analogy between measurement and simulation. The left side of this figure shows the formation of cracks at the bottom of the scratch. In the case of total material destruction, no stress can be transferred through the material and a lower stress is displayed.

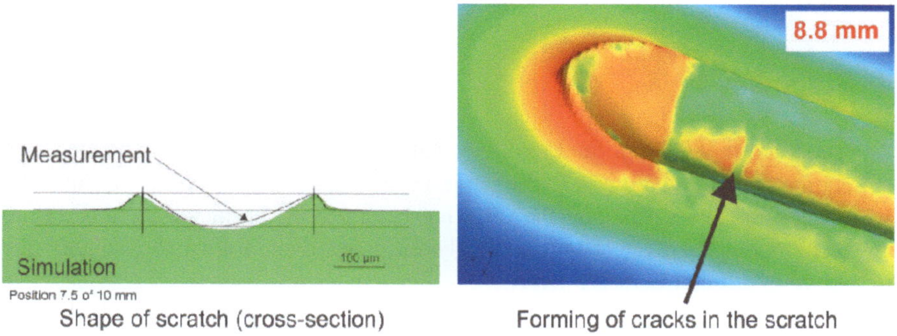

Fig. 9 Results of the simulation of a scratch test with steel (C45); comparison of cross-section (left) and stress distribution during the scratch (right)

3.2 *Simulation with Coating*

The simulation with a coating is more difficult. To be able to exactly assess the stress-strain behavior, some information about the interface between layer and base material is necessary. Fig. 10 shows the results. It is clearly visible that the direction of the curve changes at a depth of 50 μm. This change is much more obvious in the simulation than in the measurement. This is an indication that the adhesion model is not working correctly at the moment.

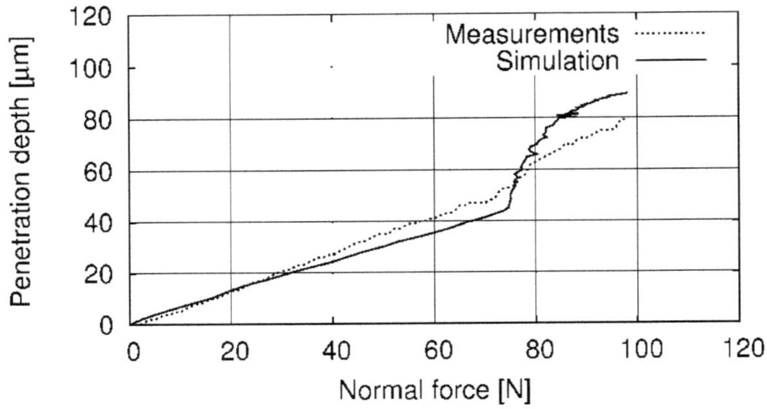

Fig. 10 Comparison of measurement and simulation of an electroless plated steel (C45) with nickel. The thickness of the nickel layer is 50 µm.

The greatest problem with the remeshing algorithm is that it still has difficulties with small coatings. Therefore, we are planning to develop our own algorithm to achieve the aim of simulating particle-reinforced coatings and multi-layer coatings.

4 Summary

The scratch test is an important tool for testing surfaces. By emulating this test in a simulation, not only the external change of the material is visible, but also the stress state in the interior and the resulting behavior of the material. In addition, material models can be tested relatively easy for their ability to simulate the scratch tests or other abrasive wear.

By modeling the various stress areas of ductile materials, the behavior of uncoated samples can be calculated very precisely. The greatest problem of the simulation of coated systems is to find and implement an applicable remeshing algorithm. This is the next step in our research. The future aim is to simulate particle-reinforced coatings and multi-layer coatings for a better understanding of these systems and the relationship with wear simulations.

References

1. Goto, H., Gotoh, M., Ejiri, S., Horimoto, Y., Hirose, Y.: Scratch test of TiCN thin films with different preferred orientation. Materials Science Forum 524, 729–734 (2006)
2. Ko, D., Lee, J., Kim, B.: Mechanical and Adhesive Properties of Plasma CVD Coatings on Various Substrates using Scratch Test. Solid State Phenomena 116(17), 304–307 (2006)

3. Xu, C., Fan, C., Zhang, Y., Abys, J.A.: Oberflächenanalyseverfahren im Überblick. Mo Metalloberfläche 54(11), 53–59 (2000)
4. Bolelli, G., Cannillo, V., Lusvarghi, L., Montorsi, M., Mantini, F.P., Barletta, M.: Microstructural and tribological comparison of HVOF-sprayed and post-treated M–Mo–Cr–Si (M= Co, Ni) alloy coatings. Wear 263(7-12), 1397–1416 (2007)
5. Burnett, P.J., Rickerby, D.S.: The relationship between hardness and scratch adhesion. Thin Solid Films 154, 403–416 (1987)
6. Bull, S.J.: Failure modes in scratch adhesion testing. Surface & Coatings Technology 50(1), 25–32 (1991)
7. Bull, S.J., Berasetegui, E.G.: An overview of the potential of quantitative coating adhesion measurement by scratch. Tribology International 39(2), 99–114 (2006)
8. Taube, K.: Qualitätssicherung an tribologischen Schichten-Eigenschaften und Messverfahren. Mat.-wiss. und Werkstofftechnik 3

Automatic Variable Noise Suppression for Laser Based Classification of Explosive Materials*

Jan Schlenke and Lars Hildebrand

Dortmund University of Technology, Computer Science Department,
Chair 1, Otto-Hahn-Str. 16, 44221 Dortmund, Germany
{Jan.Schlenke,Lars.Hildebrand}@tu-dortmund.de

Abstract. Efficient noise suppression is an important factor in every form of sensory data analysis. Data acquired outside controlled lab environment often suffer from severe noise interferences that need to be minimized in order to perform reliable interpretation and evaluation. Existing solution often employ repeated measurements to enhance signal to noise ratios in which case measurement time becomes a primary factor. Furthermore iterated measurements are difficult if the measured object is destroyed or partially consumed in the measuring process. We propose a noise reduction approach that uses wavelet transformation in combination with an automatically derived threshold function to reduce noise levels while minimizing signal degradation and line broadening. We illustrate the achieved results using measurements of explosive materials acquired using LIBS and Raman spectroscopy.

Keywords: Automatic de-noising, Wavelet transformation, Shift invariant, Thresholding, LIBS, Raman, OPTIX Project.

1 Introduction

Noise is a common source of interference occurring in a wide range, if not all of, measuring techniques. Options to reduce the influence of noise and increase the signal to noise ratio (SNR) are numerous and range from choosing the experimental setup, over optimizing hardware and calibration which affect the measurement directly to software based filters which are used to enhance the measurement after it has been recorded. A simple and effective way to reduce noise or enhance the SNR is repeated sampling. Repeated sampling works under the assumption that mean noise energy is zero and signal responses are constant, and thus reduces noise distortions by calculating the mean of the accumulated measurements.

* ework Program (FP7/2007-2013) under GrThe research leading to the results presented here has received funding from the European Community's Seventh Framant Agreement No. 218037.

When repeated sampling is not a possible option due to time or technological constrains filters are needed that can enhance single measurements. Technologies used for OPTIX include Laser-induced breakdown spectroscopy (LIBS) [1] and Raman spectroscopy [2], LIBS consumes small portions of the measured substance while Raman measurements which are usually accumulated are limited by the available acquisition time. A known drawback of filters that operates on single measurements lies in the fact that the underlying signal is altered in the filter process - often by broadening and shrinking signals or introducing unwanted oscillations - resulting in a trade-off between noise suppression and signal preservation. This makes the degree of smoothing a critical parameter as too intensive smoothing can result in some signal becoming undetectable or in introduce fake signals while weak or no noise suppression cannot remove the problems inherent to poor signal to noise ratio. In recent years several noise suppression approaches using wavelet transformation have been proposed based on the works of Donoho and Johnstone [3],[4],[5]. A general introduction to wavelets and discrete wavelet transformation can be found in [6],[7] and [8]. In this paper we propose a new approach to automatic smoothing by extending Donoho's thresholding scheme for denoising to utilize a redundant form of wavelet transformation described by Lang et.al.[9],[10] in combination with automatically derived threshold functions to handle variable noise levels within measurements.

2 Examples

Real measurement often exhibit noise intensities that are not constant through the spectral domain. Filtering such variable noise with constant filter parameters naturally leads to sub smoothing effects in areas where only a portion of the noise is sufficiently suppressed and over smoothing effects which degrade the original signal in areas that suffer from less intense noise.

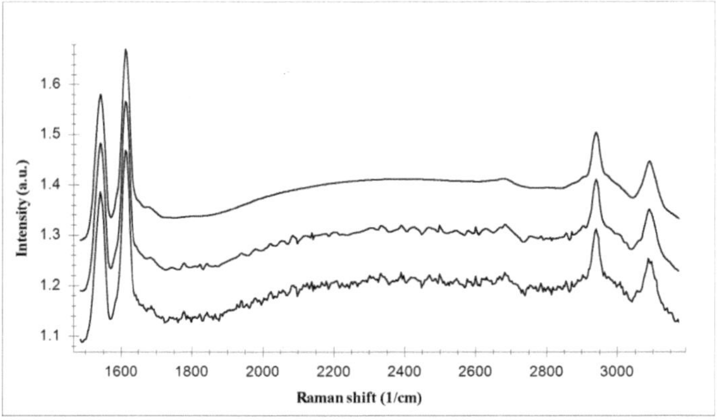

Fig. 1 Effects of different smoothing techniques tested with a high quality Raman spectrum of Explosive substance A. Bottom: Original measurement. Middle: Spectrum treated with constant universal threshold. Top: Spectrum treated with variable noise suppression technique.

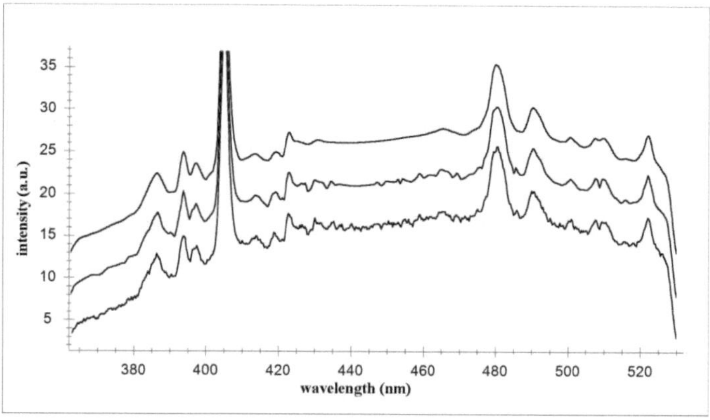

Fig. 2 Partial LIB Spectrum of Explosive substance. Bottom: Original Spectrum. Middle: Results of noise suppression using constant universal threshold. Top: Spectrum treated with variable wavelet thresholding.

Experiments with real measurements of Raman and LIB spectroscopy show the positive effects of variable thres- holds see Figure 1 and Figure 2. In both cases the constant threshold (middle) visibly reduces the noise intensity but is not able to fully remove the distortions while the variable threshold (top) creates a very smooth spectral line without major distortions to visible signals. Figure 3 illustrates the different results obtained using constant and variable threshold using a Raman spectrum suffering from higher relative noise intensities than the partial spectrum seen in Figure 1. Filter results suggest that noise intensity in this real measurement is not constant as filtered spectrum displays an increasingly irregular behavior with higher Raman shift. Please note that an edge filter was used to suppress the laser signal resulting in a strong suppression of all signals left of shift zero.

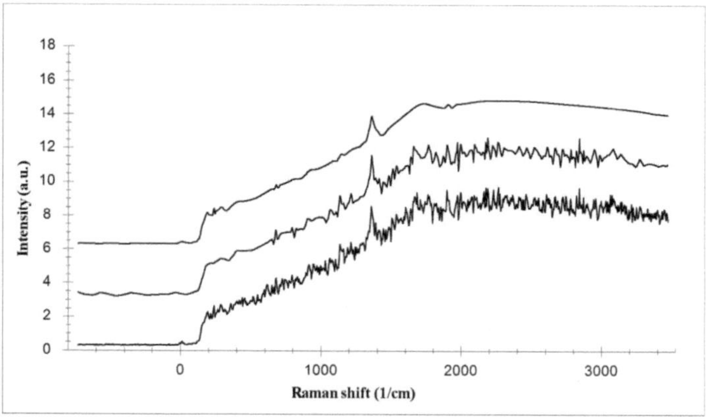

Fig. 3 Noise suppression on low quality Raman spectrum of an explosive substance. Bottom: Original Spectrum. Middle: Results of noise suppression using constant universal threshold. Top: Spectrum treated with variable wavelet thresholding.

The suppression of variable noise intensities with constant thresholds can also lead to complete signal extinction. Figure 4 illustrates this using an artificially created signal and strongly modulated noise. The original artificial spectrum contains six lorenzian peak signals of different intensities. The spectrum has been afflicted with normal distributed noise modulated with a low frequency sine function so parts of the spectral domain suffer from high intensity noise while other parts are almost noise free. The smallest of the six peak signals has been placed in the region of lowest noise intensity. Treating the noise signal with a constant threshold filter completely erases the small signal while filtering with variable threshold is able to preserve all five signals while still achieving comparable noise suppression in the regions suffering from intense noise.

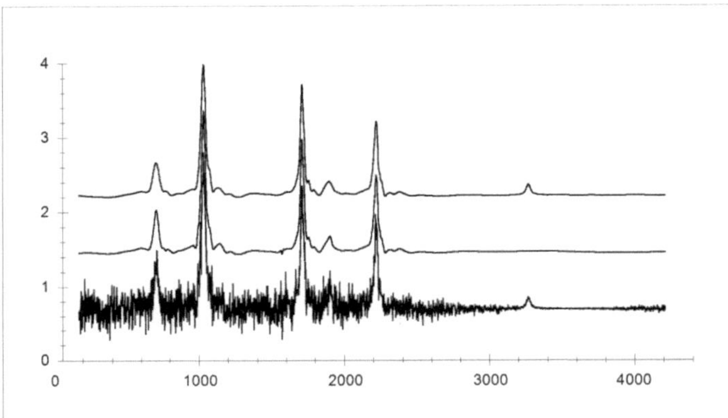

Fig. 4 Artificial spectrum filtered with constant universal threshold (middle) and variable threshold (top)

3 Method

Wavelet transformation describes a family of transformations which use basis functions, so called Wavelets, to decompose data into distinct subspace scales of different detail. Other than the Fourier transformation, wavelet transformation uses finite basis functions and thus retains a sense of local information in the transformed data. Scales are nested satisfying the multiresolution analysis requirement, meaning that the space that contains finer scales also contains larger, coarser scales.

$$V_j \subset V_{j+1} \quad \forall j \in Z \tag{1}$$

In the continuous case one has the finest scale $V_{-\infty} = 0$ and $V_{+\infty} = g$, with g representing the regarded function in its entirety. In the discrete case the finest scales usually span two sample points while the coarsest scale spans the entire signal. The wavelet transformation of a given signal uses a set of functions $\psi_{j,k}(t)$ called wavelets, which span the differences between the spaces spanned by the scaling functions. Each scaling space V_j can thus be described by the scaling space V_{j-1} and the wavelet subspace W_{j-1} which describes the differences between V_j and V_{j-1}.

$$V_j = V_{j-1} \oplus W_{j-1} \tag{2}$$

By extending this idea one can transform the relation of cascading scaling subspaces to render:

$$\mathrm{R} = \mathrm{V}_0 + W_0 + W_1 + \ldots + \mathrm{W}_{\mathrm{n\text{-}1}} + W_n \tag{3}$$

with R denoting the space of presentable functions, V_0 denoting the coarsest scaling space and W_j the wavelet space corresponding to scale j. The definition of the discrete wavelet transformation of a signal $f_k(t)$ is closely related to the above relation of subspaces and can be written as:

$$f_k(t) = c(k)\cdot \quad (\mathrm{t} - \mathrm{k}) + \sum_j d_j(k)\cdot \; 2^{j/2}\psi(2^j \mathrm{t} - \mathrm{k}) \tag{4}$$

$\varphi(k)$ is called the scaling function with the corresponding coefficients $c(k)$ while $\psi(k)$ represents the wavelet function with corresponding coefficients $d_j(k)$. In practice this means that given a specific basis function or motherwavelet a function f can be fully characterized by the coefficients given in $c(k)$ and $d_j(k)$, similar to the representation as coefficients of the discrete Fourier transformation. The stationary or shift invariant wavelet transformation variant requires a storage space of $N \log N$ and can be computed in $O(N \log N)$ increasing both factors by $\log N$ compared to normal discrete wavelet transformation [11].

Given the noise standard deviation σ Donoho and Johnson have proposed the universal threshold $\lambda = \sigma \cdot \sqrt{2\log(N)}$ as an asymptotically optimal solution to suppress noise using wavelet transformation [4]. To estimate the non constant threshold function we can use the fact that wavelet coefficients, and especially those at higher detailed scales, describe a well localized influence on the original data. Let $f_m(x)$ be the noise modulating function then the universal threshold function for modulated noise can be written as:

$$f_{\text{universal}}(x) = \sqrt{2\mathrm{Log}_e(N)} \cdot \sigma \cdot f_m(x) \tag{5}$$

In practice σ is usually unknown and has to be estimated from the measured data. Donoho and Johnstone propose the following estimation, $\sigma = median(|d_i|)/0.6475$, i $\in$ finest detail scale [3]. However this methods leads to the above mentioned problems if σ is modulated since large parts of the signal are treated with noise suppression parameters that do not fit the actual occurring noise. To approximate the universal threshold function in the case of modulated noise it is possible to extrapolate σ and $f_m(x)$ individually from a given measurement but estimate the behavior of the product $\sigma \cdot f_m(x)$ using a windowed median filter. The window-size used to describe the range of the median filter becomes an additional parameter which has to match the underlying modulation. Large window-sizes limit the responsiveness and result in slow changing threshold functions while smaller window-sizes allow for a more dynamic curve but are also more prone to error due to the smaller sample number. For window sizes that are large in relation to the noise modulation frequency the windowed median function converges towards the global median which we have already shown to be disadvantageous for non constant noise levels. To address the issue of noise passing the suppression unaltered we therefore used an alternative method to extrapolate a threshold function which converges towards an upper-bound noise intensity threshold instead of the threshold optimal for the median noise intensity. This alternative method uses upper and lower linear enveloping curves to extrapolate an approximation of the optimal threshold function.

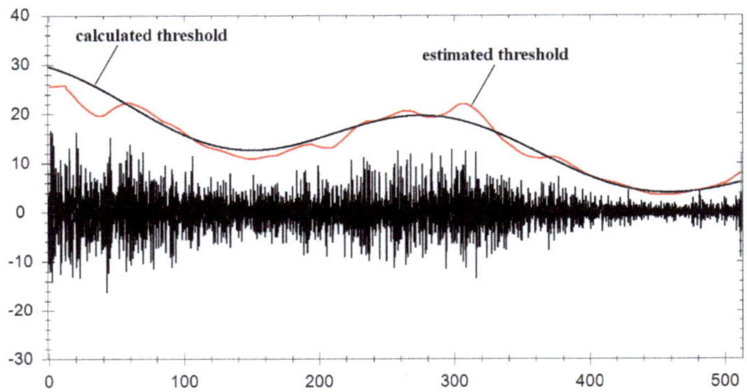

Fig. 5 Illustration of optimal variable noise threshold estimation from the finest detail scale of the wavelet transformation. The calculated threshold uses information of noise standard deviation and modulation that are usually not available in practice. The estimated threshold is based solely on the information contained in the wavelet coefficients.

The curves are constructed as interpolations between the extreme values occurring within a defined window of data-points. The upper-lower(UL) envelop is defined as the lower envelop of the upper envelop of the coefficients and the lower-upper(LU) envelop respectively as the upper envelop of the lower envelop of the coefficient values. The idea of using an aggregation of approximations is inspired by the opening and closing operators found in the field of morphology [12]. In this case the application of both operations serves the purpose to add robustness to the estimation in case fine scale wavelet coefficients contain portions of signals mixed with noise. The threshold function is calculated as the difference of UL and LU, f_{thres} = UL - LU. Note that in this variant there is no longer the need to estimate σ. The window size to calculate the enveloping functions used for our experiments was chosen as w=N/50 and has returned reliably good results for uniform and normal distributed noise in samples of N=1024, 2048, 4096 and 32768 data-points. Larger window-sizes generally result in higher threshold values and generally more rigid threshold behavior. Figure 5 illustrates the finest detail scale obtained by wavelet transformation of the artificial spectrum given in Figure 3 as well as the estimated threshold function using the cascaded enveloping functions and the calculated threshold function based on the full knowledge of modulating function and noise standard deviation.

4 Conclusions

Noise reduction via wavelet thresholding yields high quality results with regard to smoothness and signal preservation. Problems like signal degradation and line broadening are significantly less severe compared to mean or gauss smoothing. The proposed method is able to detect and suppress variable intensities of noise within a single measurement without the need for user interaction or repeated measuring. These characteristics are well suited for the automatic detection and analysis of substances in uncontrolled environments where interference intensity and variability are difficult to predict and control.

References

1. Rusak, D.A., Castle, B.C., Smith, B.W., Winefordner, J.D.: Fundamentals and applications of laser-induced breakdown spectroscopy. Critical Reviews in Analytical Chemistry 27(4), 257–290 (1997)
2. Norman, B., Colthup, L.H.: Introduction to infrared and Raman spectroscopy. Academic Press (1990)
3. Johnstone, J.M., Donoho, D.L.: Adapting to unknown smoothness via wavelet shrinkage. Journal of the American Statistical Association 90, 1200–1224 (1995)
4. Donoho, D.L., Johnstone, J.M.: Ideal spatial adaptation by wavelet shrinkage. Biometrika 81(3), 425–455 (1994)
5. Donoho, D.L., Johnstone, J.M.: Asymptotic minimaxity of wavelet estimators with sampled data, statist. Sinica, 1–32 (1999)
6. Daubechies, I.: Ten Lectures on Wavelets (CBMS-NSF Regional Conference Series in Applied Mathematics). SIAM: Society for Industrial and Applied Mathematics (June 1992)
7. Daubechies, I.: Orthonormal bases of compactly supported wavelets ii: variations on a theme. SIAM J. Math. Anal. 24, 499–519 (1993)
8. Shensa, M.J.: The discrete wavelet transform: wedding the a trous and mallat algorithms. IEEE Transactions on Signal Processing 40(10), 2464–2482 (1992)
9. Lang, M., Guo, H., Odegard, J.E., Burrus, C.S., Wells, R.O.: Nonlinear processing of a shift invariant dwt for noise reduction (1995)
10. Lang, M., Guo, H., Odegard, J.E., Burrus, C.S., Lang, M., Guo, H., Burrus, C.S., Wells, R.O.: Noise reduction using an undecimated discrete wavelet transform (1996)
11. Beylkin, G.: On the representation of operators in bases of compactly supported wavelets. SIAM J. Numer. Anal. 6(6), 1716–1740 (1992)
12. Serra, J.: Image Analysis and Mathematical Morphology. Academic Press, Inc., Orlando (1983)

Peak Detection Algorithm Based on Second Derivative Properties for Two Dimensional Ion Mobility Spectrometry Signals

Rafael Slodzinski[1], Lars Hildebrand[2], and Wolfgang Vautz[1]

[1] Leibniz-Institut für Analytische Wissenschaften - ISAS - e.V., Bunsen-Kirchhoff-Str.11, 44139 Dortmund, Germany
[2] Dortmund University of Technology, Computer Science Department, Chair 1, Otto-Hahn-Str. 16, 44221 Dortmund, Germany
{rafael.slodzinski,wolfgang.vautz}@isas.de
lars.hildebrand@udo.edu

Abstract. In this paper we propose a novel peak detection algorithm for 2-dimensional analytical data. The proposed algorithm utilizes the properties of the second derivative and curvature of (regular) surfaces to perform peak detection. Raw data used in this study for performance demonstration were obtained by a hyphenated system called gas-chromatographic column ion mobility spectrometer (GC-IMS). GC-IMS is a two stage technique for separation of gas-phased (organic) compounds. Despite the good performance of the MCC-IMS separation in general, there are still unsatisfactory cases where the substances overlap and the recorded signals nearly merge. Frequently used peak detection algorithm for 2-dimensional data, like the watershed algorithm, do not perform well in those cases. Preliminary empirical results show good peak detection performance of the proposed algorithm. Furthermore the results indicate that the proposed algorithm is capable to solve the problem of peak detection even in cases of strong peak overlapping.

Keywords: Ion Mobility Spectrometry, Peak Detection, Overlapping Peaks, Signal Processing , Mathematical Differentiation, Differential geometry.

1 Introduction

Ion mobility spectrometry (*IMS*) is an analytical method for the detection of gas-phased compounds. The technique is non-invasive and provides excellent detection limits of trace compounds from ppm_v down to ppt_v range. Moreover, it exhibits short time of analysis and is operated at ambient temperature and pressure, thus causing low technical expenditure. The introduction of *IMS* lies almost half a century back, when ion mobility spectrometers were used for military purpose (e.g. chemical war fare agents) and later on in security sensitive domains for detection of explosives and drugs [1,2]. In the course of time *IMS* found its way to civilian applications. Today ion mobility spectrometers are applied in process control [3-9], environmental and indoor air quality monitoring [10-15] and for medical and biological issues. In latter applications, the interest lies on the detection of biomarkers and metabolites for early diagnosis and for online medication control [16-19].

In early applications, the focus was on the detection (identification and quantification) of one or few specific compounds. For complex samples like human breath or the headspace of bacteria the discrimination of sample compounds along the ion mobility dimension only is not adequate. Indeed for numerous substances the ion mobility is known. However, many of them have similar or even equal ion mobility values, thus making identification difficult or even impossible. Hence it is an accepted approach for medical and biological applications to couple *IMS* with gas-chromatographic pre-separation techniques. Such a hyphenated system (a *GC-IMS*) generates two dimensional analytical signals represented in a matrix form due to both separation dimensions. Ideally, sample compounds separated by the *GC-IMS* appear within the matrix as spatially disjoint Gaussian shaped areas of high signal intensity.

An evaluation of a single measurement or even a complete measurement series in the traditional way - considering the huge amount of data – is laborious and time consuming. Hence an automated evaluation is desirable, beginning with data normalization, noise reduction, peak detection and finally pattern recognition in measurements. The stage of the peak detection may be the biggest challenge, even for automated approaches, since complex organic samples may contain hundreds of substances which can't be perfectly separated. This paper focuses on peak detection and reports preliminary results obtained by novel approach for 2-dimensional analytical data.

Numerous peak detection algorithms for data of hyphenated systems like *MCC-IMS*, *GC-MS*, *GC-GC*, *LC-MS* can be found in the literature [20-22]. Most of those algorithms utilize the "two step" approach of peak detection. In the first stage, peak detection in one dimension is performed, in subsequent stage the results are merged to obtain the peak position in both dimensions. Other classes of algorithms are based on the drain algorithm, an inversion of the watershed algorithm known from digital image processing [23-25], utilize wavelet transformations or perform peak detection for several levels of signal intensity and merge the levels for final result to uncover locations of sample compounds [26]. However, almost all algorithms lack of ability to detect superposed peaks that appear as shoulders of dominant peaks.

The approach presented here utilizes mathematical differentiation on 2-dimensional data in particular *GC-IMS* data and exploits the properties of the second derivative and curvature of peaks in similar manner as derivatives are used for the calculation of maxima of good-natured mathematical functions.

Preliminary results indicate good peak detection performance and are demonstrated for compiled mixtures of specific substances and for real world data like human breath.

2 Methods and Definitions

2.1 Multi-capillary Column Ion Mobility Spectrometry

The working principle of the classical *IMS* is based on differences in velocities of ions which move in a weak electric field in a particular drift gas. Figure 1 illustrates the design of an ion mobility spectrometer. A gas phase sample is introduced into the ionization chamber and undergoes a chemical ionization reaction. Radioactive Nickel ^{63}Ni as ionization source is usually applied. The source emits beta rays, which ionise the molecules of the carrier gas. This type of ions is referred as reactant ions. Their number per time interval is limited and depends on the type of ionization source and carrier gas as well. In this study synthetic air was used as carrier gas. When the sample molecules enter the ionization chamber, proton transfer takes place, thus producing analyte ions and a simultaneous reduction of the reactant ions. Recurrent an ion shutter opens and ions are exposed to the electric field E. The field's strength is controlled by the length l_D of the drift tube and the applied drift voltage U_D .The electric field instantly accelerates the ions through the drift tube towards the detector located on the opposite side. While the ions migrate through the drift tube, they collide several times with molecules of a neutral counter flowing gas – termed drift gas – and group in swarms of ions. Each collision reduces the velocity of involved ions, where the number of collisions depends mainly on ion's size, charge and shape. The interaction of collisions and steady acceleration leads to a specific average migration velocity v_D of the ion swarms. After a specific drift time t_D the ion swarms reach the detector and the resulting electrical signal is measured in equidistant time intervals. Since the ambient temperature and pressure essentially impact on the measured drift times, the results are normalized and reported as inverse reduced ion mobility $1/K_o$.

Due to the complexity of real world samples, a gas-chromatographic pre-separation is applied before the sample enters the reaction chamber. In this study a multi-capillary column is applied. A *MCC* contains nearly 1000 parallel capillaries, each capillary 40μm in diameter. The *MCC* is operated at constant temperature. While the sample molecules migrate through the column, they interact with the coated inner surface of the capillaries. Depending on the chemical properties of a sample compound the inner surface bound each compound for different time. The migration time from injection to elution is termed retention time of a sample compounds.

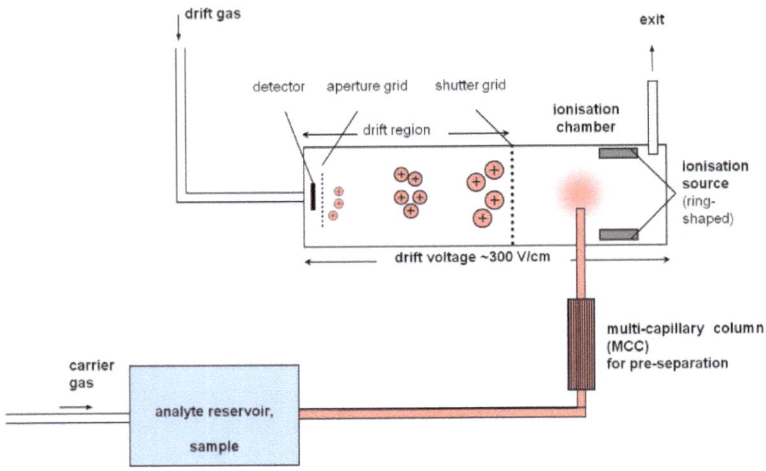

Fig. 1 Hyphenated system of an ion mobility spectrometer coupled with multi-capillary column for gas-chromatographic pre-separation

2.2 Data Format

A spectrum – the output data of a single *IMS* measurement - is a vector $S = (z_0, z_1, \dots, z_N)$ of signal intensities z_i measured in equidistant time point dt_i, i ∈ $\{1 \dots N\}$. If a *MCC* is coupled with an *IMS* one obtains an additional dimension of retention time. As a consequence the result of a *MCC-IMS* measurement is a series of *R* one dimensional *IMS* spectra recorded at equidistant retention time point rt_j, j ∈ $\{1 \dots R\}$. The series of spectra usually is represented as a matrix

$$M_{ims} = \begin{pmatrix} z_{11} & \cdots & z_{N1} \\ \vdots & \ddots & \vdots \\ z_{1R} & \cdots & z_{NR} \end{pmatrix} \tag{1}$$

Each data point z_{ij} of M_{ims} denotes the signal intensity at a specific drift time dt_i and specific retention time rt_j. Figure 2 illustrates a single spectrum taken from a breath analysis. The prominent peak at $1/K_o = 0.485$ is termed Reactant Ion Peak (*RIP*). As the name denotes, he arises from remaining reactant ions that do not take part in the chemical reaction with the sample compounds. Occurring peaks with $1/K_o > 0.485$ on the right of the *RIP* originate from exhaled sample compounds. The region from $0.1 \cdot index_{Rip}$ and $0.8 \cdot index_{Rip}$ is called signal free region *(SFR)*. This region is predominated by noise; no analyte has been observed in this region in an *MCC-IMS* measurement up to now. In subsequent figure 3 the entire measurement is displayed. In this figure the horizontal axis defines the drift time dimension while the vertical axis defines the retention time dimension; the signal intensity is coded in an arbitrary but fixed color scheme. The complete data format including header information on the experimental setup and on the sample is described elsewhere [27].

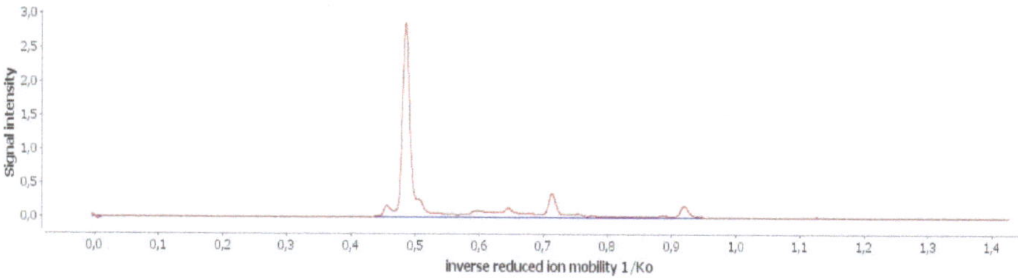

Fig. 2 Graph of a single IMS spectrum

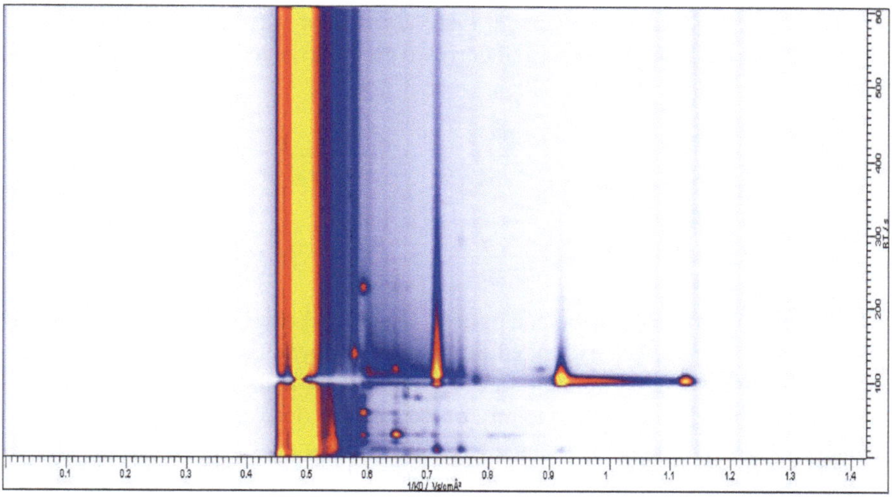

Fig. 3 Series of IMS spectra merged in a heatmap

2.3 *Partial Differentiation and Differential Geometry*

To clarify the underlying idea of the algorithm, it is suitable, to visualize the output data matrix M_{ims} as points in a three-dimensional space. The datum plane is spanned by vectors of equidistant data points in inverse reduced ion mobility dimension and retention time. The signal intensity determines the elevation of each data point above the datum plane. Connecting each point with his direct neighbors (Moore neighborhood) in terms of datum plane, a surface in a three dimensional Euclidean space is generated. The surface is commonly curved, but exhibits neither edges nor folds. It is free from self-intersections as well. In differential geometry such benign surfaces are termed regular surfaces [28-29]. Interpreting the output data in this way, allows using the resources of the differential geometry to determine peaks even in the case of strong overlap. Relevant concepts from differential geometry used in this study are the concepts of partial derivatives and the notion of curvature. Partial derivative of a function f of several variables is its derivative with respect to one of those variables, where the remaining ones are kept constant. Considering a function $f(x_1,\ldots,x_n)$ in Euclidean space $\mathbb{R}^n$ its first partial derivative in direction x_i at point $P=(p_1,\ldots p_n)$ is defined as:

$$\frac{\partial f}{\partial x_i} f(p_1,\ldots,p_n) = \lim_{h \to o} \frac{f(p_1,\ldots,p_i+h,\ldots,p_n) - f(p_1,\ldots,p_i,\ldots,p_n)}{h} \tag{2}$$

If the function f has partial derivatives $\frac{\partial f}{\partial x_i}$ with respect to each variable x_i at P the defined vector is termed gradient $\nabla(f)$.

The second partial derivative formed by the two-fold application of derivation on the function $f(x_1,\ldots,x_n)$. In general there are several ways to derive f twice. All combinations of the two-fold partial derivatives are arranged in a matrix H, termed Hessian:

$$H(f) = \begin{pmatrix} \frac{\partial^2 f}{\partial x_1^2} & \cdots & \frac{\partial^2 f}{\partial x_1 \partial x_n} \\ \vdots & \ddots & \vdots \\ \frac{\partial^2 f}{\partial x_n \partial x_1} & \cdots & \frac{\partial^2 f}{\partial x_n^2} \end{pmatrix} \tag{3}$$

The Hessian matrix $H(f)$ is the analogue to the second derivative of functions with one variable. The surface examined in this study depends on two variables due to both separation stages. The curvature of a surface at a specific point P indicates how strong the surface in the vicinity of the point P derivates from the tangential plane E in P. The tangent plane E at P may be spanned by any two linearly independent vectors T_1 and T_2; hence vectors in direction of the coordinate axes – the partial derivatives of equitation (2) - may be used. In addition the tangent vectors define a normal vector N at P by their cross product. The intersection of the examined surface with two planes that are spanned by the normal vector and each of the two tangent vectors, result in two intersection curves that cross the point P on the surface. The curvature of each curve at P describes the curvature of the surface with respect to the tangent vectors T_1 and T_2 respectively. The curvatures are called principal curvatures k_1 and k_2. The Gaussian curvature K is defined as the product of k_1 and k_2 and the mean curvature M is defined as the average of k_1 and k_2.

The literature of differential geometry provides formulas which allow direct calculation of K and M in a specific point P of an examined surface by utilizing the first and second partial derivatives (see ch.7 in [28]). Given a surface defined by the function $f(x_1,x_2)$ the Gaussian curvature K in point P is defined as

$$K(P) = \frac{\frac{\partial^2 f}{\partial x_1^2} \cdot \frac{\partial^2 f}{\partial x_2^2} - \left(\frac{\partial^2 f}{\partial x_1 \partial x_2}\right)^2}{\left(1 + \left(\frac{\partial f}{\partial x_1}\right)^2 + \left(\frac{\partial f}{\partial x_2}\right)^2\right)^2} \tag{4}$$

And the mean curvature M in point P respectively is defined as

$$M(P) = \frac{\left(1+\left(\frac{\partial f}{\partial x_1}\right)^2\right)\cdot\frac{\partial^2 f}{\partial x_2^2}+\left(2\cdot\frac{\partial f}{\partial x_1}\cdot\frac{\partial f}{\partial x_2}\cdot\frac{\partial^2 f}{\partial x_1 \partial x_2}\right)+\frac{\partial^2 f}{\partial x_1^2}\cdot\left(1+\left(\frac{\partial f}{\partial x_2}\right)^2\right)}{2\cdot\sqrt[3]{1+\left(\frac{\partial f}{\partial x_1}\right)^2+\left(\frac{\partial f}{\partial x_2}\right)^2}} \quad (5)$$

Both curvatures classify a point P on a surface as

Elliptic , if the Gaussian curvature $K(P) > 0$
Hyperbolic , if the Gaussian curvature $K(P) < 0$
Parabolic , if the Gaussian curvature $K(P) = 0$ and mean curvature $M(P) \neq 0$
Flat , if the Gaussian curvature $K(P) = 0$ and mean curvature $M(P) = 0$

The following figure 4 illustrates the surface points determined by different curvature properties.

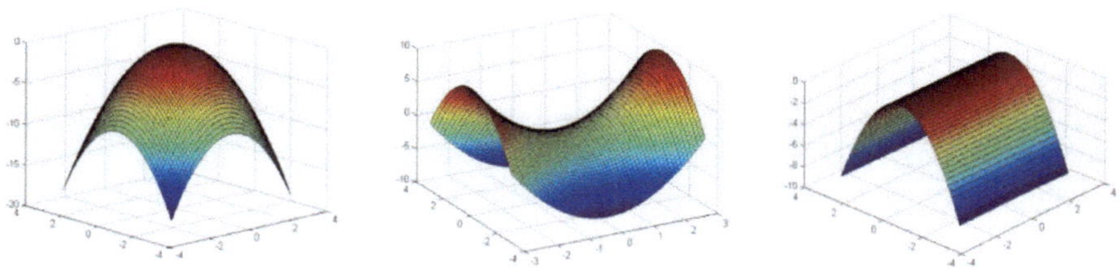

Fig. 4 Types of points on a surface from left to right: elliptic, hyperbolic and parabolic surface points

The idea of the algorithm is to analyze the Gaussian curvature K in each point of the examined surface, mark points with elliptic properties that in addition exhibit a negative second partial derivative in direction of coordinate axes. The elliptic shaped areas are assumed to contain highest levels of concentration for a specific sample compound. Moreover, the information of peak location is more accurately retained in the curvature property as in the original output data. Superposed peaks indicated as shoulders of adjacent prominent peaks become detectable using this approach. The figure 5 (a)-(d) illustrates this idea. Figure 5 displays the cross section of two *2D*-Gaussian functions which overlap to varied degrees. The red line indicates the raw output data for each point in the cross section, the blue curve denotes the curvature in the same point.

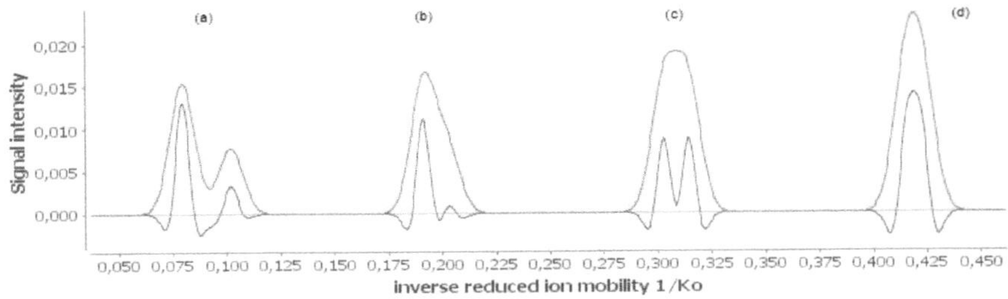

Fig. 5 Comparison of preservation of information in curvature property and raw output data for a cross section of two overlapping 2D-Gaussian functions

2.4 Algorithm

In the pre-processing stage the raw output data matrix M_{ims} is smoothed and baseline corrected. Data smoothing is performed by applying 2D-Gaussian filter [30]. Furthermore the gradient $\nabla(P)$ and the Hessian H (P) for each point P of the surface are calculated using eq. (2) and eq. (3) respectively. Basing on the obtained matrices the Gaussian curvature $K\ (P)$ for each point P is computed utilizing the equation (4).To obtain the first differentiation in peak and non-peak points, the signal free region *(SFR)* between $0.1{\cdot}index_{Rip}$ and $0.8{\cdot}index_{Rip}$ is analyzed. For each point P_{SFR} of *SFR* is tested, if its Gaussian curvature K is positive $(K(P_{SFR})>0$) and, if the first component H_{11} of its Hessian $H\ (P_{SFR})$ meet the condition $H_{11}\ (Ppoint\ _{SFR})\ <0$. If the examined point P_{SFR} fulfills both conditions, the values of K and H_{11} and the signal intensity $I\ (P_{SFR})$ of the point P are stored in designated arrays. After examination of all points in *SFR*, the median value for each parameter is determined.

This set of parameters K_{med}, H_{med}, I_{med} is used as threshold for the subsequent analysis of the entire measurement. A point P within the matrix is assumed to be a potential peak point, if the conditions $K\ (P)>K_{med}$, $H_{11}\ (P)\ <H_{med}$ and $I\ (P)>I_{med}$ hold. Examined points that fulfill these conditions, become vertices of an undirected graph G. Each vertex V of G is tagged with a coordinate pair. The tag is composed of row and column index the primary point P resides in the data matrix M_{ims}. Two vertices V_1 and V_2 are connected by an edge, if the primal points of these vertices are adjacent in the data matrix M_{ims}.

Once the construction of graph G completed, isolated vertices of G are discarded. This operation is justified, since it is implausible that true peaks appear in such narrow shaped area (here as one point only). For the remaining graph G connected components are calculated. By definition, the connected components are regions with elliptic curvature properties. They specify local maxima due to the properties of Hessian and exhibit certain level of signal intensity. Determination of the connected components of G covers the cases (a) and (b) illustrated in figure 5. To solve the case (c) of figure 5, a local drain algorithm on each connected component of G is applied. From here the connected components satisfy all mathematical (elliptic shape + local maxima) demands made to true peaks in two dimensional space. In consequence detected connected components are assumed to contain peaks of sample compounds. For final discrimination between peak and non-peak areas a score S_{cc} of each connected component CC_i of G is calculated. S_{cc} is the sum of curvature values of all points belonging to a connected component CC_i. Basing on the S_{cc} values of connected components from the signal free region *SFR* a mean score S_{mean} and the standard deviation S_δ is calculated and standardized. A connected component is assumed to be a true peak region, if its score S_{cc} meets the condition of equation (6), where λ is a user defined variable:

$$S_{cc} \geq Smean + \lambda \cdot S\delta \quad (6)$$

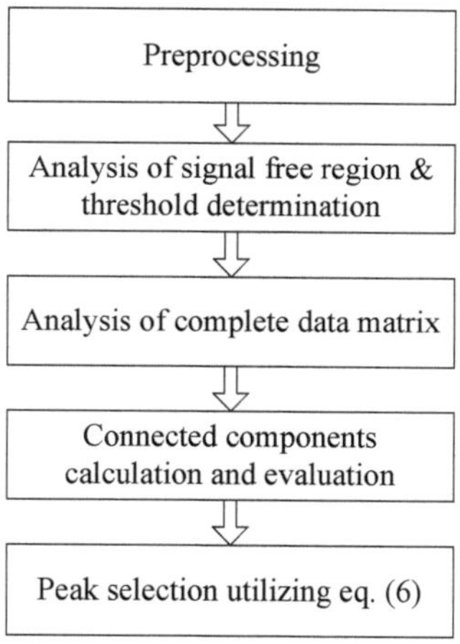

Fig. 6 Proposed algorithm outlined as flow diagram

3 Results

We demonstrate the algorithm's detection capabilities on two real world measurements. The first data matrix is obtained from a synthetic sample mixture of 16 different substances. The ion mobility and the retention time of each substance and its possible polymer are given in table 1. The ion mobility and retention time values of each analyte in table 1 were obtained by several reference measurements for each compound. The result of the proposed peak detection algorithm on this mixture is illustrated in figure 7. The figure focuses on the important region ($1/K_o$: 0.45-1.1 and retention time from 0s to 450s) of the measurement, where most of analytes appear. All anaytes of table 1 except propofol has been successfully detected. The appearance of peaks within the figure 7 is indicated by black crosses.

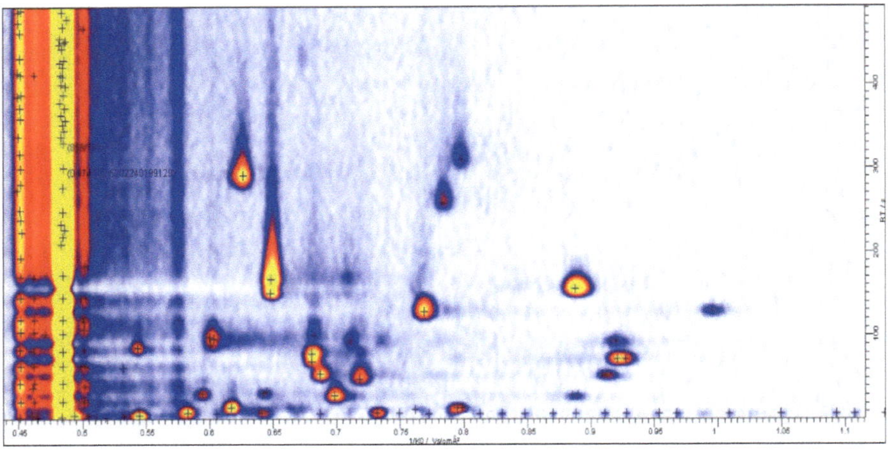

Fig. 7 Result of peak detection algorithm performed on the mixture specified by table 1

Table 1 Ion mobility and retention time of analytes in the tested sample mixture

analyte	*1/K₀*	Retention time in sec.
Acetone M	0.4960	2.5
Acetone D	0.5440	2.5
2-Hexanone M	0.5820	6.4
2-Hexanone D	0.6410	6.4
2-Heptanone M	0.7940	11.6
2-Heptanone D	0.6170	11.6
2-Octanol M	0.8900	25.9
2-Octanol D	0.6990	25.9
Limonene M	0.5930	29.8
Limonene D	0.6440	29.8
1-Octanol	0.7180	45.4
2-Nonanone M	0.6880	50.6
2-Nonanone D	0.9120	50.6
Isopulegol M	0.6820	70.1
Isopulegol D	0.9230	70.1
Naphthalene	0.5430	81.8
Menthol M	0.9170	89.4
Menthol D	0.6020	89.4
Menthol T	0.7110	89.4
Decanal M	0.7690	128.7
Decanal D	0.9990	128.7
Carvone M	0.6490	155.7
Carvone D	0.8880	155.7
1-Decanol	0.7840	259.6
Thymol	0.6270	292.0
2-Undecanol	0.7980	312.8
Propofol	0.6730	441.2

The second example addresses the issue of detection of overlapped peaks within a data matrix. For this purpose we compared the capabilities of the adapted watershed algorithm for 2-dimensional analytical data with the proposed algorithm. Figure 8 illustrates both results obtained by the adapted watershed (left part of figure 8) and the proposed algorithm (right part of figure 8). One clearly sees in the ion mobility region from 0.53 to 0.58 and retention time from 2s to 30s the watershed algorithm only detected one pronounced peak (at $1/K_o$: 0.548 and 11s retention time). Adjacent peaks in front (at $1/K_o$: 0.532) as well as in subsequent (at $1/K_o$: 0.570 and 0.579), appearing as shoulders of the dominant peak has been not detected, due to the working principle of the watershed algorithm.

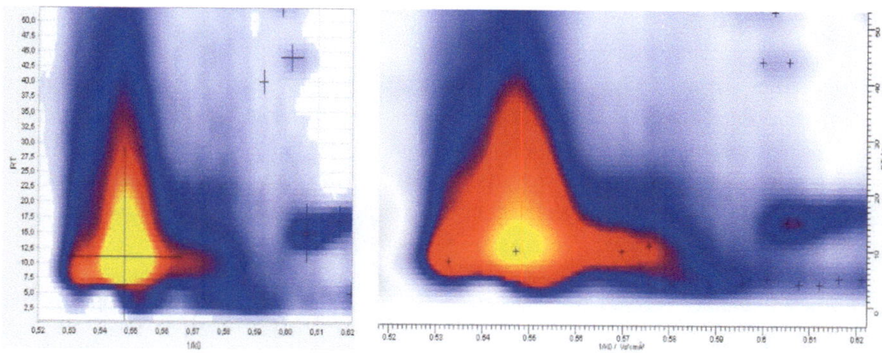

Fig. 8 Comparison of peak detection results on data matrix from breath analysis: obtained by watershed algorithm (left) and obtained by the proposed algorithm (right)

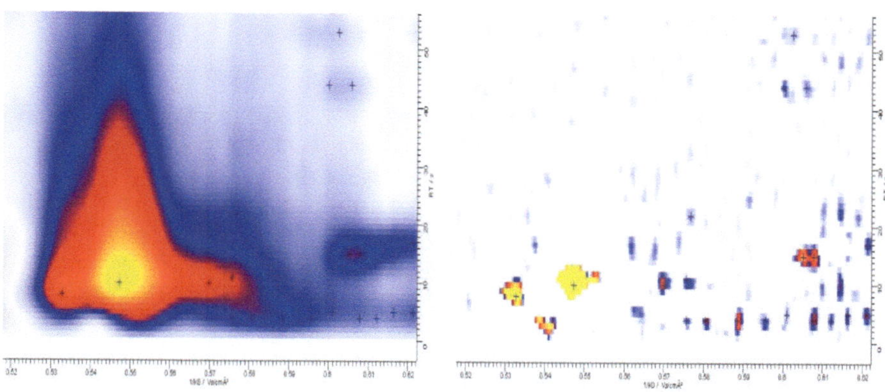

Fig. 9 Comparison of raw data matrix (left) and the corresponding curvature matrix (right)

In contrast the proposed algorithm performed well case of strong signal overlapping. Here, the analysis of the curvature property for each matrix point significantly contributed to the improved detection results. The detectability of superposed peaks at $1/K_o$: 0.532 , 0.570 and 0.579 in addition to the dominant peak at $1/K_0$: 0.548 can be easily understood by examination of curvature values of each data point calculated by eq. (4). The figure 9 (left) illustrates again the result achieved on the data matrix; in addition the corresponding matrix of curvature values is displayed in the right of figure 9. In the right part of figure 9 elliptic curvature properties that describe local extreme values are highlighted in different colors. In accordance with considerations made in section 2.3, these regions have been marked as true peaks.

4 Conclusions

In this paper we proposed a novel algorithm for peak detection on 2-dimensional analytical data. The consideration of the curvature properties of examined surfaces leads to improved detection of 2D-peaks, even in cases of strong peak overlapping. The preliminary results are promising but further examination is needed. In particular the impact of noise has to be examined in detail and the options for automated selective noise reduction explored.

Acknowledgments. This research was founded by the European Union as part of the Second Generation Locator for Urban Search and Rescue (SGL for USaR). SGL for USaR is a collaborative project (number 217967) founded under call identifier FP7-SEC-2007-1, part of the seventh framework program. Furthermore, the financial support of the Bundesministerium für Bildung und Forschung and The Ministerium für Innovation, Wissenschaft und Forschung des Landes Nordrhein-Westfalen is gratefully acknowledged.

References

1. Eiceman, G.A., Karpas, Z.: Ion Mobility Spectrometry, 2nd edn. CRC Press (2005)
2. Louis, R.H., Hill, H.H.: Ion Mobility Spectrometry in Analytical Chemistry. Critical Reviews in Analytical Chemistry 21(5), 321–355 (1990)
3. West, C., Baron, G., Minett, J.J.: Detection of Gunpowder Stabilizers with Ion Mobility Spectrometry. Forensic Science Int. 166, 91–101 (2007)
4. Raatikanien, J., Reinikainen, V., Minkkinen, P., Ritvanen, T., Muje, P., Pursiainen, J., Hiltunen, T., Hyvonen, P., Wright, A., Reinikainen, S.: Multivariate Modeling of Fish Freshness Index Based on Ion Mobility Spectrometry Measurements. Anal. Chim. Acta 544, 128–134 (2005)
5. Vautz, W., Zimmerman, D., Hartmann, M., Baumbach, I.J., Nolte, J., Jung, J.: Ion Mobility Spectrometry for Food quality and Safty. Food Additives & Contaminants 23, 1064–1073 (2005)
6. Vautz, W., Sielemann, S., Baumbach, J.I.: Determination of Terpenes in Humid Ambient Air Using Ultraviolet Ion Mobility Spectrometry. Anal. Chim. Acta 513, 393–399 (2004)

7. Vautz, W., Bauchmach, J.I., Jung, J.: Beer Fermentation Control Using Ion Mobility Spectrometry. Journal of The Institute of Brewing 112, 157–164 (2006)
8. Sielemann, S., Baumbach, J.I., Schmidt, H., Pilzecker, P.: Detection of Alcohols Using UV-ion Mobility Spectrometers. Int. Journal of Ion Mobility Spectrometry 5(3), 7–10 (2002)
9. Baumbach, J.I.: Process Analysis Using Ion Mobility Spectrometry. Anal. Bioanal. Chem. 384, 1059–1070 (2006)
10. Leonhardt, J.W., Rohrbeck, W., Bensch, H.: A High Resolution IMS for Environmental Studies. Int. Journal of Ion Mobility Spectrometry 3, 43–49 (2000)
11. Baumbach, J.I., Berger, D., Leonhart, J.W., Klockow, D.: Ion Mobility Sensor in Environmental Analytical-Chemistry – Concept and 1st Results. Int. J. Environ. Anal. Chem. 52, 189–193 (1993)
12. Borsdorf, H., Rammler, A., Schulze, D., Boadu, K.O., Feist, B., Weiss, H.: Rapid on-site Determination of Chlorobenzene in Water Samples using Ion Mobility spectrometry. Anal. Chim. Acta 440, 63–70 (2001)
13. Eiceman, G.A., Snyder, A.P., Blyth, D.A.: Monitoring of Airbone Organic Vapors using Ion Mobility Spectrometry. Int. J. Environ. Anal. Chem. 38, 415–425 (1990)
14. Eiceman, G.A., Leasure, C.S., Vandiver, V.J.: Negative-Ion Mobility Spectrometry for Selected Inorganic Pollutant and Gas-Mixtures in Air. Anal. Chem. 58, 76–80 (1986)
15. Reese, E.S., Limero, T.F.: Calibration of the Volatile Organic Analyzer for the International Space Station Unit. Int. Journal of Ion Mobility Spectrometry 4, 51–53 (2001)
16. Ruzsanyi, V.: Analyse flüchtiger Metabolite von der Ausatemluft mittels Ionenbeweglichkeitsspektrometer. Phd Thesis University of Dortmund (2005)
17. Perl, T., Carstens, E., Hirn, A., Quintel, M., Nolte, J., Jünger, M., Vautz, W.: Determination of Serum Propofol Concentrations by Breath Analysis Using Ion Mobility Spectrometry. Br. J. Anaesth 103(6), 822–827 (2009)
18. Vautz, W., Baumbach, J.I.: Exemplar Application of Multicapillary Column Ion Mobility Spectrometry for Biological and Medical Purpose. Int. Journal of Ion Mobility Spectrometry 11, 11–35 (2008)
19. Prasad, S., Schmidt, H., Lampen, P., Wang, M., Guth, R., Rao, J.M., Smith, G.B., Eiceman, G.A.: Analysis of Bacterial Strains with Pyrolysis-gas Chromatography/Differential Mobility Spectrometry. Analyst 131(11), 1216–1225 (2006)
20. Dixon, S.J., Brereton, R.G., Soini, H.A., Novotny, M.V., Penn, D.J.: An automated method for peak detection and matching in large gas chromatography-mass spectrometry data sets. J. Chemom 20, 325–340 (2006)
21. Peters, S., Vivo-Truyols, G., Marriott, P.J., Schoenmakers, P.J.: Development of an algorithm for peak detection in comprehensive two-dimensional chromatography. J. Chromatogr. A 1156, 14–24 (2007)
22. Hastings, C.A., Norton, S., Roy, S.: New algorithms for processing and peak detection in liquid chromatography/mass spectrometry data. S. Rapid Commun. Mass Spectrom 16, 462–467 (2002)
23. Reichenbach, S.E., Ni, M., Kottapalli, V., Visvanathan, A.: Information technologies for comprehensive two-dimensional gas-chromatography. Chemom. a. Intell. Lab. Sys. 71, 107–120 (2004)

24. Beucher, S., Lantuejoul, C.: Use of watersheds in contour detection. In: International Workshop on Image Processing, Real-Time Edge and Motion Detection/Estimation, pp. 17–24 (1979)
25. Riese, M.: Peak Detection on GC-IMS Data using Water Shed Transformation. Beachelor thesis. University of Bielefeld (2008)
26. Bader, S.: Identification and Quantification of Peaks in Spectrometric Data. Phd Thesis. Dortmund University of Technology (2008)
27. Vautz, W., Bödeker, B., Bader, S., Baumbach, J.I.: Recommendation of a Standard Format for Data Sets from GC/IMS with Sensor-Controlled Sampling. Intern. J. for Ion Mobility Spectrometry (11), 71–76 (2008)
28. Pressley, A.: Elementary differential Geometry. Springer (2001)
29. Carmo, M.P.: Differential Geometry of Curves and Surfaces, 1st edn. Prentice-Hall (1976)
30. Gonzales, R., Woods, R.: Digital Image Processing, 2nd edn. Prentice-Hall (2002)

Design of Semiactive Damper in Vehicle Suspension Considering the Tire Lift Off

Miloš Musil and Ferdinand Havelka

STU, Faculty of Mechanical Engineering,
Institute of Applied Mechanics and Mechatronics,
Námestie slobody 17, 812 31 Bratislava 1, Slovak Republic
{milos.musil,ferdinand.havelka}@stuba.sk

Abstract. A quarter car model with magnetorheological (MR) damper is studied in this paper. An adopted experimentally verified non – linear hysteretic mathematical model is used to represent the MR damper. Approaching road disturbances are measured by a sensor. Optimal preview control strategy for fully active suspension is derived with respect to road holding, suspension rattle space and ride comfort. Continuous inverse mathematical model of the MR damper for the use of control is derived. The effect of tire lift off is modeled using a continuous mathematical function.

Keywords: Quarter car, MR damper, Optimal preview control, Tire lift off.

1 Introduction

In recent years active and semi-active suspensions have been investigated due to their ability to adapt to various types of road excitations. Compared with passive suspension systems, which can only dissipate the energy present in the system, active suspension systems can supply the flow of energy into the system and can generate forces which are independent of the state of the system. Effective compromise between passive and active suspension systems are semi-active suspensions. Semi-active suspension systems are less expensive than the active ones, they require much less energy intensive source and even if the source of energy fails they can still operate as passive suspension systems.

In this paper a semi-active magnetorheological damper is utilized in the suspension system. An experimentally verified non-linear hysteretic mathematical model is used to represent the dynamics of the MR damper. A basic quarter car model is utilized to simulate the vertical dynamics of vehicle. For the use of control algorithm a continuous inverse mathematical model of the MR damper is derived. Dissipative force generated by the MR damper tries to match the one generated by a fictive ideal active system. Optimal preview control strategy for the fully active system with respect to road holding, suspension rattle space and ride comfort is derived such that it is supposed that approaching road disturbances are measured by a sensor and are known within a certain distance ahead. Active and semi-active suspension systems with preview are examined in vehicle traveling over a bump and in both cases the effect of tire lift off is investigated.

2 Quarter Car Model with Idealized Active Suspension and with Consideration the Tire Lift Off

To simulate the vertical dynamics of vehicle, the quarter car model is utilized – figure 1a.). The equations of motion with consideration the tire lift off problem are follows:

$$\begin{aligned} m_1\ddot{y}_1 &= -k_1(y_1 - y_2) + u - m_1 g \\ m_2\ddot{y}_2 &= -k_2(y_2 - w)\left[1 - H(y_2 - w)\right] - k_1(y_2 - y_1) - u - m_2 g \end{aligned} \tag{1}$$

where *H(-)* is the heaviside step function and u is the force generated by active control element, which dynamics is neglected. The function modeling the tire lift off problem can be rewritten into form:

$$1 - H(y_2 - w) = \frac{1}{2}\left[1 - \operatorname{sgn}(y_2 - w)\right] \cong \frac{1}{2}\left\{1 - \tanh\left[\beta_r(y_2 - w)\right]\right\} \tag{2}$$

where now the function *tanh(-)* is a continuous function and β_r is a coefficient large enough.

Noting the relation (2), equations of motion in the matrix form are

$$\mathbf{M}_\mathbf{a}\ddot{\mathbf{q}}_\mathbf{aA} + \mathbf{K}_\mathbf{aL}\mathbf{q}_\mathbf{aR} + \mathbf{K}_\mathbf{aN}\mathbf{q}_\mathbf{aR}\mathbf{e}_\mathbf{a}\tanh(\beta_r\mathbf{q}_\mathbf{aR}) = \mathbf{b}_\mathbf{ag}g + \mathbf{b}_\mathbf{u}u \tag{3}$$

Transformation from absolute to the relative coordinates is realized through the equation

$$\mathbf{q}_\mathbf{aR} = \mathbf{T}_\mathbf{aA}\mathbf{q}_\mathbf{aA} + \mathbf{T}_\mathbf{aw}w \tag{4}$$

Combining equation (3) and differentiated equation (4), state space model of system in the figure 1 a.) is obtained:

$$\dot{\mathbf{x}}_\mathbf{a} = \mathbf{A}_\mathbf{aL}\mathbf{x}_\mathbf{a} + \mathbf{A}_\mathbf{aN}\mathbf{x}_\mathbf{a}\mathbf{E}_\mathbf{a}\tanh(\beta_r\mathbf{x}_\mathbf{a}) + \mathbf{B}_\mathbf{a}u + \mathbf{G}_\mathbf{a}\mathbf{f}_\mathbf{gw} \tag{5}$$

where

$$\begin{gathered} \mathbf{A}_\mathbf{aL} = \begin{bmatrix} \mathbf{O}_\mathbf{a} & \mathbf{T}_\mathbf{aA} \\ -\mathbf{M}_\mathbf{a}^{-1}\mathbf{K}_\mathbf{aL} & \mathbf{O}_\mathbf{a} \end{bmatrix},\ \mathbf{A}_\mathbf{aN} = \begin{bmatrix} \mathbf{O}_\mathbf{a} & \mathbf{O}_\mathbf{a} \\ -\mathbf{M}_\mathbf{a}^{-1}\mathbf{K}_\mathbf{aN} & \mathbf{O}_\mathbf{a} \end{bmatrix},\ \mathbf{E}_\mathbf{a} = \begin{bmatrix} \mathbf{e}_\mathbf{a}^T \\ \mathbf{o}_\mathbf{a} \end{bmatrix}^T,\ \mathbf{B}_\mathbf{a} = \begin{bmatrix} \mathbf{o}_\mathbf{a} \\ \mathbf{M}_\mathbf{a}^{-1}\mathbf{b}_\mathbf{u} \end{bmatrix}, \\ \mathbf{G}_\mathbf{a} = \begin{bmatrix} \mathbf{G}_\mathbf{ag} & \mathbf{G}_\mathbf{aw} \end{bmatrix},\ \mathbf{G}_\mathbf{ag} = \begin{bmatrix} \mathbf{o}_\mathbf{a} \\ \mathbf{M}_\mathbf{a}^{-1}\mathbf{b}_\mathbf{ag} \end{bmatrix},\ \mathbf{G}_\mathbf{aw} = \begin{bmatrix} \mathbf{T}_\mathbf{aw} \\ \mathbf{o}_\mathbf{a} \end{bmatrix},\ \mathbf{f}_\mathbf{gw} = \begin{bmatrix} g \\ \dot{w} \end{bmatrix},\ \mathbf{x}_\mathbf{a} = \begin{bmatrix} \mathbf{q}_\mathbf{aR} \\ \dot{\mathbf{q}}_\mathbf{aA} \end{bmatrix} \\ \mathbf{M}_\mathbf{a} = \begin{bmatrix} m_1 & 0 \\ 0 & m_2 \end{bmatrix},\ \mathbf{K}_\mathbf{aL} = \begin{bmatrix} k_1 & 0 \\ -k_1 & \frac{k_2}{2} \end{bmatrix},\ \mathbf{K}_\mathbf{aN} = \begin{bmatrix} 0 & 0 \\ 0 & -\frac{k_2}{2} \end{bmatrix},\ \mathbf{e}_\mathbf{a} = \begin{bmatrix} 0 \\ 1 \end{bmatrix}^T, \mathbf{b}_\mathbf{ag} = -\begin{bmatrix} m_1 \\ m_2 \end{bmatrix}, \\ \mathbf{b}_\mathbf{u} = \begin{bmatrix} 1 \\ -1 \end{bmatrix},\ \mathbf{q}_\mathbf{aA} = \begin{bmatrix} y_1 \\ y_2 \end{bmatrix},\ \mathbf{q}_\mathbf{aR} = \begin{bmatrix} y_1 - y_2 \\ y_1 - w \end{bmatrix},\ \mathbf{T}_\mathbf{aA} = \begin{bmatrix} 1 & -1 \\ 0 & 1 \end{bmatrix},\ \mathbf{T}_\mathbf{aw} = \begin{bmatrix} 0 \\ -1 \end{bmatrix} \end{gathered} \tag{6}$$

where $\mathbf{O_a}$ and $\mathbf{o_a}$ are zero matrix and vector respectively of appropriate dimensions.

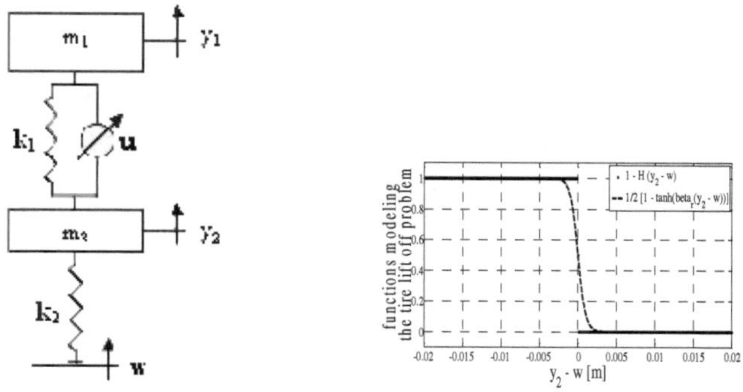

a.) b.)

Fig. 1 a.) quarter car model with active suspension b.) functions modeling the tire lift off problem

3 Optimal Linear Preview Control of Idealized Active Suspension

As was shown by many authors, e.g. (Hać & Youn, 1991; Hać, 1992; Thompson & Pearce, 1998), optimal linear control with preview, i.e. the case when approaching road disturbances are known within a certain distance ahead, reduces variances of car body acceleration, suspension travel and tire deflection at the same time compared with the no preview case. In this section, optimal linear preview control of idealized active suspension is derived. Incoming road disturbances are measured by a sensor at some distance ahead of the vehicle. The tire lift off effect is not considered in the deriving. The equations of motion for such a problem are:

$$\mathbf{M}_a\ddot{\mathbf{q}}_{aA} + \mathbf{K}_a\mathbf{q}_{aR} = \mathbf{b}_u u \tag{7}$$

Combining equation (7) and differentiated equation (4), state space model of idealized active suspension with preview and without consideration the tire lift off effect is obtained:

$$\dot{\mathbf{x}}_a = \mathbf{A}_a\mathbf{x}_a + \mathbf{B}_a u + \mathbf{G}_{aw}\dot{w} \tag{8}$$

where

$$\mathbf{A}_a = \begin{bmatrix} \mathbf{O}_a & \mathbf{T}_{aA} \\ -\mathbf{M}_a^{-1}\mathbf{K}_a & \mathbf{O}_a \end{bmatrix}, \quad \mathbf{K}_a = \begin{bmatrix} k_1 & 0 \\ -k_1 & k_2 \end{bmatrix} \tag{9}$$

The state vector $\boldsymbol{x}_a$ contains relative displacements $\mathbf{q}_{AR}$ (suspension and the tire deflections) and absolute velocities $\mathbf{q}_{AR}$ (car body and the wheel velocities) – see (6). Such a description leads to the velocity of road disturbances at the input. It is assumed that the road disturbances $w(t)$ are measured at the distance l_p in front of the vehicle, i.e. at time t the preview information about incoming road disturbances is available from the time t up to time $t + t_p$, where $t_p = l_p/v$ is the preview time and v is the vehicle velocity. The active force generator is optimized in regard to ride comfort, suspension rattle space and road holding. Corresponding variables to be minimized are car body acceleration, suspension deflection and tire deflection.

$$\mathbf{z} = \begin{bmatrix} \ddot{y}_1 & y_1 - y_2 & y_2 - w \end{bmatrix}^T \tag{10}$$

and in the state space form

$$\mathbf{z} = \mathbf{C_a x_a} + \mathbf{D_a} u + \mathbf{H_a} \dot{w} \tag{11}$$

Then the performance index involves appropriately weighted variances of optimized variables (13) that are to be minimized and weighted variance of active control force that is also to be minimized:

$$J = \lim_{T\to\infty} \frac{1}{T} \int_0^T \left[q_1 \ddot{y}_1^2 + q_2 \left(y_1 - y_2\right)^2 + q_3 \left(y_2 - w\right)^2 + Ru^2 \right] dt = \lim_{T\to\infty} \frac{1}{T} \int_0^T \left(\mathbf{z}^T \mathbf{Q} \mathbf{z} + u^T R u \right) dt \tag{12}$$

where

$$\mathbf{Q} = diag\left(\begin{bmatrix} q_1 & q_2 & q_3 \end{bmatrix}\right) \tag{13}$$

$\mathbf{Q}$ is weighting matrix and R is control cost constant. Then the Hamiltonian H for this issue is in the form

$$\mathbf{H} = \frac{1}{2}\left(\mathbf{x_a^T Q_1 x_a} + u^T R_1 u + \dot{w}^T Q_2 \dot{w} + 2\mathbf{x_a^T N_1} u + 2\mathbf{x_a^T N_2} \dot{w}\right) + \boldsymbol{\lambda}^T \left(\mathbf{A_a x_a} + \mathbf{B_a} u + \mathbf{G_{aw}} \dot{w} - \dot{\mathbf{x}}_\mathbf{a}\right) \tag{14}$$

After some manipulations, the active control force is obtained

$$u = -\mathbf{K_b x_a} - \mathbf{K_f r} \tag{15}$$

where

$$\begin{gathered} \mathbf{K_b} = R_1^{-1}\left(\mathbf{N_1^T} + \mathbf{B_a^T P}\right),\ \mathbf{K_f} = R_1^{-1}\mathbf{B_a^T},\ \mathbf{Q_1} = \mathbf{C_a^T Q C_a},\ R_1 = \mathbf{D_a^T Q D_a} + R,\ Q_2 = \mathbf{H_a^T Q H_a},\ \mathbf{N_1} = \mathbf{C_a^T Q D_a} \\ \mathbf{N_2} = \mathbf{C_a^T Q H_a},\ \mathbf{Q_n} = \mathbf{Q_1} - \mathbf{N_1} R_1^{-1} \mathbf{N_1^T},\ \mathbf{A_n} = \mathbf{A_a} - \mathbf{B_a} R_1^{-1} \mathbf{N_1^T},\ \mathbf{A_c} = \mathbf{A_n} - \mathbf{B_a} R_1^{-1} \mathbf{B_a^T P},\ \mathbf{G_r} = \mathbf{P G_{aw}} + \mathbf{N_2} \end{gathered} \tag{16}$$

where $\mathbf{P}$ is a nonnegative definite symmetric solution of the algebraic Riccati equation and $\mathbf{r}$ is calculated as follows

$$\begin{gathered} \mathbf{P A_n} + \mathbf{A_n^T P} - \mathbf{P B_a} R_1^{-1} \mathbf{B_a^T P} + \mathbf{Q_n} = \mathbf{O} \\ \mathbf{r}(t) = \int_0^{t_p} e^{\mathbf{A_c^T}\sigma} \mathbf{G_r} \dot{w}(t+\sigma) d\sigma \end{gathered} \tag{17}$$

4 MR Damper – Hydromechanical and Mathematical Model

Hydromechanical and mathematical model of the MR damper were identified by the author in publication (Úradníček, 2008), where they were designed on the basis of model of an electrorheological damper compiled in publication (Hong & Choi, 2005).

Hydromechanical parameters of the hydromechanical model of the MR damper – figure 2 a) – are:

C_1, C_2, C_4 – compressibility of the volumes front of, behind the piston and of gas storage.
A_1, A_2, A_4, A_f – cross-sectional area of the bottom, top part of the piston and area of the membrane and of the grooves.
p_1, p_2, p_4 – pressures of the MR fluid front of, behind the piston and of gas storage
I_f – inertia of the MR fluid flowing through the grooves of the piston.
R_f – hydraulic resistance of the MR fluid flowing through the groove of the piston
Δp_{MR} – pressure drop in the groove of the piston caused by friction force.
y_r – displacement of the MR fluid flowing through the groove of the piston relative to piston.
y_p – displacement of the piston of the MR damper, y_m – displacement of the membrane separating the MR fluid from gas.

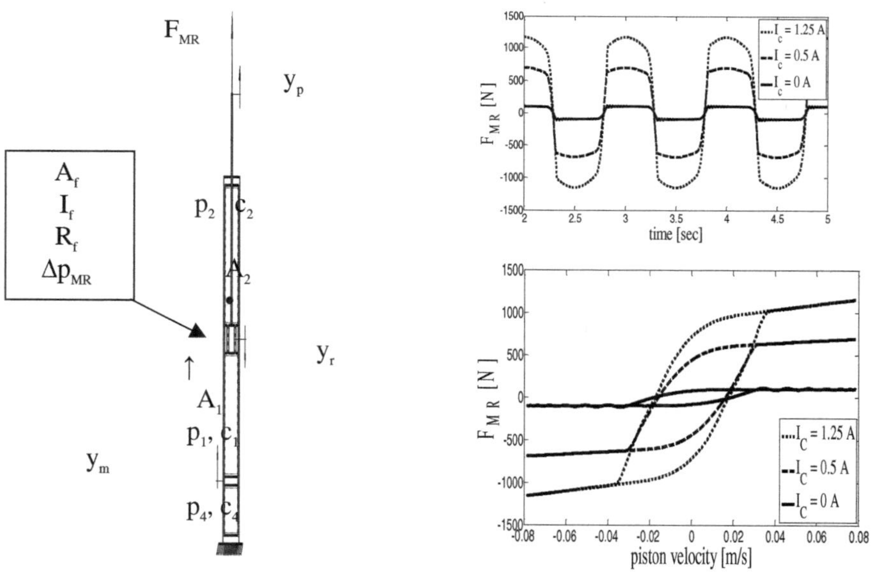

Fig. 2 a.) hydromechanical model of the MR damper b.) dynamical characteristics of the MR damper

The equation of motion of the hydromechanical model of the MR damper in the figure 2 a.) is

$$\begin{aligned} m_f \ddot{y}_r + c_f \dot{y}_r + F_y \tanh(\beta_d \dot{y}_r) + k_{f_1} y_r &= k_{f_2} y_p \\ F_{MR} &= m_f \ddot{y}_r + c_f \dot{y}_r + F_y \tanh(\beta_d \dot{y}_r) \end{aligned} \tag{18}$$

where

$$m_f = I_f A_f A_p,\ c_f = R_f A_f A_p,\ F_y = \Delta p_{MR} A_p,\ k_{f_1} = A_f\left(\frac{1}{c_1+c_4}+\frac{1}{c_2}\right)A_p,\ k_{f_2} = \left(A_p - A_f\right)\left(\frac{1}{c_1+c_4}+\frac{1}{c_2}\right) \quad (19)$$

In publication (Úradníček, 2008) author experimentally identified the parameters of the MR damper LORD RD – 1005 – 3 on the basis of mathematical model described by relations (49) and (51) as functions of electric current flowing through the coil of the MR damper (for the range of electric current 0 – 1.25 A):

$$\begin{aligned} m_f &= d_5 & c_f(I_c) &= c_1 I_c + d_1 & F_y(I_c) &= a_2 I_c^3 + b_2 I_c^2 + c_2 I_c + d_2 \\ \beta_d &= 80 & k_{f_1}(I_c) &= b_3 I_c^2 + c_3 I_c + d_3 & k_{f_2}(I_c) &= b_4 I_c^2 + c_4 I_c + d_4 \end{aligned} \quad (20)$$

5 Quarter Car Model with MR Damper and with Consideration the Tire Lift Off

The mathematical model of the semi-active MR damper described in the previous section is implemented in the quarter car model described in the second section. So the idealized fully active suspension is replaced by the MR damper. In the equations of motion (1) the control force u is replaced by the one generated by the MR damper – F_{MR}.

$$\begin{aligned} m_1\ddot{y}_1 &= -k_1\left(y_1 - y_2\right) + F_{MR} - m_1 g \\ m_2\ddot{y}_2 &= -k_2\left(y_2 - w\right)\left[1 - H\left(y_2 - w\right)\right] - k_1\left(y_2 - y_1\right) - F_{MR} - m_2 g \end{aligned} \quad (21)$$

Dynamics of the MR damper is described by the relations (18) in the previous section.

$$\begin{aligned} m_f\ddot{y}_r + c_f\dot{y}_r + F_y\tanh\left(\beta_d\dot{y}_r\right) + k_{f_1}y_r &= -k_{f_2}\left(y_1 - y_2\right) \\ F_{MR} &= -k_{f_2}\left(y_1 - y_2\right) - k_{f_1}y_r \end{aligned} \quad (22)$$

By substituting the MR damping force from the second relation of (22) into (21) the equations of motion of the quarter car model with MR damper and with consideration the tire lift off in the matrix form are

$$\mathbf{M_s}\ddot{\mathbf{q}}_{\mathbf{sA}} + \mathbf{B_{sL}}\dot{\mathbf{q}}_{\mathbf{sA}} + \mathbf{B}_{s\mathbf{N}}\tanh\left(\beta_d\dot{\mathbf{q}}_{\mathbf{sA}}\right) + \mathbf{K_{sL}}\mathbf{q_{sR}} + \mathbf{K_{sN}}\mathbf{q_{sR}}\mathbf{e_s}\tanh\left(\beta_r\mathbf{q_{sR}}\right) = \mathbf{b_{sg}}g \quad (23)$$

Transformation from absolute to the relative coordinates is realized through the equation

$$\mathbf{q_{sR}} = \mathbf{T_{sA}}\mathbf{q_{sA}} + \mathbf{T_{sw}}w \quad (24)$$

Combining equation (23) and differentiated equation (24), state space model is obtained:

$$\dot{\mathbf{x}}_{\mathrm{s}} = \mathbf{A}_{\mathrm{sL}}\mathbf{x}_{\mathrm{s}} + \mathbf{A}_{\mathrm{sr}}\mathbf{x}_{\mathrm{s}}\mathbf{E}_{\mathrm{s}} \tanh\left(\beta_r \mathbf{x}_{\mathrm{s}}\right) + \mathbf{A}_{\mathrm{sd}} \tanh\left(\beta_d \mathbf{x}_{\mathrm{s}}\right) + \mathbf{G}_{\mathrm{s}}\mathbf{f}_{\mathrm{gw}} \tag{25}$$

where

$$\mathbf{A}_{\mathrm{sL}} = \begin{bmatrix} \mathbf{O}_{\mathrm{s}} & \mathbf{T}_{\mathrm{sA}} \\ -\mathbf{M}_{\mathrm{s}}^{-1}\mathbf{K}_{\mathrm{sL}} & -\mathbf{M}_{\mathrm{s}}^{-1}\mathbf{B}_{\mathrm{sL}} \end{bmatrix},\ \mathbf{A}_{\mathrm{sr}} = \begin{bmatrix} \mathbf{O}_{\mathrm{s}} & \mathbf{O}_{\mathrm{s}} \\ -\mathbf{M}_{\mathrm{s}}^{-1}\mathbf{K}_{\mathrm{sN}} & \mathbf{O}_{\mathrm{s}} \end{bmatrix},\ \mathbf{A}_{\mathrm{sd}} = \begin{bmatrix} \mathbf{O}_{\mathrm{s}} & \mathbf{O}_{\mathrm{s}} \\ \mathbf{O}_{\mathrm{s}} & -\mathbf{M}_{\mathrm{s}}^{-1}\mathbf{B}_{\mathrm{sN}} \end{bmatrix},\ \mathbf{E}_{\mathrm{s}} = \begin{bmatrix} \mathbf{e}_{\mathrm{s}}^{T} \\ \mathbf{o}_{\mathrm{s}} \end{bmatrix}^{T},$$
$$\mathbf{G}_{\mathrm{s}} = \begin{bmatrix} \mathbf{G}_{\mathrm{sg}} & \mathbf{G}_{\mathrm{sw}} \end{bmatrix},\ \mathbf{G}_{\mathrm{sg}} = \begin{bmatrix} \mathbf{o}_{\mathrm{s}} \\ \mathbf{M}_{\mathrm{s}}^{-1}\mathbf{b}_{\mathrm{sg}} \end{bmatrix},\ \mathbf{G}_{\mathrm{sw}} = \begin{bmatrix} \mathbf{T}_{\mathrm{sw}} \\ \mathbf{o}_{\mathrm{s}} \end{bmatrix},\ \mathbf{x}_{\mathrm{s}} = \begin{bmatrix} \mathbf{q}_{\mathrm{sR}} \\ \dot{\mathbf{q}}_{\mathrm{sA}} \end{bmatrix}, \mathbf{T}_{\mathrm{sA}} = \begin{bmatrix} 1 & -1 & 0 \\ 0 & 1 & 0 \\ 0 & 0 & 1 \end{bmatrix},\ \mathbf{T}_{\mathrm{sw}} = \begin{bmatrix} 0 \\ -1 \\ 0 \end{bmatrix}, \tag{26}$$

where $\mathbf{O}_{\mathrm{s}}$ and $\mathbf{o}_{\mathrm{s}}$ are zero matrix and vector respectively of appropriate dimensions.

6 Continuous Inverse Mathematical Model of the MR Damper

Mathematical model of the MR damper works such – see relations (22) – that for given electric current and for kinematic variables the damping force F_{MR} is calculated. An inverse model of the MR damper designed for the use of control should calculate the control electric current for given kinematic variables and for required damping force.

In section 3 the optimal preview control and corresponding active control force were derived. In this section as required damping force F_{MR} that is trying to be matched by the MR damper the active control force u from section 3 is taken. In some situations the active force is physically unable to be achieved by the MR damper. This problem is solved below. As it was pointed out the active control force that would be generated by fully active system is trying to be matched by the one generated by the MR damper, so with respect to the second relation of (56) it can be written

$$\begin{aligned} F_{MR} &= u \\ -k_{f_2}\left(y_1 - y_2\right) - k_{f_1} y_r &= u \end{aligned} \tag{27}$$

For the use of control it is appropriate to replace the variables $k_{f1}(I_c)$ and $k_{f2}(I_c)$ from relations (20) by linear functions $k_{f1r}(I_r)$ and $k_{f2r}(I_r)$ and for given range of control electric current to optimize their parameters by the least squares method – see figure (3)

$$k_{f_{1r}}\left(I_r\right) = c_{3r} I_r + d_{3r}, \qquad k_{f_{2r}}\left(I_r\right) = c_{4r} I_r + d_{4r} \tag{28}$$

These relations (28) are used only in the inverse model of the MR damper for calculating the required electric current.

Fig. 3 Comparison of the functions used in mechanical and in inverse model of the MR damper – see relations (20) and (28)

After mentioned replacement the second relation of (27) now has the form

$$-k_{f_{2r}}\left(y_1 - y_2\right) - k_{f_{1r}} y_r = u \tag{29}$$

By substituting (28) and (15) – relation for calculating the active control force u – into (29) we obtain

$$-\left(c_{4r} I_r + d_{4r}\right)\left(y_1 - y_2\right) - \left(c_{3r} I_r + d_{3r}\right) y_r = -K_{b_1}\left(y_1 - y_2\right) - K_{b_2}\left(y_2 - w\right) - K_{b_3}\dot{y}_1 - K_{b_4}\dot{y}_2 - \mathbf{K_f r} \tag{30}$$

After some manipulations the required control electric current is obtained

$$I_r = \frac{\left(\mathbf{K}_{\mathbf{I_n}} + \mathbf{K}_{\mathbf{b_s}}\right)\mathbf{x_s} + \mathbf{K_f r}}{\mathbf{K}_{\mathbf{I_d}}\mathbf{x_s}} \tag{31}$$

If this fictive required electric current I_r flew through the coil of the MR damper, the ideal control active force would be achieved. The required electric current I_r calculated from the relation (31) can be any real number (also negative, which is physically impossible). But there are some restrictions of electric control current. Working range of the electric current is limited to *0 – 1 A*. Instead of commonly used saturation a continuous function is utilized for the calculation of theoretical control electric current for used MR damper (Havelka, 2010).

$$I_{ct} = \frac{1}{2}\left\{1 + \tanh\left[\alpha\left(I_r - \frac{1}{2}\right)\right]\right\} \tag{32}$$

The parameter α was optimized by the least squares method to match the commonly used saturation.

Since the response time of an actual MR damper to the theoretical required control electric current I_{ct} is not immediate but time-delayed, this effect can be included into the model using a first order filter

$$\dot{I}_c(t) = -\frac{1}{T_{MR}}\left[I_c(t) - I_{ct}(t)\right] \tag{33}$$

where T_{MR} is the time constant of the MR damper and I_c is the actual electric current applied to the model (20) of the MR damper. It further means that the system matrices $\mathbf{A}_{\mathbf{sL}}$ and $\mathbf{A}_{\mathbf{sd}}$ of the state space model (25) are also dependent on the applied electric current I_c.

7 Simulations and Results

To significantly demonstrate the benefits of preview control the vehicle model was let to travel over a bump and further it was supposed that all the state variables are measured. The bump is described by equation

$$w(t)=\left\{H\left(t-t_{bs}\right)-H\left[t-\left(t_{bs}+t_{b}\right)\right]\right\}\frac{b_h}{2}\left\{1-\cos\left[2\pi\frac{1}{t_b}\left(t-t_{bs}\right)\right]\right\} \tag{34}$$

where t_{bs} is the "starting" time of the bump, b_h is the height of the bump and $t_b = b_l / v$ is the duration time of the bump where v is the vehicle velocity.

The quarter car equipped with fully active idealized suspension was let to travel over a bump at velocity *4 m/s* –figure 4. The preview distance in front of the front wheel was set to *1.6 m*. This implies the preview time t_p = *0.4 sec.*

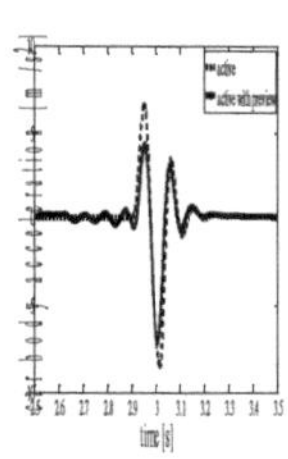

a.) car body acceleration

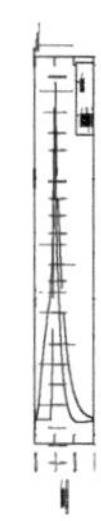

b.) car body displacement

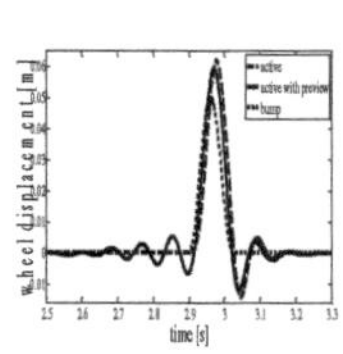

c.) wheel displacement

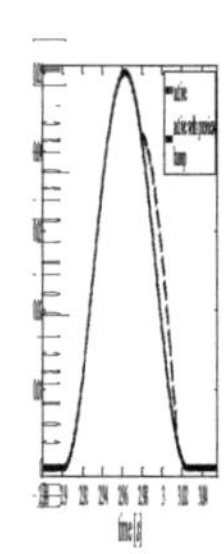

d.) tire contact point diplacement

Fig. 4 Responses of the quarter car model traveling over a bump equipped with fully active system controlled by "active" and "active with preview" control strategies

In Figure 4 car body acceleration and car body, wheel and tire contact point displacements of fully active quarter car model traveling over a bump controlled by "active" and "active with preview" control strategies are shown. In all cases the active preview control strategy provides lower amplitudes and also smaller variances. As shown in Figure 4 c.) – the preview controlled active suspension acts the wheel before the bump excitation comes to smoothly lift it over the bump and avoids the tire lift off the road compared with the no preview case when the undesired tire lift off effect occurs – Figure 4 d.).

Then the quarter car equipped with the MR damper in suspension was let to travel over the bump at velocity *3.5 m/s* – figure 5. The preview distance in front of the front wheel was set to *1.6 m*. This implies the preview time $t_p = 0.457\ sec$.

Figure 5 shows that the semi-active MR damper with the preview case provides some small improvement in car body displacement compared with the no preview case, but in terms of the wheel displacement no difference between the preview and no preview case can be observed. This is because the semi-active MR damper is unable to supply energy into the system and cannot generate forces when there is no changing suspension deflection, i.e. the MR damper cannot act the system before the excitation comes – see figure 6 b.).

The control electric current is trying to change the MR fluid properties to achieve the required active control force – see figure 6 b.) from the time *2.5 sec* up to time *3 sec* – but during this time period there is no changing suspension deflection so the MR damper can produce no force – see F_{MR} in the figure 6 b.) during this period.

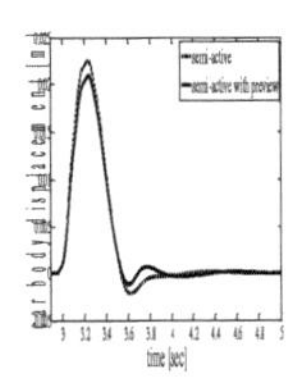

a.) car body displacement

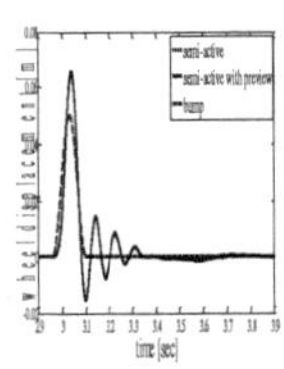

b.) wheel displacement

Fig. 5 Responses of the quarter car model traveling over a bump equipped with MR damper controlled with "no preview" and "preview" control strategies

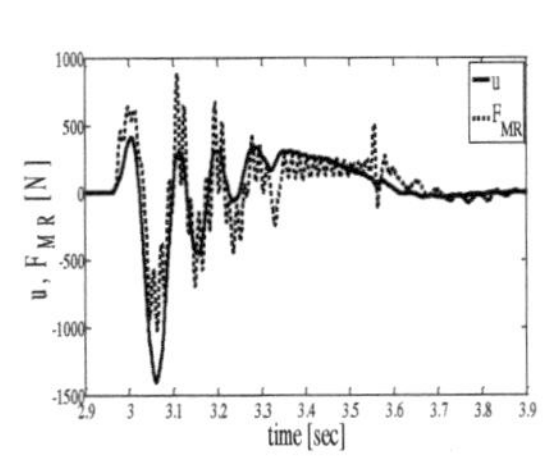

a.) no preview case

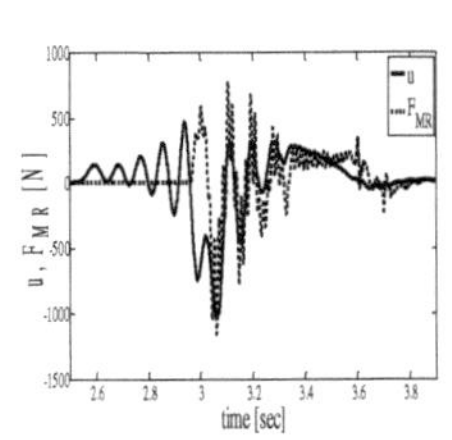

b.) preview case

Fig. 6 Required active force "u" and the actual control force of the MR damper "F_{MR}" in the quarter car model traveling over a bump controlled with" no preview" and "preview" control strategies

8 Conclusions

The preview controlled active suspension acts the vehicle before the excitation comes, i.e. prepares the vehicle for approaching road disturbances to smoothly travel over them and reduces the probability of undesired tire lift off effect. Active suspension with preview compared with the no preview case provides lower maximum amplitudes and smaller variances of car body acceleration, suspension travel and tire deflection at the same time.

In the case of utilizing the semi-active MR damper in suspension difference between the preview and no preview control strategies almost diminishes. This is because MR damper is only able to dissipate the energy present in the system and cannot generate independent forces when there is no changing suspension deflection.

Acknowledgments. This work was supported by grant Vega 1/0197/12.

References

1. Hać, A.: Optimal Linear Preview Control of Active Vehicle Suspension. Vehicle System Dynamics 21, 167–195 (1992)
2. Hać, A., Youn, I.: Optimal Semi-Active Suspension with Preview Based on a Quarter Car Model. In: American Control Conference, Boston, MA, USA, pp. 433–438 (1991)
3. Havelka, F., Zuščík, M., Musil, M.: Návrh spojitého inverzného matematického modelu magnetoreologického tlmiča pre riadenie. In: Noise and Vibration in Practice: Proceedings of the 15th International Acoustic Conference, Kočovce, Slovensko, pp. 53–56 (2010)
4. Hong, S.R., Choi, S.B.: A hydro-mechanical model for hysteretic damping force prediction of ER damper: experimental verification. Journal of Sound and Vibration 285, 1180–1188 (2005)
5. Prabakar, R.S., Sujatha, C., Narayanan, S.: Optimal semi-active preview control response of a half car vehicle model with magnetorheological damper. Journal of Sound and Vibration 326(3-5), 400–420 (2009)
6. Thompson, A.G., Pearce, C.E.M.: Physically Realisable Feedback Controls for a Fully Active Preview Suspension Applied to a Half-Car Model. Vehicle System Dynamics 30, 17–35 (1998)
7. Úradníček, J.: Multidisciplinárna optimalizácia modelu odpruženia vozidla vybaveného semiaktívnym magnetoreologickým tlmičom. Kandidátska dizertačná práca. STU v Bratislave SjF, Bratislava (2008)
8. Musil, M.: Iterative model correction based on transfer function. In: Engineering Mechanics 2004. Institute of Thermomechanics Academy of Sciences of the Czech Republic, Praha (2004) ISBN 80-85918-88-9

Author Index

MIX
Papier aus verantwortungsvollen Quellen
Paper from responsible sources
FSC® C105338

If you have any concerns about our products,
you can contact us on
ProductSafety@springernature.com

In case Publisher is established outside the EU,
the EU authorized representative is:
Springer Nature Customer Service Center GmbH
Europaplatz 3, 69115 Heidelberg, Germany

Printed by Libri Plureos GmbH
in Hamburg, Germany